Ferromagnetic and Ferroelectric Materials: Synthesis, Applications, and Techniques

Ferromagnetic and Ferroelectric Materials: Synthesis, Applications, and Techniques

Guest Editor

Dana-Georgeta Popescu

MDPI

Basel • Beijing • Wuhan • Barcelona • Belgrade • Novi Sad • Cluj • Manchester

Guest Editor
Dana-Georgeta Popescu
Surface and Interface Science
National Institute of Materials
Physics
Magurele
Romania

Editorial Office
MDPI AG
Grosspeteranlage 5
4052 Basel, Switzerland

This is a reprint of the Special Issue, published open access by the journal *Materials* (ISSN 1996-1944), freely accessible at: https://www.mdpi.com/journal/materials/special_issues/ferroelectricity_perovskite_oxides.

For citation purposes, cite each article independently as indicated on the article page online and as indicated below:

Lastname, A.A.; Lastname, B.B. Article Title. *Journal Name* **Year**, *Volume Number*, Page Range.

ISBN 978-3-7258-3739-7 (Hbk)
ISBN 978-3-7258-3740-3 (PDF)
https://doi.org/10.3390/books978-3-7258-3740-3

Contents

About the Editor

Popescu Dana-Georgeta

Dr. Dana-Georgeta Popescu is a researcher specializing in surface and interface science. She earned her Engineer Diploma in 2009 and her Ph.D. in Engineering in 2014 from the Politehnica University of Bucharest, where her doctoral research focused on photonic crystal technology.

Since 2010, Dr. Popescu has been part of the surface and interface science group at the National Institute of Materials Physics (NIMP), working under the guidance of Dr. Cristian-Mihail Teodorescu. Her research has explored semiconducting surfaces and their interfaces with magnetic metals, contributing to the development of systems exhibiting unusual magnetism at room temperature. She further expanded her expertise in electron spectroscopy techniques during a three-month research stay at CEA in France, working with Dr. N. Barret. Additionally, she has conducted advanced experiments at synchrotron facilities in Italy, Switzerland, and Poland. Her recent research focuses on ferroelectric Rashba materials.

Dr. Popescu has an extensive academic portfolio, including 39 published articles, five book chapters, and four national patents. She has successfully led two young research team projects (2015, 2021) and secured multiple research grants, including the prestigious *L'Oréal–UNESCO "For Women in Science"* award (2021). She is also serving as the academic editor of two books: *Silicon Photonics–From Fundamentals to Devices* and *Mastering Optoelectronics–Fundamentals, Applications, and Innovations*.

materials

MDPI

Article

Improved Energy Storage Density and Efficiency of Nd and Mn Co-Doped $Ba_{0.7}Sr_{0.3}TiO_3$ Ceramic Capacitors Via Defect Dipole Engineering

Hyunsu Choi [1,†], Srinivas Pattipaka [1,†], Yong Hoon Son [1], Young Min Bae [1], Jung Hwan Park [2], Chang Kyu Jeong [3], Han Eol Lee [3], Sung-Dae Kim [1], Jungho Ryu [4] and Geon-Tae Hwang [1,*]

1 Department of Materials Science and Engineering, Pukyong National University, 45, Yongso-Ro, Nam-Gu, Busan 48513, Republic of Korea; sky5021184@pukyong.ac.kr (H.C.); cnuphy444@gmail.com (S.P.); mike4009@pukyong.ac.kr (Y.H.S.); dud1560@pukyong.ac.kr (Y.M.B.); sdkim@pknu.ac.kr (S.-D.K.)
2 Department of Mechanical Engineering, Department of Aeronautics, Mechanical and Electronic Convergence Engineering, Kumoh National Institute of Technology, 61 Daehak-Ro, Gumi 39177, Republic of Korea; parkjh1151@kumoh.ac.kr
3 Division of Advanced Materials Engineering, Jeonbuk National University, Jeonju 54896, Republic of Korea; ckyu@jbnu.ac.kr (C.K.J.); haneol@jbnu.ac.kr (H.E.L.)
4 School of Materials Science and Engineering, Yeungnam University, 280 Daehak-Ro, Gyeongsan-si 38541, Republic of Korea; jhryu@ynu.ac.kr
* Correspondence: gthwang@pknu.ac.kr
† These authors contributed equally to this work.

Abstract: In this paper, we investigate the structural, microstructural, dielectric, and energy storage properties of Nd and Mn co-doped $Ba_{0.7}Sr_{0.3}TiO_3$ [$(Ba_{0.7}Sr_{0.3})_{1-x}Nd_xTi_{1-y}Mn_yO_3$ (BSNTM) ceramics (x = 0, 0.005, and y = 0, 0.0025, 0.005, and 0.01)] via a defect dipole engineering method. The complex defect dipoles $(Mn''_{Ti} - V^{\cdot\cdot}_O)^{\cdot}$ and $(Mn''_{Ti} - V^{\cdot\cdot}_O)$ between acceptor ions and oxygen vacancies capture electrons, enhancing the breakdown electric field and energy storage performances. XRD, Raman, spectroscopy, XPS, and microscopic investigations of BSNTM ceramics revealed the formation of a tetragonal phase, oxygen vacancies, and a reduction in grain size with Mn dopant. The BSNTM ceramics with x = 0.005 and y = 0 exhibit a relative dielectric constant of 2058 and a loss tangent of 0.026 at 1 kHz. These values gradually decreased to 1876 and 0.019 for x = 0.005 and y = 0.01 due to the Mn^{2+} ions at the Ti^{4+}- site, which facilitates the formation of oxygen vacancies, and prevents a decrease in Ti^{4+}. In addition, the defect dipoles act as a driving force for depolarization to tailor the domain formation energy and domain wall energy, which provides a high difference between the maximum polarization of P_{max} and remnant polarization of P_r (ΔP = 10.39 μC/cm^2). Moreover, the complex defect dipoles with optimum oxygen vacancies in BSNTM ceramics can provide not only a high ΔP but also reduce grain size, which together improve the breakdown strength from 60.4 to 110.6 kV/cm, giving rise to a high energy storage density of 0.41 J/cm^3 and high efficiency of 84.6% for x = 0.005 and y = 0.01. These findings demonstrate that defect dipole engineering is an effective method to enhance the energy storage performance of dielectrics for capacitor applications.

Keywords: ceramic capacitors; donor–acceptor complex; defect dipole engineering; dielectric and ferroelectric properties; energy storage density and efficiency

Citation: Choi, H.; Pattipaka, S.; Son, Y.H.; Bae, Y.M.; Park, J.H.; Jeong, C.K.; Lee, H.E.; Kim, S.-D.; Ryu, J.; Hwang, G.-T. Improved Energy Storage Density and Efficiency of Nd and Mn Co-Doped $Ba_{0.7}Sr_{0.3}TiO_3$ Ceramic Capacitors Via Defect Dipole Engineering. *Materials* **2023**, *16*, 6753. https://doi.org/10.3390/ma16206753

Academic Editor: David Marrero Lopez

Received: 13 September 2023
Revised: 15 October 2023
Accepted: 16 October 2023
Published: 18 October 2023

1. Introduction

Dielectric capacitors are key components of pulsed power applications, and are extensively used in microwave communications, electromagnetic devices, hybrid electric vehicles, and high-frequency inverters [1–5]. Notably, dielectric capacitors display ultrahigh power density, ultrafast charge–discharge rates, excellent fatigue resistance, and thermal stability as compared to batteries [6–8]. However, their energy storage density performance is lower than that of batteries because of their low breakdown strength (BDS), which limits their applications in energy storage devices [9–11]. It is thus necessary to develop new dielectric capacitors with high energy storage density and high energy efficiency to meet the increasing demands for energy storage devices.

The key parameters for energy storage in dielectric capacitors, such as the total energy storage density (W_{tot}), recoverable energy density (W_{rec}), and energy efficiency (η) can be calculated by the following equations [8,9,12]:

$$W_{tot} = \int_{0}^{P_{max}} E\,dP \tag{1}$$

$$W_{rec} = \int_{P_r}^{P_{max}} E\,dP \tag{2}$$

$$\eta = \frac{W_{rec}}{W_{rec+}W_{loss}} \times 100\% \tag{3}$$

where E is the applied electric field, P is induced polarization, P_{max} is maximum polarization, P_r is remnant polarization, and W_{loss} is hysteresis loss (Figure 1). According to these equations, W_{rec} and η can be improved by increasing the difference between P_{max} and P_r ($\Delta P = P_{max} - P_r$) and the BDS/breakdown electric field (E_{BD}), which means that energy storage mostly depends on the ΔP and E_{BD} parameters, hence a larger E_{BD} is the cause of high energy storage density. Normally, high-dielectric-constant materials with a large P_{max} display high dielectric loss, which leads to low BDS and W_{rec} [13]. Researchers have sought to enhance BDS by modifying extrinsic properties, such as reducing the thickness of dielectric capacitors [14,15], porosity [16,17], and grain size [18,19], and adopting a core–shell structure [20,21]. They have also modified intrinsic properties, including enhancing the bandgap energy [9,22], tailoring electrical homogeneity, and reducing electrical conductivity [23].

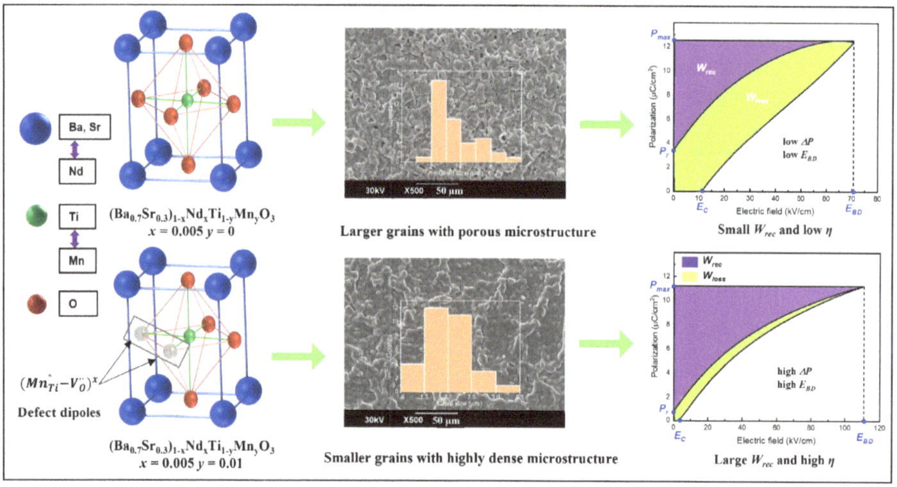

Figure 1. Schematic illustration for energy storage performance of Nd and Mn co-doped BST ceramics. Defect dipoles $(Mn''_{Ti} - V_O^{\cdot\cdot})^x$ between acceptor ions and oxygen vacancies capture electrons, reduce grain size, and provide a high difference between P_{max} and P_r, which improve the breakdown electric field with Mn, resulting in a high energy storage density and high energy efficiency in BSNTM ceramics.

In recent years, lead-free dielectric capacitors have received significant attention, and a great deal of research has been carried out to enhance energy storage properties due to lead toxicity and environmental issues. Lead-free dielectrics, such as $BaTiO_3$ (BT) [15,24,25],

$Bi_{0.5}Na_{0.5}TiO_3$ (BNT) [13,26–29], $BiFeO_3$ (BFO) [1,23,30], and $K_{0.5}Na_{0.5}NbO_3$ (KNN) [31,32]-based materials/composites, afford improved energy storage performance and energy efficiency for energy storage applications. In particular, BT-based ceramics are potential candidates and are widely used for capacitor applications due to their high polarization, high dielectric constant, and low Curie temperature (T_C) [33,34]. Few oxide materials (Al_2O_3, SiO_2, and MgO) are used as additives to improve the BDS and energy storage properties of BT-based ceramics [35–37]. Kovbasiuk et al. [38] investigated the thermophysical properties of PbO–ZnO–B_2O_3 with the doping of Al_2O_3, SiO_2, and BaO oxides for dielectric layers on film-heating elements. Rafik et al. [39] reported Sr substitution at the A-site of BT ($Ba_{0.7}Sr_{0.3}TiO_3$) ceramics and improved dielectric properties. However, oxygen vacancies and conduction electrons can occur during the sintering process of BT-based ceramics at high temperatures, resulting in a high dielectric loss [40].

Aliovalent doping is an effective method for tailoring the electrical properties of oxide materials. The use of donor dopants, such as La and Nd, is an effective approach to compensate for the formation of oxygen vacancies to improve the dielectric properties of BT ceramics. Morison and Shaikh et al. [41,42] reported La- and Nd-doped BT ceramics with a high dielectric constant of 25,000 and 13,000 at T_C, respectively. On the other hand, acceptor (Mn^{2+} at Ti^{4+}-site) doping in BT ceramics promotes the formation of oxygen vacancies and minimizes the decrease in Ti^{4+} during the sintering process in low-oxygen atmospheres. Therefore, Mn-doped BT decreases dielectric loss [43–45]. Recently, Yueshun et al. [46] demonstrated defect dipoles via oxygen vacancies in acceptor-doped (specifically Fe) $Sr_2Bi_4Ti_{(5-x)}Fe_xO_{18}$ (x = 0.04–0.12), and enhanced E_{BD} and energy storage properties.

In this paper, we present a defect dipole engineering method to improve the breakdown strength and energy storage performance by co-doping Nd and Mn in $Ba_{0.7}Sr_{0.3}TiO_3$ (BST) ceramics that have been prepared via the traditional solid-state reaction method. Nd-doped BST [$(Ba_{0.7}Sr_{0.3})_{1-x}Nd_xTiO_3$, BSNT] ceramics can compensate for the formation of oxygen vacancies, improving the dielectric constant of BSNT ceramics. In contrast, Mn-doped BSNT ceramics [$(Ba_{0.7}Sr_{0.3})_{1-x}Nd_xTi_{1-y}Mn_yO_3$, BSNTM] facilitate the formation of oxygen vacancies, prevent a decrease in Ti^{4+}, and yield low dielectric loss. Therefore, simultaneously, a high dielectric constant and low dielectric loss can be expected with Nd and Mn co-dopants in BST. Moreover, complex defect dipoles with uniform and small-grained microstructure provide a high difference between P_{max} and P_r (ΔP~10.39 μC/cm^2); these show the improved breakdown strength of 110.6 kV/cm with Nd and Mn, which results in a high energy storage density of 0.41 J/cm^3 and high efficiency of 84.6% in BSNTM ceramics, as schematically shown in Figure 1.

2. Materials and Methods

$(Ba_{0.7}Sr_{0.3})_{1-x}Nd_xTi_{1-y}Mn_yO_3$ (BSNTM) (x = 0, 0.005, and y = 0, 0.0025, 0.005, and 0.01) lead-free ceramics were synthesized using the traditional solid-state reaction method. The raw materials $BaCO_3$ (Sigma-Aldrich, St. Louis, MO, USA, 99%), $SrCO_3$ (Sigma-Aldrich, 98%), Nd_2O_3 (Sigma-Aldrich, 99.9%), TiO_2, (Sigma-Aldrich, 99%), and MnO_2 (Sigma-Aldrich, 99%) were weighed in stoichiometric proportions and ball-milled for 24 h. After drying the slurry, the BSNTM powder was calcined at 1150 °C for 3 h to obtain the phase of BSNTM. Further, 3 wt.% of Li_2CO_3 (Junsei, London, UK, 99%) powder was added to this calcined powder as a sintering aid, and the powder was again ball-milled for 12 h to reduce the sintering temperature and increase its bulk density. Subsequently, 5 wt.% of polyvinyl alcohol (Sigma-Aldrich, 99%) was added, and the powder was pressed into pellets with dimensions of 10 mm in diameter and 0.5 mm in thickness at a pressure of 10 MPa, followed by sintering at 1050 °C for 2 h. Finally, silver paste (ELCOAT, Electroconductives) was applied on both surfaces of the prepared pellets of BSNTM to carry out electrical characterizations.

The crystal structure of the BSNTM samples was tested using an X-ray diffractometer (Rigaku, Tokyo, Japan, Ultima IV) with Cu-Kα radiation (λ = 1.5406 Å) and a Raman spectrometer (JASCO, Tokyo, Japan, NRS-5100) with 532 nm excitation. A scanning electron microscope (SEM, TESCAN, Brno—Kohoutovice, Czech Republic, VEGA II LSU) equipped with an energy-dispersive spectrometer (EDS), and SE (secondary electron) ET (Everhart–Thornley)-type detector (YAG crystal) was used to examine the surface morphology, composition mapping, and elemental distribution at an accelerating voltage of 30 kV. Room-temperature (RT) dielectric properties were measured in the frequency range of 100 Hz–100 kHz using an impedance analyzer (Hewlett Packard, Palo Alto, CA, USA, 4294A). Ferroelectric properties (*P-E* loops) were measured using a ferroelectric tester (Aix ACT, TF Analyzer 2000). The chemical states of the BSNTM samples were measured using X-ray photoelectron spectroscopy (XPS; KRATOS Analytical Ltd., Manchester, UK, AXIS SUPRA).

3. Results and Discussion

3.1. Phase Formation and Crystal Structure

Figure 2a–e show the Rietveld refinement X-ray diffraction (XRD) patterns of BSNTM ceramics for $x = 0$ and 0.005, and $y = 0$, 0.0025, 0.005, and 0.01, in the 2θ range of 20–80°. The Rietveld refinement XRD analysis was carried out to confirm the phase and peak profiles by fitting the pseudo-Voigt function using full-proof software. All the samples exhibited a tetragonal BSNTM phase (P4 mm) with a small secondary phase of triclinic Li_2O (P1). The phase fraction of the BSNTM phase initially decreased from 94.41 to 87.73% for $x = 0.005$ and $y = 0.0025$, and further increased to 95.16% with the Mn concentration (Table 1). At RT, $Ba_{1-x}Sr_xTiO_3$ ceramics exhibit a tetragonal crystal structure for $x = 0.3$, as reported by Rafik et al. [39]. From the XRD results, there is no peak splitting/merging observed with the substitution of Nd and Mn into BST ceramics due to the lack of changes in the tetragonal crystal structure and the very low doping concentration of Nd and Mn ($x = 0.005$ and $y = 0.0025$–0.01). In Figure 2f, it can be seen that the position of the predominant (101) diffraction peak shifted towards higher angles with Nd for $x = 0.005$ and $y = 0$, and it shifted back to lower angles with Mn into BST for $x = 0.005$ and $y = 0.0025$–0.01. The shift towards lower and higher angles in the diffraction peak demonstrates an increase and decrease in the lattice cell parameters due to the incorporation of Nd and Mn in the BST system, respectively. The determined lattice cell parameters and lattice volume of all the samples are listed in Table 1. For $x = 0.005$ and $y = 0$, the Nd^{3+} (1.27 Å) ions can be occupied at the A-site of Ba^{2+} (1.61 Å) and Sr^{2+} (1.12 Å), whereas Mn^{2+} (0.66 Å) ions can occupied at the B-site of the Ti^{4+} (0.60 Å) site of the BST system for $x = 0.005$ and $y = 0.0025$–0.01, due to their mismatch of ionic radii and valences [39,47].

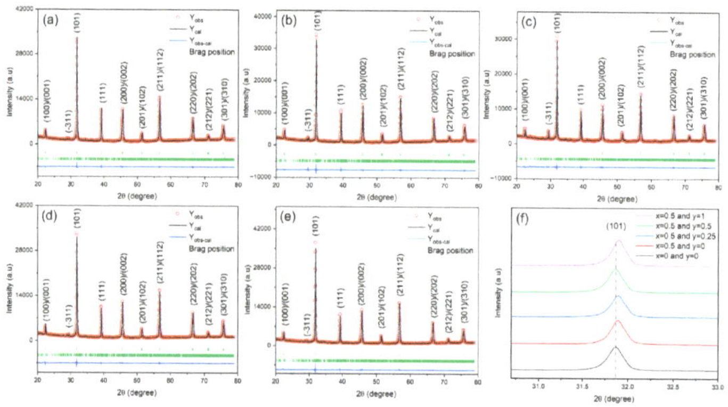

Figure 2. Rietveld refined XRD patterns of $(Ba_{0.7}Sr_{0.3})_{1-x}Nd_xTi_{1-y}Mn_yO_3$ ceramics for (**a**) $x = 0$ and $y = 0$, (**b**) $x = 0.005$ and $y = 0$, (**c**) $x = 0.005$ and $y = 0.0025$, (**d**) $x = 0.005$ and $y = 0.005$, and (**e**) $x = 0.005$ and $y = 0.01$. Figure (**f**) shows a shift in the (101) diffraction peak in the 2θ range from 30.7 − 33°.

Table 1. Rietveld refined XRD parameters of $(Ba_{0.7}Sr_{0.3})_{1-x}Nd_xTi_{1-y}Mn_yO_3$ ceramics (Note: Tetra: tetragonal and Tri: triclinic).

Composition	Lattice Parameters (Å)		V (Å^3)	Phase Fraction (%)	χ^2
	Tetragonal (P4 mm)	Triclinic (P1)			
x = 0 and y = 0	a = b = 3.97377 ± 0.00007 c = 3.98820 ± 0.00023 α = β = γ = 90°	a = 9.74235 ± 0.00045 b = 10.06442 ± 0.00025 c = 10.13491 ± 0.00020 α = 91.43°, β = 91.31° and γ = 96.38°	V_{Tetra} = 62.977 ± 0.004 V_{Tri} = 986.928 ± 0.055	Tetra = 94.41 Tri = 5.59	1.95
x = 0.005 and y = 0	a = b = 3.97989 ± 0.00022 c = 3.94835 ± 0.00064 α = β = γ = 90°	a = 9.71450 ± 0.00034 b = 10.045162 ± 0.00022 c = 10.12407 ± 0.00022 α = 91.41°, β = 91.22° and γ = 96.44°	V_{Tetra} = 62.510 ± 0.011 V_{Tri} = 981.114 ± 0.046	Tetra = 92.92 Tri = 7.08	2.67
x = 0.005 and y = 0.0025	a = b = 3.97339 ± 0.00017 c = 3.97732 ± 0.00025 α = β = γ = 90°	a = 9.72015 ± 0.00035 b = 10.06498 ± 0.00026 c = 10.13280 ± 0.00017 α = 91.46°, β = 91.28° and γ = 96.36°	V_{Tetra} = 62.793± 0.005 V_{Tri} = 984.566± 0.047	Tetra = 87.73 Tri = 12.27	2.05
x = 0.005 and y = 0.005	A = b = 3.97485 ± 0.00016 C = 3.96880 ± 0.00032 A = β = γ = 90°	a =9.71171±0.00039 b = 10.06801±0.00023 c = 10.13123±0.00022 α = 91.46°, β = 91.26° and γ = 96.38°	V_{Tetra} = 62.705± 0.006 V_{Tri} = 983.839± 0.050	Tetra = 94.03 Tri = 5.97	2.02
x = 0.005 and y = 0.01	a = b=3.97102 ± 0.00026 c = 3.97123 ± 0.00023 α = β = γ =90°	a = 9.70581 ± 0.00027 b = 10.06558 ± 0.00018 c = 10.12693 ± 0.00017 α = 91.45°, β = 91.25° and γ = 96.38°	V_{Tetra} = 62.622 ± 0.007 V_{Tri} = 982.577 ± 0.037	Tetra = 95.16 Tri = 4.84	1.77

Raman spectra of BSNTM ceramics in the range of 100–1000 cm^{-1} are shown in Figure 3. The Raman bands of all samples indicate the tetragonal phase of the perovskite structure in BST ceramics, which is similar to that in BST-based reports [39,47]. The spectral parameters of the Raman modes, such as the Raman shift of the central position of each peak and corresponding full width at half maxima (FWHM), are calculated by fitting the Gaussian function. A total of nine Raman active modes were observed. The modes that appeared around 135 and 168 cm^{-1} are associated with the vibration of A-site cations (A-O); 213, 271, and 351 cm^{-1} are related to the vibrations of B-O; 510, 539, and 565 cm^{-1} are related to the vibrations of BO_6; and 740 cm^{-1} corresponds to the A_1 + E (LO) overlapping modes [39]. The mode at 135 cm^{-1} is slightly shifted to a higher wavenumber of 138 cm^{-1} with Nd substitution for x = 0.005 and y = 0. This is caused by an *A*-site disorder, which is attributed to the incorporation of Nd^{3+} at Ba^{2+} and Sr^{2+} ions. The modes around 271 and 539 cm^{-1} shifted towards a lower wavenumber with an increasing Mn concentration from x = 0.005 and y = 0.0025 to 0.01. This is due to an increase in the *B*-site disorder in the BSNTM related to the creation of lattice tensile stress due to lattice expansion [48]. These results are well supported via XRD, dielectric, and ferroelectric properties.

3.2. Microstructural Properties

FESEM images of the BSNTM ceramics are shown in Figure 4. The x = 0.005 and y = 0.01 sample shows a uniform microstructure and has a more compact grain size distribution compared to that of pure BST and other samples of BSNTM (x = 0.005 and y < 0.01). The density of BSNTM ceramics was estimated using the Archimedes principle to confirm a dense and uniform microstructure. The estimated relative density was found to be in the range of 91% to 98% of the theocratical density, thus verifying that all the samples had a highly dense and uniform microstructure. The average grain size of the BSNTM (x = 0 and y = 0) was found to be 3.59 μm and was reduced to 1.99 μm with the substitution of Nd and Mn co-dopants in BSNTM for x = 0.005 and y = 0.01. The reduction in grain size with a uniform microstructure was due to the formation of oxygen vacancies caused by Mn^{2+} occupying Ti^{4+}. Soo and Qiaoli et al. [47,49] reported that Sm and Yb, and Nd and Mn co-doped BT ceramics with donor/donor–acceptor defect complexes via charge compensation/oxygen vacancy exhibited a uniform and small-grained microstructure. Smaller grains with uniform and dense microstructures can resist higher voltages, which results in a high BDS and enhanced energy storage properties [50,51].

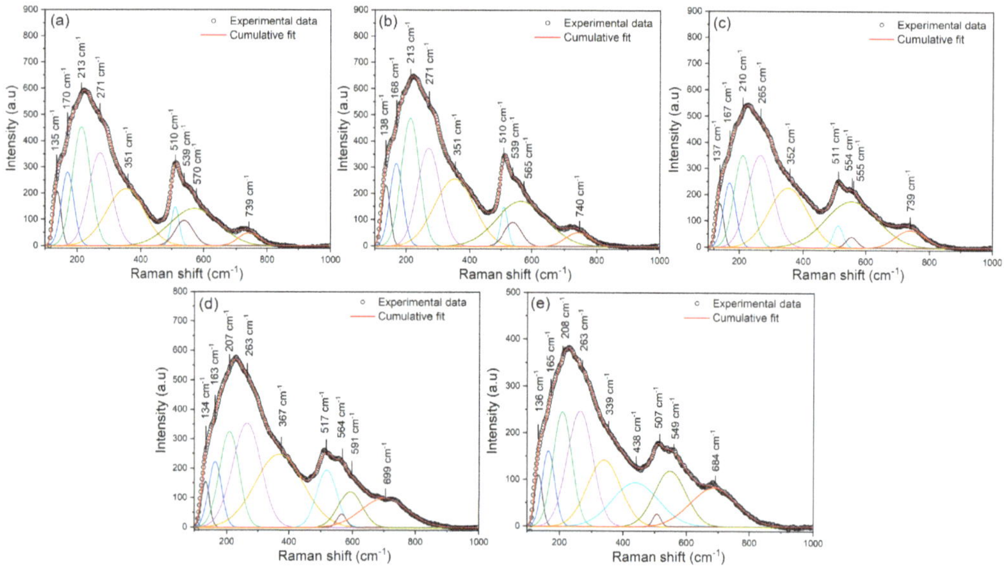

Figure 3. Raman spectra of $(Ba_{0.7}Sr_{0.3})_{1-x}Nd_xTi_{1-y}Mn_yO_3$ ceramics for (**a**) $x = 0$ and $y = 0$, (**b**) $x = 0.005$ and $y = 0$, (**c**) $x = 0.005$ and $y = 0.0025$, (**d**) $x = 0.005$ and $y = 0.005$, and (**e**) $x = 0.005$ and $y = 0.01$.

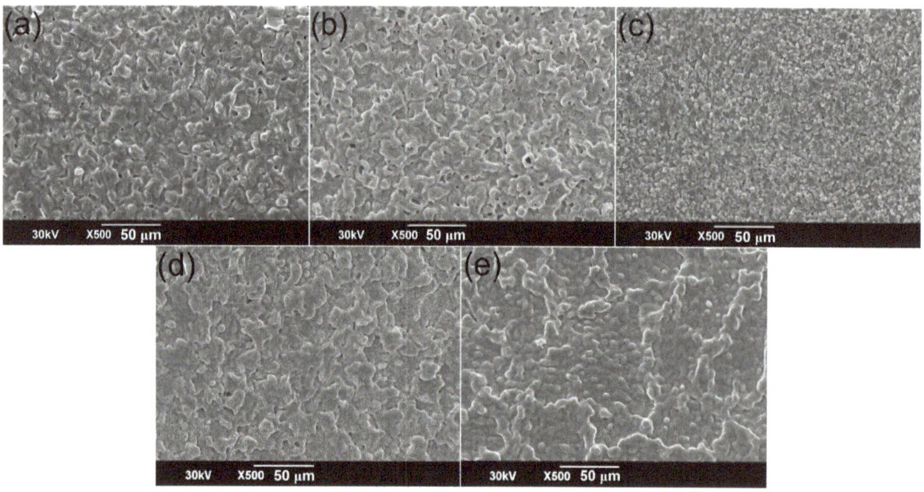

Figure 4. FESEM images of $(Ba_{0.7}Sr_{0.3})_{1-x}Nd_xTi_{1-y}Mn_yO_3$ ceramics for (**a**) $x = 0$ and $y = 0$, (**b**) $x = 0.005$ and $y = 0$, (**c**) $x = 0.005$ and $y = 0.0025$, (**d**) $x = 0.005$ and $y = 0.005$, and (**e**) $x = 0.005$ and $y = 0.01$.

SEM–energy-dispersive X-ray spectroscopy (EDX) elemental mapping was analyzed to show the incorporation and distribution of doping elements. Figure 5 shows the typical microstructure (a–e) and elemental mapping (a1–a4, b1–b5, c1–c6, d1–d6, and e1–e6) of $(Ba_{0.7}Sr_{0.3})_{1-x}Nd_xTi_{1-y}Mn_yO_3$ ceramics for all samples. The elemental mapping results show the existence of elements Ba, Sr, Nd, Ti, Mn, and O, suggesting that Nd and Mn elements enter grains and are distributed homogeneously. The measured experimental volume fraction of BSNTM composition is approximately equal to the theoretical volume fractions, confirming the stoichiometry of BSNTM (Figure 6).

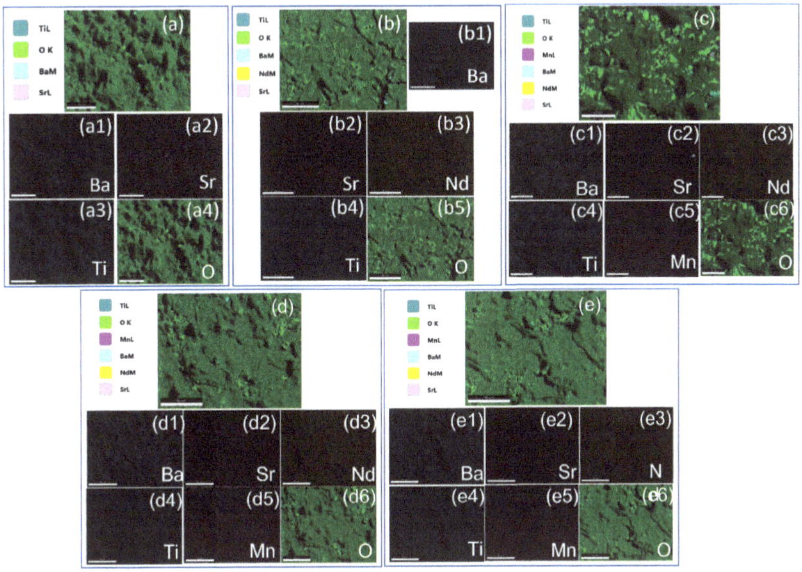

Figure 5. Typical microstructure (**a–e**) and elemental mapping (a1–a4, b1–b5, c1–c6, d1–d6, and e1–e6) of $(Ba_{0.7}Sr_{0.3})_{1-x}Nd_xTi_{1-y}Mn_yO_3$ ceramics for (**a**) $x = 0$ and $y = 0$, (**b**) $x = 0.005$ and $y = 0$, (**c**) $x = 0.005$ and $y = 0.0025$, (**d**) $x = 0.005$ and $y = 0.005$, and (**e**) $x = 0.005$ and $y = 0.01$.

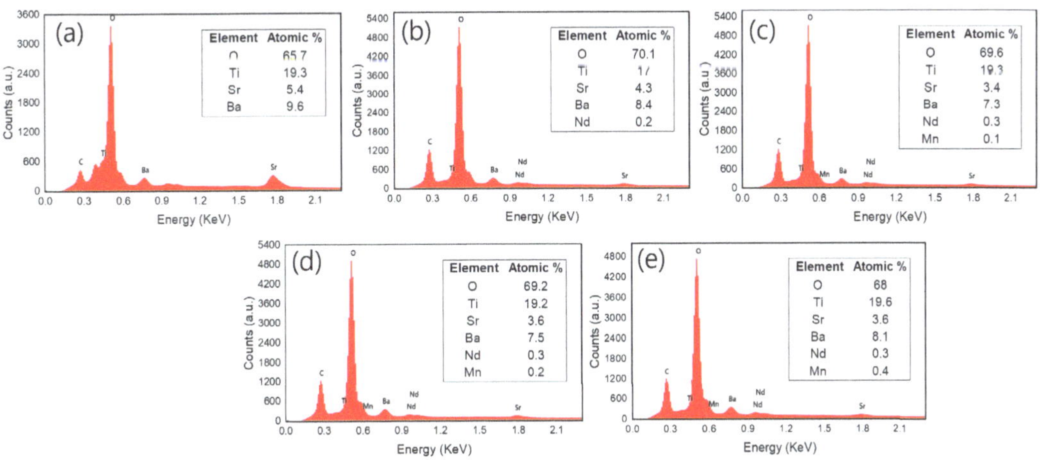

Figure 6. EDX spectra of $(Ba_{0.7}Sr_{0.3})_{1-x}Nd_xTi_{1-y}Mn_yO_3$ ceramics for (**a**) $x = 0$ and $y = 0$, (**b**) $x = 0.005$ and $y = 0$, (**c**) $x = 0.005$ and $y = 0.0025$, (**d**) $x = 0.005$ and $y = 0.005$, and (**e**) $x = 0.005$ and $y = 0.01$.

3.3. Dielectric Properties

The relative dielectric constant (ε_r) and the loss tangent ($tan\delta$) of BSNTM ceramic capacitors measured as a function frequency at RT from 100 Hz to 100 kHz are shown in Figure 7. Pure BST (x = 0 and y = 0) exhibited a ε_r of 1868 and $tan\delta$ of 0.0218 at 1 kHz, which increased to 2058 and 0.0266 with Nd substitution for x = 0.005 and y = 0. Further, these values gradually decreased to 1876 and 0.0191 with Mn substitution into BSNTM for x = 0.005 and y = 0.01. The Nd and Mn co-dopants in the BST matrix favored the formation of donor–acceptor complexes. Nd^{3+} ions in a BST system can compensate for the formation of oxygen vacancies, leading to the enhancement of the dielectric properties (ε_r and $tan\delta$) of BSNT ceramics. On the other hand, Mn^{2+} ions in the BSNT system facilitate the formation of oxygen vacancies, prevent a decrease in Ti^{4+}, and yield a low ε_r and $tan\delta$ [47].

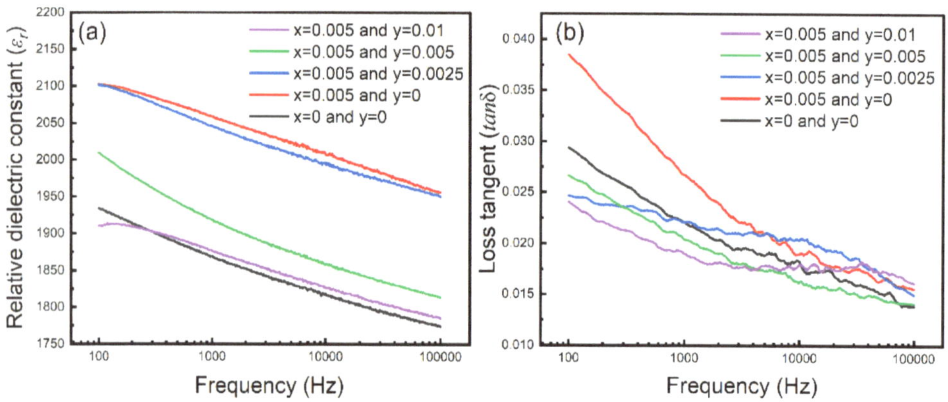

Figure 7. (**a**) Relative dielectric constant and (**b**) loss tangent as a function of the frequency of $(Ba_{0.7}Sr_{0.3})_{1-x}Nd_xTi_{1-y}Mn_yO_3$ ceramics.

3.4. P-E Loops and Energy Storage Performance

P-E loops of BSNTM ceramics measured at RT under different electric fields at a frequency of 10 Hz are shown in Figure 8. The ferroelectric BSNTM (x = 0.005 and y = 0) ceramics displayed a large maximum polarization, P_{max}, of 12.5 μC/cm^2, a small remnant polarization, P_r, of 3.35 μC/cm^2 (i.e., ΔP = 9.15 μC/cm^2), and a high coercive field, E_c, of 11.2 kV/cm. The P_{max}, P_r, and E_c values gradually reduced, ΔP and E_{BD} values increased from 9.15 to 10.39 μC/cm^2, and 70.6 to 110.6 kV/cm from x = 0.005 and y = 0 to x = 0.005 and y = 0.01, as shown in Table 2. The BSNTM sample for x = 0.005 and y = 0.01 (Figure 8e) exhibits a slim saturated *P-E* loop, and the improved E_{BD} is attributed to the decrease in grain size and defect dipoles generated with the incorporation of Mn at the Ti site of the BST host lattice; this can be understood via Kroger–Vink notation as follows [46,52]:

$$Mn_2O_3 \overset{(2TiO_2)}{\rightarrow} 2Mn'_{Ti} + 3O_o + V_o^{\cdot\cdot} \tag{4}$$

$$MnO \overset{(TiO_2)}{\longrightarrow} Mn'_{Ti} + O_o + V_o^{\cdot\cdot} \tag{5}$$

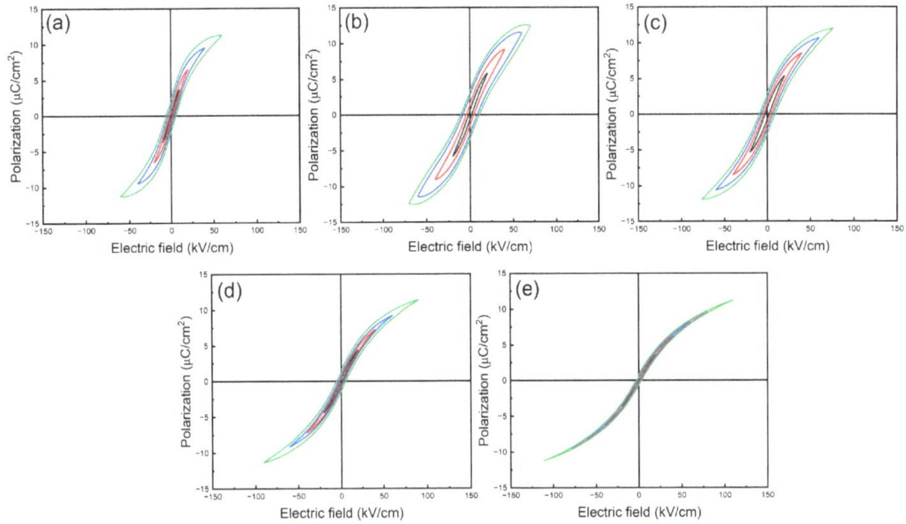

Figure 8. RT bipolar *P-E* loops of $(Ba_{0.7}Sr_{0.3})_{1-x}Nd_xTi_{1-y}Mn_yO_3$ ceramics measured under different electric fields at 10 Hz for (**a**) $x = 0$ and $y = 0$, (**b**) $x = 0.005$ and $y = 0$, (**c**) $x = 0.005$ and $y = 0.0025$, (**d**) $x = 0.005$ and $y = 0.005$, and (**e**) $x = 0.005$ and $y = 0.01$.

Table 2. Ferroelectric and energy storage parameters of BSNTM ceramics.

Composition	P_r (μC/cm²)	P_{max} (μC/cm²)	$\Delta P = P_{max} - P_r$	E_c (kV/cm)	E_{BD} (kV/cm)	W_{rec} (J/cm³)	η (%)
$x = 0$ and $y = 0$	2.22	11.2	8.98	5.94	60.4	0.15	48.5
$x = 0.005$ and $y = 0$	3.35	12.5	9.15	11.2	70.6	0.19	38.8
$x = 0.005$ and $y = 0.0025$	2.87	11.9	9.03	9.34	76.6	0.22	48.9
$x = 0.005$ and $y = 0.005$	1.57	11.4	9.83	5.45	90.5	0.3	69.4
$x = 0.005$ and $y = 0.01$	0.81	11.2	10.39	3.09	110.6	0.41	84.6

This shows that the oxygen vacancies are generated by Mn^{3+} and Mn^{2+} replacing Ti^{4+} at the B-site. In Equations (4) and (5), $2Ti^{4+}$ needs four lattice oxygen O_o to maintain charge neutrality, whereas $2Mn^{3+}$ requires $3O_o$. When $2Mn^{3+}$ substitutes at $2Ti^{4+}$, $1O_o$ is released as $\frac{1}{2}$ O_2, generating oxygen vacancies, $V_o^{\cdot\cdot}$, with two positive charges. Thus, Mn^{2+} replaces Ti^{4+} (Equation (5)) [46].

The W_{rec} and η values of BSNTM ceramic capacitors were derived from *P-E* loops using Equations (2) and (3), as shown in Table 2. The W_{rec} and η values gradually increased with the increasing Mn concentration, and the sample with $x = 0.005$ and $y = 0.01$ exhibited a high energy density of 0.41 J/cm³ at an E_{BD} of 110.6 kV/cm, and a high energy efficiency of 84.6%, as shown in Figure 9e. The enhancement in the energy storage properties is realized using defect dipole engineering via the co-doping of Nd and Mn in BST (mostly governed by Mn). $(Mn''_{Ti} - V_O^{\cdot\cdot})^{\cdot}$ and $(Mn''_{Ti} - V_O^{\cdot\cdot})$ defect dipoles between acceptor ions and oxygen vacancies can capture electrons and improve BDS. In addition, the defect dipoles act as a driving force for depolarization, making it possible to design domain formation energy and domain wall energy, which provides a high difference between P_{max} and P_r ($\Delta P = 10.39$ μC/cm²) [46]. Moreover, complex defect dipoles with optimum oxygen vacancies can provide not only a high ΔP, but also reduce grain size, which together improve breakdown strength with Mn and lead to a high energy storage density and high energy efficiency in BSNTM ceramics. It is well known that ΔP and E_{BD} are key factors for energy storage performance, i.e., higher ΔP and E_{BD} values account for huge energy storage density and efficiency [53].

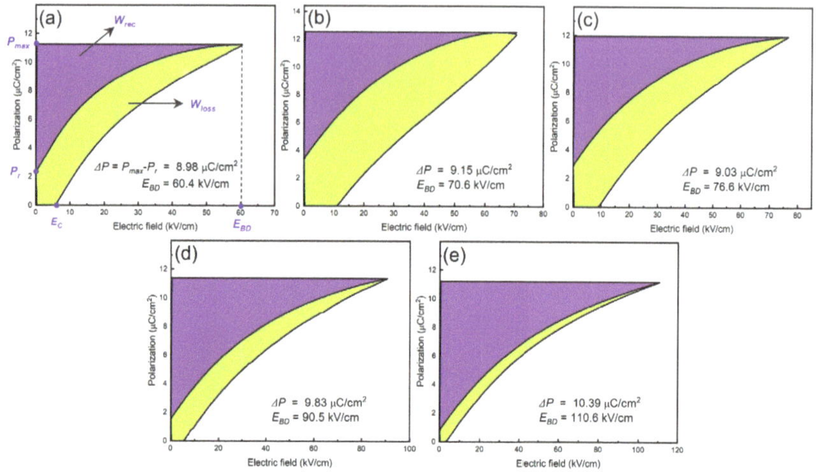

Figure 9. RT unipolar *P-E* loops of $(Ba_{0.7}Sr_{0.3})_{1-x}Nd_xTi_{1-y}Mn_yO_3$ ceramics measured at 10 Hz for (**a**) $x = 0$ and $y = 0$, (**b**) $x = 0.005$ and $y = 0$, (**c**) $x = 0.005$ and $y = 0.0025$, (**d**) $x = 0.005$ and $y = 0.005$, and (**e**) $x = 0.005$ and $y = 0.01$.

X-ray photoelectron spectroscopy (XPS) measurement was carried out to show the chemical states of Nd- and Mn-doped BST ceramics. Figure 10 shows the XPS spectra of BSNTM ceramics. Twelve peaks in terms of binding energy were detected, which correspond to those of Ba, Sr, Ti, O and C elements. Among these, C may arise from carbon pollution in the air and instruments. However, no peaks were detected for Nd and Mn elements due to the very low doping concentration ($x = 0.005$ and $y = 0.0025$–0.01). The inset of Figure 10 shows the high-resolution XPS spectra of O 1s, which was divided into two distinct peaks: the first peak (O_I) found at 529.08 eV indicates an oxygen lattice, and the second peak (O_{II}) at 531.16 eV is related to oxygen vacancies [54]. The relative intensity of the BSNTM ceramics increases with Mn doping, and the $x = 0.005$ and $y = 0.01$ sample shows much a stronger relative intensity than that of other samples, which clearly indicates that the sample processes high oxygen vacancies (inset Figure 10) [55,56].

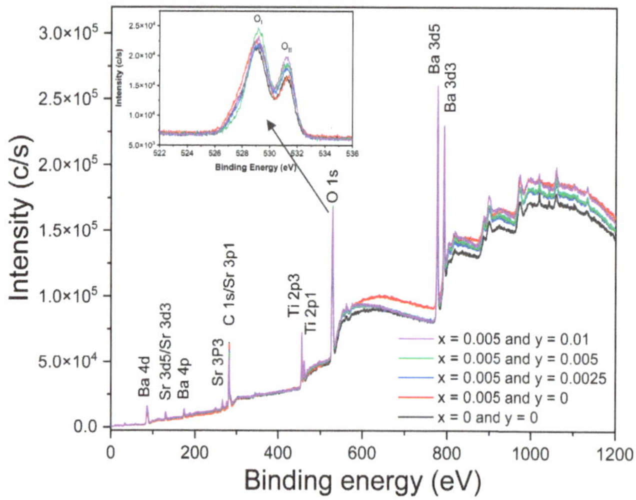

Figure 10. XPS spectra of $(Ba_{0.7}Sr_{0.3})_{1-x}Nd_xTi_{1-y}Mn_yO_3$ ceramics. Inset figure represents high-resolution XPS spectra of O 1s.

4. Conclusions

In summary, we demonstrated a defect dipole engineering method to improve the breakdown strength and energy storage properties by co-doping Nd and Mn in BST ceramics, which are fabricated via a traditional solid-stated reaction method. The XRD and Raman spectra of all samples revealed a tetragonal crystal structure. FESEM images of BSNTM ceramics exhibit a uniform and dense microstructure, whereas the average grain size decreases with an increasing Mn concentration. In addition, the dielectric properties decreased with Mn due to the formation of oxygen vacancies, which were confirmed via XPS analysis. Moreover, the complex defect dipoles with smaller grain sizes and lower dielectric losses provided a high difference between P_{max} and P_r, and improved the breakdown strength with Mn, leading to high energy density and efficiency in the BSNTM ceramics. These features suggest that defect dipole engineering is an effective approach to enhance energy storage performance for pulsed-power capacitor applications.

Author Contributions: Measurement, data curation, investigation, and visualization: H.C. Formal analysis, writing—original draft preparation, review, and editing: S.P. Data curation, and investigation: Y.H.S. Data curation, and investigation: Y.M.B. Visualization, writing—review and editing: J.H.P. Investigation, writing—review and editing: C.K.J. Validation, writing—review and editing: H.E.L. Measurement, writing—review and editing: S.-D.K. Conceptualization, validation, writing—review and editing: J.R. Conceptualization, validation, project administration, supervision, writing—review and editing: G.-T.H. All authors have read and agreed to the published version of the manuscript.

Funding: The work at PKNU was supported by the National Research Foundation of Korea (NRF-2022R1A2C4001497) grant funded by the Ministry of Science and ICT (MSIT) and the Basic Science Research Program through the National Research Foundation of Korea (NRF) funded by the Ministry of Education (gRS-2023-00245515). The work conducted at YU was supported by the National Research Foundation of Korea (NRF) grant funded by the Korean government (MSIT) (No. 2023R1A2C2005864).

Institutional Review Board Statement: Not applicable.

Informed Consent Statement: Not applicable.

Data Availability Statement: Not applicable.

Conflicts of Interest: The authors declare no conflict of interest.

References

1. Pan, H.; Li, F.; Liu, Y.; Zhang, Q.; Wang, M.; Lan, S.; Zheng, Y.; Ma, J.; Gu, L.; Shen, Y.; et al. Ultrahigh–Energy Density Lead-Free Dielectric Films via Polymorphic Nanodomain Design. *Science* **2019**, *365*, 578–582. [CrossRef] [PubMed]
2. Yao, F.-Z.; Yuan, Q.; Wang, Q.; Wang, H. Multiscale Structural Engineering of Dielectric Ceramics for Energy Storage Applications: From Bulk to Thin Films. *Nanoscale* **2020**, *12*, 17165–17184. [CrossRef] [PubMed]
3. Li, Q.; Yao, F.-Z.; Liu, Y.; Zhang, G.; Wang, H.; Wang, Q. High-Temperature Dielectric Materials for Electrical Energy Storage. *Annu. Rev. Mater. Res.* **2018**, *48*, 219–243. [CrossRef]
4. Zhao, P.; Cai, Z.; Chen, L.; Wu, L.; Huan, Y.; Guo, L.; Li, L.; Wang, H.; Wang, X. Ultra-High Energy Storage Performance in Lead-Free Multilayer Ceramic Capacitors via a Multiscale Optimization Strategy. *Energy Env. Sci.* **2020**, *13*, 4882–4890. [CrossRef]
5. Hu, Y.; Rabelo, M.; Kim, T.; Cho, J.; Choi, J.; Fan, X.; Yi, J. Ferroelectricity Based Memory Devices: New-Generation of Materials and Applications. *Trans. Electr. Electron. Mater.* **2023**, *24*, 271–278. [CrossRef]
6. Kumar, N.; Ionin, A.; Ansell, T.; Kwon, S.; Hackenberger, W.; Cann, D. Multilayer Ceramic Capacitors Based on Relaxor $BaTiO_3$-$Bi(Zn_{1/2}Ti_{1/2})O_3$ for Temperature Stable and High Energy Density Capacitor Applications. *Appl. Phys. Lett.* **2015**, *106*, 252901. [CrossRef]
7. Li, J.; Li, F.; Xu, Z.; Zhang, S. Multilayer Lead-Free Ceramic Capacitors with Ultrahigh Energy Density and Efficiency. *Adv. Mater.* **2018**, *30*, 1802155. [CrossRef]
8. Zhu, L.-F.; Zhao, L.; Yan, Y.; Leng, H.; Li, X.; Cheng, L.-Q.; Xiong, X.; Priya, S. Composition and Strain Engineered $AgNbO_3$-Based Multilayer Capacitors for Ultra-High Energy Storage Capacity. *J. Mater. Chem. A Mater.* **2021**, *9*, 9655–9664. [CrossRef]
9. Cao, W.; Lin, R.; Chen, P.; Li, F.; Ge, B.; Song, D.; Zhang, J.; Cheng, Z.; Wang, C. Phase and Band Structure Engineering via Linear Additive in NBT-ST for Excellent Energy Storage Performance with Superior Thermal Stability. *ACS Appl. Mater. Interfaces* **2022**, *14*, 54051–54062. [CrossRef]
10. Zhao, L.; Liu, Q.; Gao, J.; Zhang, S.; Li, J. Lead-Free Antiferroelectric Silver Niobate Tantalate with High Energy Storage Performance. *Adv. Mater.* **2017**, *29*, 1701824. [CrossRef]

11. Zheng, T.; Yu, Y.; Lei, H.; Li, F.; Zhang, S.; Zhu, J.; Wu, J. Compositionally Graded KNN-Based Multilayer Composite with Excellent Piezoelectric Temperature Stability. *Adv. Mater.* **2022**, *34*, 2109175. [CrossRef] [PubMed]
12. Hu, Q.; Tian, Y.; Zhu, Q.; Bian, J.; Jin, L.; Du, H.; Alikin, D.O.; Shur, V.Y.; Feng, Y.; Xu, Z.; et al. Achieve Ultrahigh Energy Storage Performance in $BaTiO_3$–$Bi(Mg_{1/2}Ti_{1/2})O_3$ Relaxor Ferroelectric Ceramics via Nano-Scale Polarization Mismatch and Reconstruction. *Nano Energy* **2020**, *67*, 104264. [CrossRef]
13. Qi, H.; Zuo, R. Linear-like Lead-Free Relaxor Antiferroelectric $(Bi_{0.5}Na_{0.5})TiO_3$-$NaNbO_3$ with Giant Energy-Storage Density/Efficiency and Super Stability against Temperature and Frequency. *J. Mater. Chem. A Mater.* **2019**, *7*, 3971–3978. [CrossRef]
14. Wang, D.; Fan, Z.; Zhou, D.; Khesro, A.; Murakami, S.; Feteira, A.; Zhao, Q.; Tan, X.; Reaney, I.M. Bismuth Ferrite-Based Lead-Free Ceramics and Multilayers with High Recoverable Energy Density. *J. Mater. Chem. A Mater.* **2018**, *6*, 4133–4144. [CrossRef]
15. Zhao, P.; Wang, H.; Wu, L.; Chen, L.; Cai, Z.; Li, L.; Wang, X. High-Performance Relaxor Ferroelectric Materials for Energy Storage Applications. *Adv. Energy Mater.* **2019**, *9*, 1803048. [CrossRef]
16. Inam, F.; Yan, H.; Jayaseelan, D.D.; Peijs, T.; Reece, M.J. Electrically Conductive Alumina–Carbon Nanocomposites Prepared by Spark Plasma Sintering. *J. Eur. Ceram. Soc.* **2010**, *30*, 153–157. [CrossRef]
17. Lu, D.-Y.; Yin, S.; Cui, S.-Z. A Fine-Grained and Low-Loss X8R $(Ba_{1-x}Dy_x)(Ti_{1-x/2}Ca_{x/2})O_3$ Ceramic. *J. Alloys Compd.* **2018**, *762*, 282–288. [CrossRef]
18. Song, Z.; Liu, H.; Zhang, S.; Wang, Z.; Shi, Y.; Hao, H.; Cao, M.; Yao, Z.; Yu, Z. Effect of Grain Size on the Energy Storage Properties of $(Ba_{0.4}Sr_{0.6})TiO_3$ Paraelectric Ceramics. *J. Eur. Ceram. Soc.* **2014**, *34*, 1209–1217. [CrossRef]
19. Lee, H.Y.; Cho, K.H.; Nam, H.-D. Grain Size and Temperature Dependence of Electrical Breakdown in $BaTiO_3$ Ceramic. *Ferroelectrics* **2006**, *334*, 165–169. [CrossRef]
20. Su, X.; Riggs, B.C.; Tomozawa, M.; Nelson, J.K.; Chrisey, D.B. Preparation of $BaTiO_3$ /Low Melting Glass Core–Shell Nanoparticles for Energy Storage Capacitor Applications. *J. Mater. Chem. A* **2014**, *2*, 18087–18096. [CrossRef]
21. Meng, L.; Zheng, L.; Cheng, L.; Li, G.; Huang, L.; Gu, Y.; Zhang, F. Synthesis of Novel Core-Shell Nanocomposites for Fabricating High Breakdown Voltage ZnO Varistors. *J. Mater. Chem.* **2011**, *21*, 11418. [CrossRef]
22. Shi, L.N.; Ren, Z.H.; Jain, A.; Jin, R.H.; Jiang, S.S.; Zhou, H.Z.; Chen, F.G.; Wang, Y.G. Enhanced Energy Storage Performance Achieved in $Na_{0.5}Bi_{0.5}TiO_3$–$Sr_{0.7}Bi_{0.2}TiO_3$ Ceramics via Domain Structure and Bandgap Width Tuning. *Ceram. Int.* **2023**, *49*, 12822–12831. [CrossRef]
23. Wang, G.; Li, J.; Zhang, X.; Fan, Z.; Yang, F.; Feteira, A.; Zhou, D.; Sinclair, D.C.; Ma, T.; Tan, X.; et al. Ultrahigh Energy Storage Density Lead-Free Multilayers by Controlled Electrical Homogeneity. *Energy Environ. Sci.* **2019**, *12*, 582–588. [CrossRef]
24. Yuan, Q.; Li, G.; Yao, F.-Z.; Cheng, S.-D.; Wang, Y.; Ma, R.; Mi, S.-B.; Gu, M.; Wang, K.; Li, J.-F.; et al. Simultaneously Achieved Temperature-Insensitive High Energy Density and Efficiency in Domain Engineered $BaTiO_3$-$Bi(Mg_{0.5}Zr_{0.5})O_3$ Lead-Free Relaxor Ferroelectrics. *Nano Energy* **2018**, *52*, 203–210. [CrossRef]
25. Asbani, B.; Gagou, Y.; Ben Moumen, S.; Dellis, J.-L.; Lahmar, A.; Amjoud, M.; Mezzane, D.; El Marssi, M.; Rozic, B.; Kutnjak, Z. Large Electrocaloric Responsivity and Energy Storage Response in the Lead-Free $Ba(Ge_xTi_{1-x})O_3$ Ceramics. *Materials* **2022**, *15*, 5227. [CrossRef] [PubMed]
26. Yin, J.; Zhang, Y.; Lv, X.; Wu, J. Ultrahigh Energy-Storage Potential under Low Electric Field in Bismuth Sodium Titanate-Based Perovskite Ferroelectrics. *J. Mater. Chem. A Mater.* **2018**, *6*, 9823–9832. [CrossRef]
27. Lu, Y.; Zhang, H.; Yang, H.; Fan, P.; Samart, C.; Takesue, N.; Tan, H. SPS-Prepared High-Entropy $(Bi_{0.2}Na_{0.2}Sr_{0.2}Ba_{0.2}Ca_{0.2})TiO_3$ Lead-Free Relaxor-Ferroelectric Ceramics with High Energy Storage Density. *Crystals* **2023**, *13*, 445. [CrossRef]
28. Jiang, Y.; Niu, X.; Liang, W.; Jian, X.; Shi, H.; Li, F.; Zhang, Y.; Wang, T.; Gong, W.; Zhao, X.; et al. Enhanced Energy Storage Performance in $Na_{0.5}Bi_{0.5}TiO_3$-Based Relaxor Ferroelectric Ceramics via Compositional Tailoring. *Materials* **2022**, *15*, 5881. [CrossRef]
29. Pattipaka, S.; Choi, H.; Lim, Y.; Park, K.-I.; Chung, K.; Hwang, G.-T. Enhanced Energy Storage Performance and Efficiency in $Bi_{0.5}(Na_{0.8}K_{0.2})_{0.5}TiO_3$-$Bi_{0.2}Sr_{0.7}TiO_3$ Relaxor Ferroelectric Ceramics via Domain Engineering. *Materials* **2023**, *16*, 4912. [CrossRef]
30. Yang, Z.; Wang, B.; Li, Y.; Hall, D.A. Enhancement of Nonlinear Dielectric Properties in $BiFeO_3$–$BaTiO_3$ Ceramics by Nb-Doping. *Materials* **2022**, *15*, 2872. [CrossRef]
31. Chen, Q.; Gao, T.; Lang, R.; Tan, Z.; Xing, J.; Zhu, J. Achieving Outstanding Temperature Stability in KNN-Based Lead-Free Ceramics for Energy Storage Behavior. *J. Eur. Ceram. Soc.* **2023**, *43*, 2442–2451. [CrossRef]
32. Yang, Z.; Du, H.; Qu, S.; Hou, Y.; Ma, H.; Wang, J.; Wang, J.; Wei, X.; Xu, Z. Significantly Enhanced Recoverable Energy Storage Density in Potassium–Sodium Niobate-Based Lead Free Ceramics. *J. Mater. Chem. A Mater.* **2016**, *4*, 13778–13785. [CrossRef]
33. Li, X.; Chen, X.; Sun, J.; Zhou, M.; Zhou, H. Novel Lead-Free Ceramic Capacitors with High Energy Density and Fast Discharge Performance. *Ceram. Int.* **2020**, *46*, 3426–3432. [CrossRef]
34. Hu, D.; Pan, Z.; Tan, X.; Yang, F.; Ding, J.; Zhang, X.; Li, P.; Liu, J.; Zhai, J.; Pan, H. Optimization the Energy Density and Efficiency of $BaTiO_3$-Based Ceramics for Capacitor Applications. *Chem. Eng. J.* **2021**, *409*, 127375. [CrossRef]
35. Liu, B.; Wang, X.; Zhao, Q.; Li, L. Improved Energy Storage Properties of Fine-Crystalline $BaTiO_3$ Ceramics by Coating Powders with Al_2O_3 and SiO_2. *J. Am. Ceram. Soc.* **2015**, *98*, 2641–2646. [CrossRef]
36. Huang, Y.H.; Wu, Y.J.; Qiu, W.J.; Li, J.; Chen, X.M. Enhanced Energy Storage Density of $Ba_{0.4}Sr_{0.6}TiO_3$–MgO Composite Prepared by Spark Plasma Sintering. *J. Eur. Ceram. Soc.* **2015**, *35*, 1469–1476. [CrossRef]

37. Zhou, L.; Vilarinho, P.M.; Baptista, J.L. Dependence of the Structural and Dielectric Properties of Ba1-XSrxTiO_3 Ceramic Solid Solutions on Raw Material Processing. *J. Eur. Ceram. Soc.* **1999**, *19*, 2015–2020. [CrossRef]
38. Kovbasiuk, T.M.; Duriagina, Z.A.; Mierzwinski, D.; Kulyk, V.V. Thermophysical Properties of Glass-Ceramic Coatings of PbO–ZnO–B_2O_3 System Doped with Al_2O_3, SiO_2, and BaO Oxides. *Metallofiz. Noveishie Tekhnol.* **2021**, *43*, 1313–1323. [CrossRef]
39. Moussi, R.; Bougoffa, A.; Trabelsi, A.; Dhahri, E.; Graça, M.P.F.; Valente, M.A.; Barille, R.; Rguiti, M. Investigation of the Effect of Sr-Substitution on the Structural, Morphological, Dielectric, and Energy Storage Properties of $BaTiO_3$-Based Perovskite Ceramics. *Inorg. Chem. Commun.* **2022**, *137*, 109225. [CrossRef]
40. Batllo, F.; Duverger, E.; Jules, J.-C.; Niepce, J.-C.; Jannot, B.; Maglione, M. Dielectric and E.P.R. Studies of Mn-Doped Barium Titanate. *Ferroelectrics* **1990**, *109*, 113–118. [CrossRef]
41. Morrison, F.D.; Sinclair, D.C.; West, A.R. Doping Mechanisms and Electrical Properties of La-Doped $BaTiO_3$ Ceramics. *Int. J. Inorg. Mater.* **2001**, *3*, 1205–1210. [CrossRef]
42. Shaikh, A.S.; Vest, R.W. Defect Structure and Dielectric Properties of Nd_2O_3-Modified $BaTiO_3$. *J. Am. Ceram. Soc.* **1986**, *69*, 689–694. [CrossRef]
43. Jayanthi, S.; Kutty, T.R.N. Dielectric Properties of 3d Transition Metal Substituted $BaTiO_3$ Ceramics Containing the Hexagonal Phase Formation. *J. Mater. Sci. Mater. Electron.* **2008**, *19*, 615–626. [CrossRef]
44. Wang, X.; Gu, M.; Yang, B.; Zhu, S.; Cao, W. Hall Effect and Dielectric Properties of Mn-Doped Barium Titanate. *Microelectron. Eng.* **2003**, *66*, 855–859. [CrossRef]
45. Čeh, M.; Kolar, D. Solubility of Calcium Oxide in Barium Titanate. *Mater. Res. Bull.* **1994**, *29*, 269–275. [CrossRef]
46. Zhao, Y.; Yang, B.; Liu, Y.; Zhou, Y.; Wu, Q.; Zhao, S. Capturing Carriers and Driving Depolarization by Defect Engineering for Dielectric Energy Storage. *ACS Appl. Mater. Interfaces* **2022**, *14*, 6547–6559. [CrossRef] [PubMed]
47. Liu, Q.; Liu, J.; Lu, D.; Zheng, W.; Hu, C. Structural Evolution and Dielectric Properties of Nd and Mn Co-Doped $BaTiO_3$ Ceramics. *J. Alloys Compd.* **2018**, *760*, 31–41. [CrossRef]
48. Rini, E.G.; Paul, A.; Nasir, M.; Amin, R.; Gupta, M.K.; Mittal, R.; Sen, S. Correlation of Octahedral Distortion with Vibrational and Electronic Properties of $LaFe_{1-x}Ti_xO_3$ Nanoparticles. *J. Alloys Compd.* **2020**, *830*, 154594. [CrossRef]
49. Jo, S.K.; Park, J.S.; Han, Y.H. Effects of Multi-Doping of Rare-Earth Oxides on the Microstructure and Dielectric Properties of $BaTiO_3$. *J. Alloys Compd.* **2010**, *501*, 259–264. [CrossRef]
50. Huang, Y.; Zhao, C.; Wu, B.; Wu, J. Multifunctional $BaTiO_3$-Based Relaxor Ferroelectrics toward Excellent Energy Storage Performance and Electrostrictive Strain Benefiting from Crossover Region. *ACS Appl. Mater. Interfaces* **2020**, *12*, 23885–23895. [CrossRef]
51. Sun, Z.; Wang, Z.; Tian, Y.; Wang, G.; Wang, W.; Yang, M.; Wang, X.; Zhang, F.; Pu, Y. Progress, Outlook, and Challenges in Lead-Free Energy-Storage Ferroelectrics. *Adv. Electron. Mater.* **2020**, *6*, 1900698. [CrossRef]
52. Cha, S.H.; Han, Y.H. Effects of Mn Doping on Dielectric Properties of Mg-Doped $BaTiO_3$. *J. Appl. Phys.* **2006**, *100*, 104102. [CrossRef]
53. Li, J.; Shen, Z.; Chen, X.; Yang, S.; Zhou, W.; Wang, M.; Wang, L.; Kou, Q.; Liu, Y.; Li, Q.; et al. Grain-Orientation-Engineered Multilayer Ceramic Capacitors for Energy Storage Applications. *Nat. Mater.* **2020**, *19*, 999–1005. [CrossRef]
54. Ko, Y.-J.; Han, M.H.; Lim, C.; Yu, S.-H.; Choi, C.H.; Min, B.K.; Choi, J.-Y.; Lee, W.H.; Oh, H.-S. Unveiling the Role of Ni in Ru-Ni Oxide for Oxygen Evolution: Lattice Oxygen Participation Enhanced by Structural Distortion. *J. Energy Chem.* **2023**, *77*, 54–61. [CrossRef]
55. Gan, J.; Lu, X.; Wu, J.; Xie, S.; Zhai, T.; Yu, M.; Zhang, Z.; Mao, Y.; Wang, S.C.I.; Shen, Y.; et al. Oxygen Vacancies Promoting Photoelectrochemical Performance of In_2O_3 Nanocubes. *Sci. Rep.* **2013**, *3*, 1021. [CrossRef]
56. Chen, X.; Xie, Y.; Zhang, H.; Zhang, F.; Xue, Y.; Chang, A. Resistance-Temperature Characteristics of a New High-Temperature Thermistor Ceramics of Mn-Doping Ba–Ca–Zr–Ti–O System. *J. Mater. Sci. Mater. Electron.* **2021**, *32*, 25094–25107. [CrossRef]

materials

MDPI

Article

Advantages of Ferroelectrics as a Component of Heterostructures for Electronic Purposes: A DFT Insight

Irina Piyanzina [1,2], Alexander Evseev [1], Kirill Evseev [2], Rinat Mamin [2], Oleg Nedopekin [1], Dmitrii Tayurskii [1] and Viktor Kabanov [3,*]

1 Institute of Physics, Kazan Federal University, 420008 Kazan, Russia; i.piyanzina@gmail.com (I.P.); alexander-alexandrovich-evseev@mail.ru (A.E.); nedopekin@gmail.com (O.N.); dtayursk@gmail.com (D.T.)
2 Zavoisky Physical-Technical Institute, FRC Kazan Scientific Center of RAS, 420029 Kazan, Russia; ekv97@mail.ru (K.E.); rf_mamin@yahoo.com (R.M.)
3 Department for Complex Matter, Jozef Stefan Institute, 1000 Ljubljana, Slovenia
* Correspondence: viktor.kabanov@ijs.si

Abstract: The main advantage of using ferroelectric materials as a component of complex heterostructures is the ability to tune various properties of the whole system by means of an external electric field. In particular, the electric field may change the polarization direction within the ferroelectric material and consequently affect the structural properties, which in turn affects the electronic and magnetic properties of the neighboring material. In addition, ferroelectrics allow the electrostriction phenomenon to proceed, which is promising and can be used to affect the magnetic states of the interface state in the heterostructure through a magnetic component. The interfacial phenomena are of great interest, as they provide extended functionality useful for next-generation electronic devices. Following the idea of utilizing ferroelectrics in heterostructural components in the present works, we consider 2DEG, the Rashba effect, the effect of magnetoelectric coupling, and magnetostriction in order to emphasize the advantages of such heterostructures as components of devices. For this purpose, model systems of $LaMnO_3/BaTiO_3$, $La_2CuO_4/BaTiO_3$, $Bi/BaTiO_3$, and $Bi/PbTiO_3$, $Fe/BaTiO_3$ heterostructures are investigated using density functional theory calculations.

Keywords: ferroelectric; heterostructure; density functional theory; 2DEG; ME coupling

Citation: Piyanzina, I.; Evseev, A.; Evseev, K.; Mamin, R.; Nedopekin, O.; Tayurskii, D.; Kabanov, V. Advantages of Ferroelectrics as a Component of Heterostructures for Electronic Purposes: A DFT Insight. *Materials* **2023**, *16*, 6672. https://doi.org/10.3390/ma16206672

Academic Editor: Dana Popescu

Received: 20 September 2023
Revised: 9 October 2023
Accepted: 11 October 2023
Published: 13 October 2023

1. Introduction

The presence of ferroelectric materials as component in heterostructures provides outstanding new functionality which can be used in electronic devices. It is well established that the appearance of a two-dimensional gas (2DEG) or liquid is possible due to the presence of internal electrical polarization. Such polarisation, for instance in $LaAlO_3/SrTiO_3$, arises due to the charge sequence in $LaAlO_3$ atomic layers. However, 2DEG can occur even without charged atomic layers thanks to the presence of spontaneous polarisation in ferroelectric thin films [1–6]. This means that the electronic properties of the arising state can be tuned by an external field by changing the direction of the ferroelectric dipoles.

Another property which can be useful for electronic applications is magnetoelectric coupling. This property is associated with the possibility of controlling the ferromagnetic ordering at the interface due to interactions of spins through conduction electrons, leading to multiferroic properties on the part of the whole heterostructure. Multiferroic materials are compounds in which at least two order parameters coexist in the same phase. One very important and extremely rare group is that of ferroelectric ferromagnets, which have recently stimulated increasing research activity due to their scientific uniqueness and application in novel multifunctional devices. Magnetoelectric materials are mainly interesting due to the possibility of controlling the magnetic properties through an external electric field [7–11]. Due to the extraordinary challenges involved in creating multiferroic compounds, it is essential to create superlattice multicomponent materials, which consist of a magnetic insulator that supports spin-polarized 2DEG, and ferroelectric materials, which can aid in manipulating the magnetic state using an electric field, i.e., to realize a converse magnetoelectric (ME) effect [12].

Moreover, the presence of an electrostatic field in ferroelectrics due to spin–orbit (SO) coupling allows for control of SO splitting via ferroelectric polarization, which is a desirable property for spintronic applications. Natural materials demonstrating both the gigantic and ideal Rashba states are extremely rare [13,14]; computer simulations could help with this problem by allowing for the investigation of various combinations of materials.

A third promising effect for electronic devices purposes is magnetostriction in combination with ferroelectricity. In particular, within the superlattice approach it is possible to combine two materials with different features; by changing the linear sizes of the ferroelectric material by applying an external electric field the lattice parameters of a neighboring ferromagnet, the latter changes as well, leading to a change in the magnetic moments. This possibility promises significant advantages in the development of next-generation electronic devices. For instance, magnetic tunnel junctions (MTJs) are of great interest to both the experimental and theoretical communities due to their applications in magnetic random-access memory (MRAM) devices. Multiferroic materials are suitable for spin filter purposes as well [15,16]. Indeed, previous research has demonstrated that a heterostructure based on iron and a classical ferroelectric material (Fe/$BaTiO_3$) can demonstrate ideal crystallinity and heteroepitaxial growth [17]. Even more significant is that changing the polarization direction is able to affect the magnetization inside a ferromagnetic film [8]. The Fe/$BaTiO_3$ heterostructure has been widely investigated as a simple example of a model system of ferromagnetic/ferroelectric combination [5,8,17–22].

Within the last thirty years, a significant breakthrough in computational methods has been achieved thanks to the success of computer sciences. This has made possible various calculations of electronic and magnetic properties of sufficiently large and complex systems. In particular, the popular density functional theory has been implemented in a wide range of codes. Indeed, the possibilities of the approaches based on this method are very extensive.

Therefore, the present research is dedicated to an ab initio study heterostructures having a ferroelectric material as one of the components within the DFT approach. Our aim is to investigate the arising electronic and magnetic states and the possibilities for controlling the interfacial properties (2DEG, Rashba effect, ME coupling, magnetostriction) via ferroelectric polarization reversal. For this purpose, in this research we investigated $LaMnO_3$/$BaTiO_3$, La_2CuO_4/$BaTiO_3$, Bi/$BaTiO_3$, Bi/$PbTiO_3$, and Fe/$BaTiO_3$ heterostructures in order to demonstrate the effect of polarization switching onto the electronic and magnetic states, as well as the Rashba effect.

2. Materials and Methods

In the present research, calculations of structural, electronic, and magnetic properties were realized within the framework of density functional theory [23]. Exchange and correlation effects were accounted for using generalized gradient approximation (GGA-PBE) [24]. The Kohn–Sham equations [25] were solved using projectively extended wave potentials and wave functions [26]. All calculations were carried out using the VASP-6.3 (Vienna Ab initio Simulation Package) program [27] built into the MedeA computational software [28]. The cut-off of the plane wave was taken to be 400 eV, the convergence criterion for atomic relaxation was 0.02 eV/Å, and the convergence condition for self-consistent calculations was the invariance of the total energy of the system, with an accuracy of 10^{-5} eV. The Brillouin zones were sampled using Monkhorst–Pack grids [29–31], including $7 \times 7 \times 1$ and $5 \times 5 \times 1$ **k**-points, depending on the particular heterostructure under study. The Gaussian smearing was 0.05 eV. A set of calculations was carried out with a simplified $+U$ correction applied [32], which was used to ensure a better description of the electronic properties of strongly correlated electrons; an additional U value was applied to electrons of the d and f orbitals, following [33]. In particular, we applied U = 4.4, 4, 4, 4.6 eV for

3*d* orbitals of Ti, Cu, Mn, and Fe, respectively, and 8 ev for 4*f* orbitals of La. The use of additional correction is essential to ensure a better description of the magnetic state and band gap. The necessity of adding the U parameters while treating the transition metals has been widely discussed in previous works, for example [33–35]. In addition, La 4*f* states are usually shifted up in the energy scale [35].

The model of the heterostructures was constructed in such a way that the BTOs served as an overlayer for $LaMnO_3/BaTiO_3$, $La_2CuO_4/BaTiO_3$, and $LaMnO_4/BaTiO_3$; a vacuum region was added in order to imitate real heterostructures with both interface and surface regions. In the $Bi/BaTiO_3$ and $Bi/PbTiO_3$ heterostructures, the results are presented for structures with no vacuum region in order to provide a comparison with previous research [36]. Lastly, the model of $Fe/BaTiO_3$ was the same as in [18], that is, a superlattice with no added vacuum, in order to focus on the interface and avoid surface impact. It should be noted that before investigating the interface, the energy of formation was checked in order to deal with the most stable among possible surface stakings. The all-slabs terminations at the interface were the most stable among those possible.

3. Results

In this section, the effects of the presence of ferroelectric material on the interfacial conducting state, the magnetic state, the size of Rashba-type splitting, and the reverse magnetostriction and magnetoelectric couplings are presented separately.

3.1. 2DEG

The area of perovskite-based heterostructures was investigated based on the appearance of a two-dimensional conducting state (2DEG) at the interface. The conductivity at the interface occurs due to either the polar nature of one of the components or to the presence of defects [37]. Later, was been shown that 2DEG can be created at the interface of non-polar oxides, one of which is ferroelectric [1]. The main advantage of using ferroelectrics is the possibility of switching the polarization on and off, allowing for control of the properties of the electron system. Moreover, ferroelectrics have a range of other outstanding properties which might expand the scope of applications in nanoscale electronic devices, including spontaneous polarization switching, high dielectric permeability, dielectric nonlinearity, piezo- and pyro- activity, and linear and quadratic electro-optical effects.

There are two systems studied in this frame within the present paper: the heterostructure of antiferromagnet/ferroelectric, i.e., $LaMnO_3/BaTiO_3$ (LMO/BTO), and ferroelectrics with high-temperature superconductors, such as as $La_2CuO_4/BaTiO_3$ (LCO/BTO). Indeed, the creation of a 2DEG is possible when the electrostatic field along the slab is present in the system. This is possible due to the alternating charges in the atomic layers or thanks to the ferroelectric polarization being directed normally to the interface plane.

In both studied systems, the bulk components are insulators; LMO is A-type antiferromagnetic, LCO is a ferromagnetic insulator, and BTO is an insulator [38,39].

The unit cell for the LCO/BTO heterostructure is presented in Figure 1a, which consists of a middle slab of LCO as a substrate and BTO as an overlayer on both sides to make the cell symmetrical with respect to the central layer. It can be seen from the density of states (DOS) in Figure 1d that a conduction state arises, which is provided mostly by oxygen-polarized electrons. These conducting electrons are located mostly within the interfacial CuO layer (Figure 1b,c). In fact, the DOS at the Fermi level gradually increases from the surface of the ferroelectric towards the interfacial CuO layer and monotonously decreases towards the center of the LCO slab. The calculated number of the charge carrier dependence is rather discrete when carriers are present only within one atomic layer, indicating a 2D conducting character.

The other investigated heterostructure was LMO/BTO; it was constructed in the same way, with the antiferromagnet LMO surrounded by BTO overlayers on both sides, as depicted in Figure 2a. The optimization of the cell led to structural distortions associated with buckling within atomic layers. The most pronounced displacements were found close to the interface. These out-of-plane shifts contributed to potential build-up along the BTO overlayer, giving rise to the internal field; however, the resulting electrostatic field was not sufficiently large to promote significant electronic reconstruction and conductivity in the system, as can be seen in Figure 2d, where the DOS at the Fermi-level is zero.

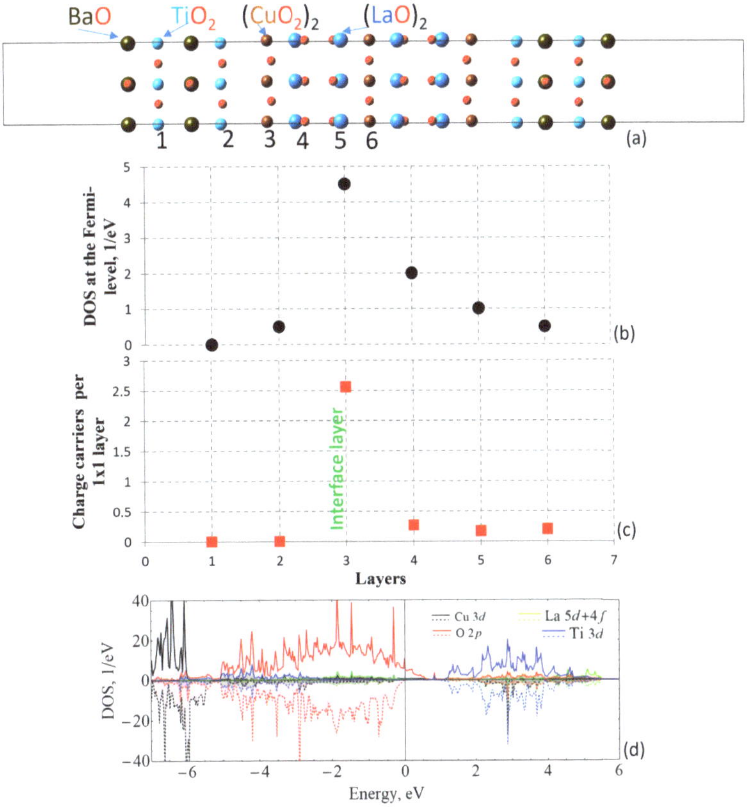

Figure 1. (**a**) The unit cell of the $La_2CuO_4/BaTiO_3$ (LCO/BTO) heterostructure; (**b**) density of states (DOS) per atomic layers as denoted in (**a**); (**c**) charge carriers per 1×1 layer; (**d**) atom-resolved DOS, where the spin components are presented at the upper and bottom parts of the graph. The red squares and black dots correspond to charge carriers and DOS at the Fermi-level respectively.

In the same way, increased polarization due to the artificial displacement of positive ions with respect to the negative oxygen ions immediately results in increased charge carriers at the surface (holes) and interface (electrons), as depicted in Figure 2b. In terms of energy state, the increased polarization shifts the Fermi level upwards (Figure 2d,e), meaning that the Ti and Mn 3*d* states become closer to the Fermi level. Contrarily, polarization towards the interface leads to an opposite situation, with holes located near the interface and electrons located near the both surfaces, as depicted in Figure 2c. The presence or absence of conducting state is mentioned in Table 1.

To sum up, the presence of the electrostatic field of the ferroelectric material is favorable in the systems where 2DEG is a desirable property. There are at least two advantages of such components: first, the arising conducting phase might be switched by external field stimulus, and second, the field within the ferroelectric material is an intrinsic feature, meaning that the requirement of an ideal interface is not indispensable here, which makes growth easier. Consequently, both features are preferred for electronic purposes.

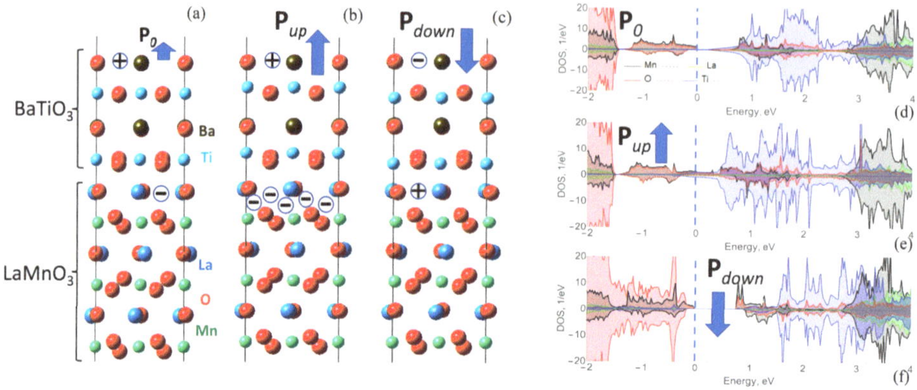

Figure 2. Half unit cells of the $LaMnO_3/BaTiO_3$ (LMO/BTO) heterostructure (**a**) fully optimized and with imposed polarization directed (**b**) towards the surface and (**c**) towards the interface (**d–f**) with corresponding atom-resolved density of states (DOS), where the spin components are presented at the upper and bottom parts of the graph. $\mathbf{P}_0$ denotes the initial polarization, $\mathbf{P}_{up}$ is the polarization directed towards the surface, $\mathbf{P}_{down}$ is the polarization directed towards the surface the interface, and the plus and minus symbols respectively correspond to the positive and negative charge distributions.

Table 1. Conductivity and distribution of magnetic moments per Mn ions over atomic layers within the $LaMnO_3$ slab of the $LaMnO_3/BaTiO_3$ heterostructure; the + and − symbols in the conductivity row respectively denote the presence and absence of the conducting state in the investigated systems.

	$\mathbf{P}_0$	$\mathbf{P}_{down}$	$\mathbf{P}_{up}$
Conductivity	−	−	+
Interfacial layer	−3.725	−4.165	−3.710
Second layer	3.670	3.635	3.705
Middle layer	−4.170	−3.91	−3.710
Second layer	3.675	3.635	3.710
Interface layer	−3.725	−4.165	−3.630

3.2. Reverse Magnetoelectric (ME) Coupling

As mentioned in the introduction, multiferroics are of a great scientific interest due to the wide range of their physical properties. Furthermore, this class of materials has a great potential for applications in switches, magnetic field sensors, and memory devices [40]. However, pure multiferroics are rare, necessitating searching among those multicomponent superlattices mainly constructed from ferroelectrics and magnets.

In the previous section, it was clearly demonstrated that the change in ferroelectric polarization direction switches the conductivity on and off. To test the possibilities, the distribution of magnetic moments within the antiferromagnetic slab of LMO was checked. It has to be noted that while LMO is a pure antiferromagnet in the bulk, due to the odd number of five MnO layers in the slab in the slab geometry used in the present work, the order is that of an unsaturated antiferromagnetic in total. This is different from the situation described in [4], for instance, where an infinite cell without a vacuum region was used for investigation. In the model constructed within the present research, the effect of changed polarization direction was observed. As listed in Table 1, the initial optimized LMO/BTO heterostructure possesses alternating magnetic moments directed along and opposite to the *z*-axis. This order is preserved for other considered cases as well, however, the magnitude changes. While these changes are insignificant, the situation might change in the superlattice geometry, which needs to be checked in further investigations.

To conclude, the ab initio observation of magnetic moments switching through the reverse magnetoelectric coupling requires an adjustment to the geometry of the heterostructure. In particular, in the LMO/BTO structure considered here, the change in polarization direction does not change the direction of the magnetic moments of Mn ions, though it does change the amplitude.

3.3. Rashba Effect

The Rashba effect constitutes the splitting of the electron conduction band along the spin due to the spin–orbit interaction. The effect is observed in structures where an effective electric field is present. Due to the presence of spin-orbit interaction, these internal electric fields lead to a splitting of the electronic states along the axis of the wave vectors. As a result, two dispersion surfaces are formed, which are connected at one Dirac point. A large and ideal Rashba-type splitting is desirable for applications in spintronic devices. There are a number of approaches that exist for enhancing this splitting. In particular, introduction of heavy elements as components of heterostructure may lead to an increase in the strength of the spin–orbit (SO) coupling [41]. Another approach is to use polar semiconductors as a substrate for the heterostructure [42,43]. This avoids mixing of the Rashba states and the spin-degenerate substrate states, allowing for the creation of so-called ideal Rashba states. The last and most promising approach is to use ferroelectric materials to enhance the electric field across the heterostructure [44]. Furthermore, the use of ferroelectrics allows for tuning the polarization, which in turn may lead to changes in the strength of SO coupling.

In the present research, two heterostructures were considered, $BaTiO_3$/Bi and $PbTiO_3$/Bi, both consisting of a ferroelectric substrate and heavy metal monolayer. The $BaTiO_3$/Bi heterostructure was investigated previously in [36,45]. Here, we followed the same heterostructure model except with the addition of a vacuum region. The structures of modeled BTO/Bi and PTO/Bi cells, along with corresponding band structures, are presented in Figure 3. One band with splitting is presented, which is the same for the S_x and S_y components; as the out of plane component S_z is negligibly small, the spins are located and split within the interfacial plane. Two path directions in the Brillouin Zone were found, namely, X–G and G–M, where G is a gamma point.

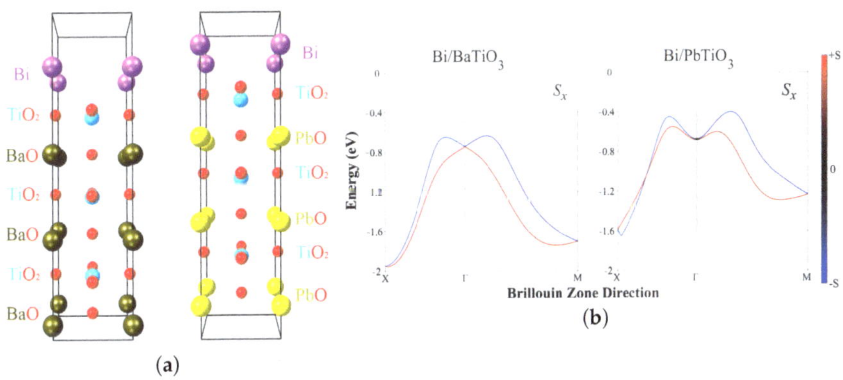

Figure 3. (**a**) The unit cells and (**b**) the corresponding band structures of the investigated BTO/Bi and PTO/Bi heterointerfaces. Only the S_x component of the split band is presented here, as the S_y component is the same and the S_z component is zero. The red and blue colors correspond to the spin-up and spin-down components.

Dispersions of surface states with characteristic features of Rashba-type splitting (shifting of the energy E_R and Rashba wave vector k_0) were found for both the $BaTiO_3$/Bi and $PbTiO_3$/Bi heterointerfaces. All data are collected in Table 2 along with effective masses and ab initio data from [45].

All obtained results for Bi/BTO are in qualitative agreement with previously published data; the difference might be due to differences in the computational parameters. The replacement of BTO by PTO increases the Rashba parameter by a factor of $\approx$1.7, which is a significant increase. Indeed, in PTO the ferroelectric polarization associated with displacements of positively charged Ti ions out of negatively charged oxygen planes is significantly higher, as can be seen in Figure 3. Thus, the electrostatic field along the cell and perpendicular to the interface plane is higher, leading to more splitting.

Table 2. Calculated Rashba splitting parameters, where E_R is the Rashba energy, k_0 is the momentum offset, α_R is the Rashba parameter, m* is the effective mass, and X–G and G–M denote the path in the Brillouin Zone. The results are shown for both fully optimized heterostructures of $BaTiO_3$/Bi and $PbTiO_3$/Bi with optimized polarization directed towards the interface.

Path	X–G				G–M			
	E_R, eV	k_0, Å^{-1}	α_R, eVÅ	m*, m_e	E_R, eV	k_0, Å^{-1}	α_R, eVÅ	m*, m_e
Bi/BTO	0.1	0.19	1.05	1.35	0.11	0.22	1.01	1.65
Bi/PTO	0.23	0.27	1.72	1.2	0.28	0.35	1.63	1.61
Bi/BTO [45]	0.16	0.22	1.45	1.14	0.18	0.25	1.42	1.36

Further, the Bi monolayer was checked separately in order to ensure that the splitting occurs only when both spin–orbit coupling and a field perpendicular to the interface plane coexist in the heterostructure. The Bi monolayer with corresponding band structure is presented in Figure 4, where the splitting is present but is not of Rashba-type.

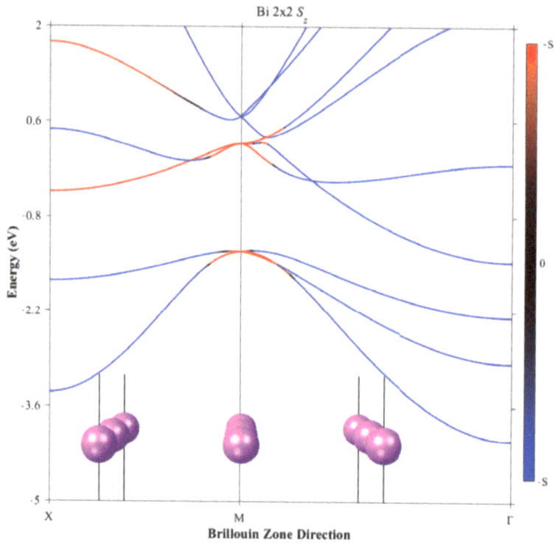

Figure 4. The unit cells and corresponding band structure of a 2 × 2 Bi mono-layer with no Rashba-type splitting. The red and blue colors correspond to the spin-up and spin-down components.

Finally, we checked the influence of applying a reverse field by shifting in the opposite direction with respect to the oxygen planes and freezing the atoms in the ferroelectric slab. This resulted in similar values to those listed in Table 2. However, the increase in ferroelectric polarization might increase the splitting. This could be a subject of further investigations, along with other combinations of heavy elements and different ferroelectrics.

3.4. *Reverse Magnetostriction Effect*

We chose the $Fe/BaTiO_3$ model system was to study the effect of linear compression sizes of the ferroelectric material on the magnetic properties of the ferromagnet. The unit cell used in the calculations is depicted in Figure 5a, where BTO acts as a ferroelectric and iron acts as a ferromagnetic overlayer. In this case, the unit cell was constructed without a vacuum region in order to exclude the impact of the surface and concentrate on the ferroelectric slab properties. The system was constructed to include a total of eighteen atomic layers in the periodic structure, seven of which were layers of bcc iron.

The initial value of the lattice parameter was equal to 3.9 Å, corresponding to the BTO substrate (the lattice mismatch was 1.36%). We calculated the values of the magnetic moments of iron atoms in the BTO/Fe heterostructure without striction, taking into account spin–orbit interactions; the data are shown in Figure 5b. After that, the entire heterostructure was simultaneously compressed along the *x*-axis by 0.2 Å and expanded along the *y*-axis by the same value. After an optimization process taking into account spin–orbit interactions, the magnetic moments of the iron atoms in each layer of the ferromagnet were calculated; these data are presented in Figure 5c. Taking spin–orbit interactions into account during the calculation makes it possible to obtain magnetic moment values that have different directions, and not just the total magnetization value of the Fe layers.

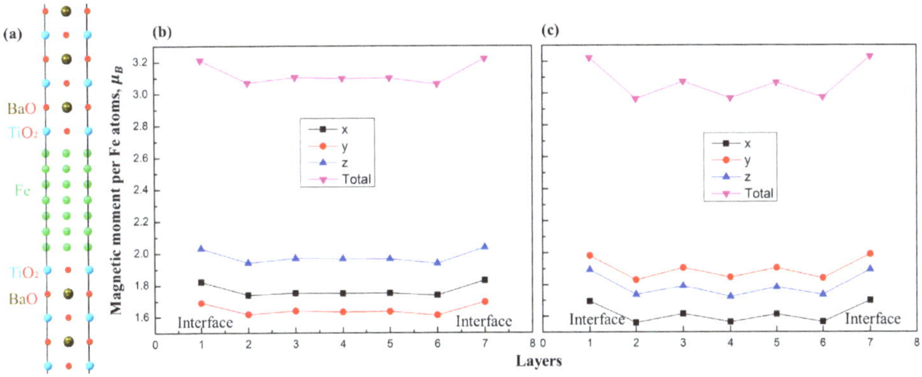

Figure 5. (**a**) The unit cell of Fe/BaTiO$_3$ heterostucture used in calculations: (**b**) magnetic moments of Fe atoms in the heterostructure without applied striction and (**c**) corresponding distribution of the magnetic moments of Fe atoms calculated for each layer of Fe/BTO heterostructure within the applied in-plane striction along the *x*-axis and extension along the *y*-axis. Each curve corresponds to the resulted magnetisation magnitude for the *x*, *y*, and *z* components of the magnetization vector and total magnetic moment in the scalar norm.

The Figure 5b shows that all curves are symmetrical with respect to the middle of the layer and have the same character of a slight increase near the interfaces. The maximum value of the total magnetic moment was found to be $\approx 3.22\,\mu_B$, which is $0.23\,\mu_B$, higher than the initial magnetic moment of iron ($2.99\,\mu_B$) calculated per Fe ion. The magnetic moments are predominantly directed along the *x*- and *z*-axes; the magnitude of magnetization directed along the *z*-axis is located in the region from $\approx 1.94\,\mu_B$ to $2.04\,\mu_B$, along *x*-axis from $\approx 1.74\,\mu_B$ to $1.83\,\mu_B$, and along the *y*-axis from $\approx 1.61\,\mu_B$ to $1.69\,\mu_B$.

In the case of a heterostructure with compression of Fe/BTO, the magnetic moment curves have the same character. The maximum value of the total magnetic moment does not change, though the minimum value does; it is observed in layer 2 and is equal to $\approx 2.96\,\mu_B$. The locations of the magnetization distribution curves along the axes have changed; in the case of the applied striction, the *y*-axis becomes predominant, the values of which range from $\approx 1.83\,\mu_B$ to $1.99\,\mu_B$, while along the *z*-axis the values range from $\approx 1.72\,\mu_B$ to $1.89\,\mu_B$ and along the *x* axis—from $\approx 1.56\,\mu_B$ to $1.70\,\mu_B$.

In order to check whether the change in magnetization is due to the shape change of BTO and corresponding influence or just from the change of the lattice parameters of Fe, we carried out a similar spin–orbit calculation only for pure bulk iron. The structure of bulk Fe is shown in Figure 6a, and was constructed in such a way that its lattice parameters (*x* and *y*) are equal to the corresponding lattice parameters of BTO, mimicking the substrate conditions. We used seven atomic layers of bcc iron, as in the heterostructure depicted in Figure 5. Figure 6b shows the results of the magnetization distribution over the atomic layers for that cell. We found all the distributions in Figure 6b to be constant, with the x and y components equal in terms of magnitude while the z-component is slightly higher (less than 0.005 μ_B). The total magnetization is same as in the heterostructure (Figure 5). Then, the cell was compressed along the x-axis and extended along y-axis, as in the previous case. Similar to the heterostructure case, the magnetization vector turned towards the extended y direction, as depicted in Figure 6c. Consequently, the z-component now coincides with the x component and the y component is 0.01 μ_B higher.

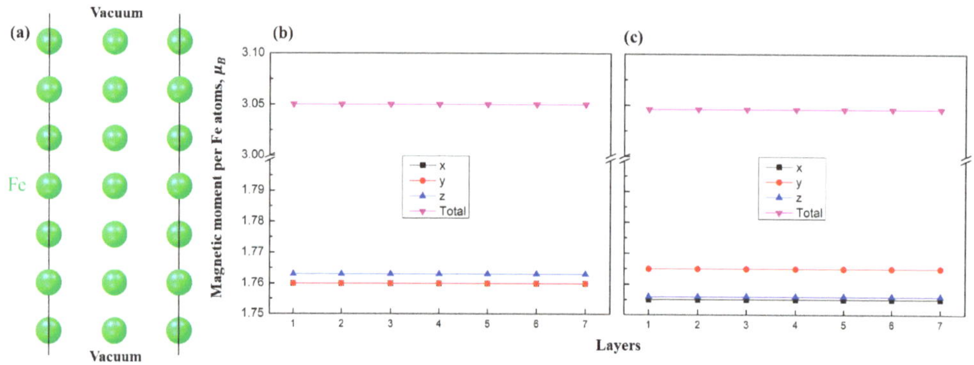

Figure 6. (**a**) The unit cell of the Fe structure used in the calculation, (**b**) the corresponding distribution of the magnetic moments of Fe atoms calculated for each layer of the structure with in-plane striction, as was in the case for the BTO substrate (1.36% mismatch), and (**c**) the corresponding distribution within the applied in-plane striction along the x-axis and extension along the y-axis in accordance with Figure 5. Each curve corresponds to the resulting magnetisation magnitude for the x, y, and z components of the magnetization vector and total magnetic moment in the scalar norm.

Overall, in the case of the heterostructure the effect is more pronounced, the magnitudes of all components are similar, and the total magnetization is roughly the same; however, there are a few significant differences. First, the components are more distant from each other, meaning that it may be possible to distinguish them easier. In the heterostructure, the difference between the x and y components is $\approx$0.1 μ_B, while the z component is higher by $\approx$0.2 μ_B. The change in magnetic moments along the y direction is $\approx$0.3 μ_B due to extending the lattice by 0.2 Å. Moreover, this extension and compression lead to increments of difference in the x and y components from 0.1 μ_B to 0.4 μ_B, whereas in the bulk the difference is only 0.01 μ_B. Thus, our results show that compression along one axis and expansion along the other can affect the direction of magnetization such that the magnetic moment turns predominantly along the axis with expanded length. In the heterostructure geometry, this effect is significant and could find interesting practical applications.

4. Conclusions

In this paper, we have demonstrated that the incorporation of ferroelectrics as a component of the heterostructure may enhance the desirable properties of the heterostructure.

In the case of the creation of a interfacial conducting state, the presence of a material with intrinsic ferroelectric polarization is favorable due to the opportunity to toggle the conductivity by means of an external field. Using the example of two heterostructures, $LaMnO_3/BaTiO_3$ and $La_2CuO_4/BaTiO_3$, we have seen that even without charged layers, as in $LaAlO_3/SrTiO_3$, the 2DEG might arise within the interfacial layers. Furthermore, another advantage concerns the growth conditions of the ferroelectric overlayer. Indeed, the alternation of charges in $LaAlO_3/SrTiO_3$-type heterostructures and the pure interface without defects are substantial conditions for a 2DEG, as the field arises in the ferroelectric material anyway.

The second very important feature desirable for electronic applications, especially for memory devices, is reverse magnetoelectric coupling. Systems with the ability to change the magnetic ordering by means of electric stimulus are of particular interest due to higher energy storage density. In the present paper, we observed the effect of the ferroelectric polarisation on the magnitude of magnetic moments in the $LaMnO_3/BaTiO_3$.

The presence of ferroelectric material as a component of a heterostructure along with a heavy metal such as Bi may enhance splitting thanks to ferroelectric polarization. In addition, the external electric field may increase the internal electrostatic field and slightly increase the splitting. The change in the ferroelectric polarization direction does not significantly change the Rashba parameter.

Lastly, the impact of anisotropic striction of ferroelectric $BaTiO_3$ on the magnetic moments of ferromagnetic bcc iron was investigated. Striction in one direction (x-axis) and corresponding extension in the other direction (y-axis) resulted in magnetization redistribution; in particular, the y component of the magnetization vector becomes predominant, whereas in the initial heterostructure the z component has the highest magnitude. Furthermore, comparison to bulk Fe with same lattice parameters suggests the conclusion that BTO has additional impact on the distribution of the magnetic moments. More precisely, the magnetization vectors in the heterostructure components are more pronounced, i.e., there is a $\approx 0.2\,\mu_B$ difference between each component without applied striction; moreover, the applied anisotropic striction makes the difference between the x and y components equal to $0.3\,\mu_B$, which is significant and might find interesting applications.

Overall, the use of ferroelectrics opens up new possibilities when combining materials with different properties within one device, and can make possible to tune certain properties of the interfacial state, namely, conductivity, Rashba splitting, and magnetization. This can provide an opportunity to tune the magnetic moments via the magnetoelectric coupling and striction. The present research provides an overview of areas where ferroelectrics might be useful and should be considered as possible components in a multilayer structure. In other words, it provides examples of phenomena which arise in heterostructures with ferroelectrics and which might be promising avenues for further investigation. Lastly, the research presents a general methodology used for treating heterostructures via ab initio calculations.

Author Contributions: Conceptualization, I.P.; methodology, O.N. and D.T.; software, I.P.; validation, I.P., A.E. and O.N.; formal analysis, I.P.; investigation, I.P., A.E. and K.E.; resources, I.P.; data curation, R.M.; writing—original draft preparation, I.P.; writing—review and editing, O.N., D.T. and R.M.; visualization, I.P.; supervision, V.K.; project administration, V.K.; funding acquisition, R.M. and V.K. All authors have read and agreed to the published version of the manuscript.

Funding: The work of I.P. and R.M. concerning LMO/BTO was funded by the Russian Scientific Foundation according to research project No. 21-12-00179. The part concerning the magnetostriction effect was funded by the Russian Scientific Foundation according to research project No. 21-72-10178. The part concerning LCO/BTO was supported by the government assignment for the FRC Kazan Scientific Center of RAS. V.K. acknowledges financial support from the Slovenian Research Agency Program No. P1-0040.

Institutional Review Board Statement: Not applicable.

Informed Consent Statement: Not applicable.

Data Availability Statement: The data presented in this study are available on request from the corresponding author.

Acknowledgments: Computational resources were provided by the Laboratory for Computer-Aided Design of New Materials and Machine Learning of Kazan Federal University.

Conflicts of Interest: The authors declare no conflict of interest.

References

1. Fredrickson, K.D.; Demkov, A.A. Switchable conductivity at the ferroelectric interface: Nonpolar oxides. *Phys. Rev. B* **2015**, *91*, 115126. [CrossRef]
2. Niranjan, M.K.; Wang, Y.; Jaswal, S.S.; Tsymbal, E.Y. Prediction of a switchable two-dimensional electron gas at ferroelectric oxide interfaces. *Phys. Rev. Lett.* **2009**, *103*, 016804. [CrossRef]
3. Liu, X.; Tsymbal, E.Y.; Rabe, K.M. Polarization-controlled modulation doping of a ferroelectric from first principles. *Phys. Rev. B* **2018**, *97*, 094107. [CrossRef]
4. Weng, Y.; Niu, W.; Huang, X.; An, M.; Dong, S. Ferroelectric control of a spin-polarized two-dimensional electron gas. *Phys. Rev. B* **2021**, *103*, 214101. [CrossRef]
5. Cao, C.; Chen, S.; Deng, J.; Li, G.; Zhang, Q.; Gu, L.; Ying, T.P.; Guo, E.J.; Guo, J.G.; Chen, X. Two-Dimensional Electron Gas with High Mobility Forming at $BaO/SrTiO_3$ Interface. *Chin. Phys. Lett.* **2022**, *39*, 047301. [CrossRef]
6. Kabanov, V.V.; Piyanzina, I.I.; Lysogorskiy, Y.V.; Tayurskii, D.A.; Mamin, R.F. Ab initio investigation of electronic and magnetic properties of antiferromagnetic/ferroelectric $LaMnO_3/BaTiO_3$ interface. *Mater. Res. Express* **2020**, *7*, 055020. [CrossRef]
7. Niranjan, M.K.; Burton, J.D.; Velev, J.P.; Jaswal, S.S.; Tsymbal, E.Y. Magnetoelectric effect at the $SrRuO_3/BaTiO_3$ (001) interface: An ab initio study. *Appl. Phys. Lett.* **2009**, *95*, 052501. [CrossRef]
8. Duan, C.G.; Jaswal, S.S.; Tsymbal, E.Y. Predicted magnetoelectric effect in $Fe/BaTiO_3$ multilayers: Ferroelectric control of magnetism. *Appl. Phys. Lett.* **2006**, *97*, 047201. [CrossRef]
9. Niranjan, M.K.; Velev, J.P.; Duan, C.G.; Jaswal, S.S.; Tsymbal, E.Y. Magnetoelectric effect at the $Fe_3O_4/BaTiO_3$ (001) interface: A first-principles study. *Phys. Rev. B* **2008**, *78*, 104405. [CrossRef]
10. Demkov, A.A.; Niu, Q.; Li, Z.; Shi, J.; Wang, E. Magnetoelectric coupling and electric control of magnetization in ferromagnet/ferroelectric/normalmetal superlattices. *Phys. Rev. B* **2009**, *80*, 140415. [CrossRef]
11. Baloni, M.; Sharma, R.C.; Singh, H.; Khan, B.; Singh, M.K.; Sati, P.C.; Thakur, V.N.; Kotnala, R.K.; Kumar, A. Energy storage and magnetoelectric coupling in neodymium (Nd) doped $BiFeO_3$-$PbTiO_3$ solid solution. *J. Alloys Compd.* **2023**, *946*, 169333. [CrossRef]
12. Piyanzina, I.I.; Mamin, R.F. Toward the ferroelectric field-effect transistor on $BaTiO_3/LaMnO_3$ heterostructure: DFT investigation. *J. Mater. Sci.* **2022**, *57*, 21620–21629. [CrossRef]
13. Chen, M.; Liu, F. Prediction of giant and ideal Rashba-type splitting in ordered alloy monolayers grown on a polar surface. *Natl. Sci. Rev.* **2021**, *8*, nwaa241. [CrossRef]
14. Bychkov, Y.A.; Rashba, E.I. Properties of a 2D electron gas with lifted spectral degeneracy. *Sov. Phys.-JETP Lett.* **1984**, *39*, 78–81.
15. Borek, S.; Braun, J.; Ebert, H.; Minár, J. Multiferroic heterostructures for spin filter applications: An ab initio study. *Phys. Rev. B* **2015**, *92*, 174408. [CrossRef]
16. Fechner, M.; Maznichenko, I.V.; Ostanin, S.; Ernst, A.; Henk, J.; Bruno, P.; Mertig, I. Magnetic phase transition in two-phase multiferroics predicted from first principles. *Phys. Rev. B* **2008**, *78*, 212406. [CrossRef]
17. Lu, N.; Zhang, P.; Zhang, Q.; Qiao, R.; He, Q.; Li, H.B.; Wang, Y.; Guo, J.; Zhang, D.; Duan, Z.; et al. Electric-field control of tri-state phase transformation with a selective dual-ion switch. *Nature* **2017**, *546* , 124–128. [CrossRef] [PubMed]
18. Piyanzina, I.; Evseev, K.; Kamashev, A.; Mamin, R. Ab Initio Characterization of Magnetoelectric Coupling in $Fe/BaTiO_3$, $Fe/SrTiO_3$, $Co/BaTiO_3$ and $Co/SrTiO_3$ Heterostructures. *Magnetism* **2023**, *3*, 215–225. [CrossRef]
19. Leksin, P.V.; Garifyanov, N.N.; Garifullin, I.A.; Schumann, J.; Vinzelberg, H.; Kataev, V.; Klingeler, R.; Schmidt, O.G.; Buchner, B. Full spin switch effect for the superconducting current in a superconductor/ferromagnet thin film heterostructure. *Appl. Phys. Lett.* **2010**, *97*, 102505. [CrossRef]
20. Zhao, Y.; Peng, R.; Guo, Y.; Liu, Z.; Dong, Y.; Zhao, S.; Li, Y.; Dong, G.; Hu, Y.; Zhang, J.; et al. Ultraflexible and malleable $Fe/BaTiO_3$ multiferroic Heterostructures for functional devices. *Adv. Funct. Mater.* **2021**, *31*, 2009376. [CrossRef]
21. Velev, J.P.; Duan, C.G.; Belashchenko, K.D.; Jaswal, S.S.; Tsymbal, E.Y. Effects of ferroelectricity and magnetism on electron and spin transport in $Fe/BaTiO_3/Fe$ multiferroic tunnel junctions. *J. Appl. Phys.* **2008**, *103*, 07A701. [CrossRef]
22. Duan, C.G.; Sahoo, S.; Srinivas, P.; Jaswal, S.S.; Tsymbal, E.Y.; Binek, C. Ferroelectric control of magnetism in $BaTiO_3/Fe$ heterostructures via interface strain coupling. *Phys. Rev. B* **2007**, *76*, 092108. [CrossRef]
23. Kresse, G.; Furthmüller, J. Efficiency of ab-initio total energy calculations for metals and semiconductors using a plane-wave basis set. *Comput. Mater. Sci.* **1996**, *6*, 15–50. [CrossRef]
24. Perdew, J.P.; Burke, K.; Ernzerhof, M. Generalized gradient approximation made simple. *Phys. Rev. Lett.* **1996**, *77*, 3865. [CrossRef]
25. Kohn, W.; Sham, L.J. Self-consistent equations including exchange and correlation effects. *Phys. Rev.* **1965**, *140*, A1133. [CrossRef]
26. Blöchl, P.E. Projector augmented-wave method. *Phys. Rev. B* **1994**, *50*, 17953. [CrossRef]
27. Kresse, G.; Furthmüller, J. Efficient iterative schemes for ab initio total-energy calculations using a plane-wave basis set. *Phys. Rev. B* **1996**, *54*, 11169. [CrossRef]
28. *MedeA Version 3.7*; MedeA Is a Registered Trademark of Materials Design, Inc., San Diego, USA. Available online: https://www.materialsdesign.com/ (accessed on 20 September 2023).
29. Monkhorst, H.J.; Pack, J.D. Special points for Brillouin-zone integrations. *Phys. Rev.* **1976**, *13*, 125188. [CrossRef]
30. Blöchl, P.E.; Jepsen, O.; Andersen, O.K. Improved tetrahedron method for Brillouin-zone integrations. *Phys. Rev.* **1994**, *49*, 16223. [CrossRef]
31. Methfessel, M.; Paxton, A.T. High-precision sampling for Brillouin-zone integration in metals. *Phys. Rev.* **1989**, *40*, 3616. [CrossRef]

32. Dudarev, S.L.; Botton, G.A.; Savrasov, S.Y.; Humphreys, C.J.; Sutton, A.P. Electron-energy-loss spectra and the structural stability of nickel oxide: An LSDA+U study. *Phys. Rev. B* **1998**, *57*, 1505–1509. [CrossRef]
33. Calderon, C.E.; Plata, J.J.; Toher, C.; Oses, C.; Levy, O.; Fornari, M.; Natan, A.; Mehl, M.J.; Hart, G.; Nardelli, M.B.; et al. The AFLOW standard for high-throughput materials science calculations. *Comput. Mater. Sci.* **2015**, *108*, 233–238. [CrossRef]
34. Wang, L.; Maxisch, T.; Ceder, G. Oxidation energies of transition metal oxides within the GGA+*U* framework. *Phys. Rev. B* **2006**, *73*, 195107. [CrossRef]
35. Piyanzina, I.I.; Kopp, T.; Lysogorskiy, Y.V.; Tayurskii, D.A.; Eyert, V. Electronic properties of $LaAlO_3/SrTiO_3$ n-type interfaces: A GGA+*U* study. *J. Phys. Condens. Matter* **2017**, *29*, 095501. [CrossRef]
36. Evseev, A.A.; Piyanzina, I.I.; Nedopekin, O.V.; Tayurskii, D.A. Ab initio investigation of $Bi/BaTiO_3$ and $Bi/PbTiO_3$ heterostructures for spintronic applications. *Ferroelectrics* **2023**, *605*, 27–35. [CrossRef]
37. Piyanzina, I.I.; Eyert, V.; Lysogorskiy, Y.V.; Tayurskii, D.A.; Kopp, T. Oxygen vacancies and hydrogen doping in $LaAlO_3/SrTiO_3$ heterostructures: Electronic properties and impact on surface and interface reconstruction. *J. Phys. Condens. Matter.* **2019**, *31*, 295601. [CrossRef]
38. Piyanzina, I.I.; Mamin, R.F. Electronic and Magnetic Properties of the $BaTiO_3/LaMnO_3$ Interface: A DFT Study. *J. Supercond. Nov. Magn.* **2022**, *35*, 2225–2229. [CrossRef]
39. Kabanov, V.V.; Piyanzina, I.I.; Tayurskii, D.A.; Mamin, R.F. Towards high-temperature quasi-two-dimensional superconductivity. *Phys. Rev. B* **2018**, *98*, 094522. [CrossRef]
40. Ramesh, R.; Spaldin, N. Multiferroics: Progress and prospects in thin films. *Nat. Mater.* **2007**, *6*, 21–29. [CrossRef]
41. Ast, C.R.; Henk, J.; Ernst, A.; Moreschini, L.; Falub, C.M.; Pacilé, D.; Bruno, P.; Kern, K.; Grioni, M. Giant spin splitting through surface alloying. *Phys. Rev. Lett.* **2007**, *98*, 186807. [CrossRef]
42. Singh, S.; Romero, A.H. Giant tunable Rashba spin splitting in a two-dimensional BiSb monolayer and in BiSb/AlN heterostructures. *Phys. Rev. B* **2017**, *95*, 165444. [CrossRef]
43. Eremeev, S.V.; Nechaev, I.A.; Koroteev, I.A.; Echenique, P.M.; Chulkov, E.V. Ideal two-dimensional electron system with a giant Rashba-type spin splitting in real materials: Surfaces of bismuth tellurohalides. *Phys. Rev. Lett.* **2012**, *108*, 246802. [CrossRef] [PubMed]
44. Lutz, P.; Figgemeier, T.; Abd El-Fattah, Z.M.; Bentmann, H.; Reinert, F. Large Spin Splitting and Interfacial States in a $Bi/BaTiO_3$ Rashba Ferroelectric Heterostructure. *Phys. Rev. Appl.* **2017**, *7*, 044011. [CrossRef]
45. Mirhosseini, H.; Maznichenko, I.V.; Abdelouahed, S.; Ostanin, S.; Ernst, A.; Mertig, I.; Henk, J. Toward a ferroelectric control of Rashba spin-orbit coupling: Bi on $BaTiO_3$ (001) from first principles. *Phys. Rev. B* **2010**, *81*, 073406. [CrossRef]

materials

MDPI

Article

Anisotropic Properties of Epitaxial Ferroelectric Lead-Free $0.5[Ba(Ti_{0.8}Zr_{0.2})O_3]$-$0.5(Ba_{0.7}Ca_{0.3})TiO_3$ Films

Nicholas Cucciniello [1,2], Alessandro R. Mazza [2], Pinku Roy [2], Sundar Kunwar [2], Di Zhang [2], Henry Y. Feng [1], Katrina Arsky [3], Aiping Chen [2,*] and Quanxi Jia [1,*]

[1] Department of Materials Design & Innovation, University at Buffalo, The State University of New York, Buffalo, NY 14260, USA; ncucciniello@lanl.gov or ngcuccin@buffalo.edu (N.C.); henryfeng1@berkeley.edu (H.Y.F.)

[2] Center for Integrated Nanotechnologies (CINT), Los Alamos National Laboratory, Los Alamos, NM 87545, USA; armazza@lanl.gov (A.R.M.); pinkur@lanl.gov (P.R.); sundar@lanl.gov (S.K.); dizhang@lanl.gov (D.Z.)

[3] Department of Materials Science & Engineering, University of Illinois Urbana, Urbana, IL 61801, USA

* Correspondence: apchen@lanl.gov (A.C.); qxjia@buffalo.edu (Q.J.)

Abstract: As the energy demand is expected to double over the next 30 years, there has been a major initiative towards advancing the technology of both energy harvesting and storage for renewable energy. In this work, we explore a subset class of dielectrics for energy storage since ferroelectrics offer a unique combination of characteristics needed for energy storage devices. We investigate ferroelectric lead-free $0.5[Ba(Ti_{0.8}Zr_{0.2})O_3]$-$0.5(Ba_{0.7}Ca_{0.3})TiO_3$ epitaxial thin films with different crystallographic orientations grown by pulsed laser deposition. We focus our attention on the influence of the crystallographic orientation on the microstructure, ferroelectric, and dielectric properties. Our results indicate an enhancement of the polarization and strong anisotropy in the dielectric response for the (001)-oriented film. The enhanced ferroelectric, energy storage, and dielectric properties of the (001)-oriented film is explained by the coexistence of orthorhombic-tetragonal phase, where the disordered local structure is in its free energy minimum.

Keywords: ferroelectric; anisotropy; dielectric; energy storage; morphotropic phase boundary

Citation: Cucciniello, N.; Mazza, A.R.; Roy, P.; Kunwar, S.; Zhang, D.; Feng, H.Y.; Arsky, K.; Chen, A.; Jia, Q. Anisotropic Properties of Epitaxial Ferroelectric Lead-Free $0.5[Ba(Ti_{0.8}Zr_{0.2})O_3]$-$0.5(Ba_{0.7}Ca_{0.3})TiO_3$ Films. *Materials* **2023**, *16*, 6671. https://doi.org/10.3390/ma16206671

Academic Editor: Dana Popescu

Received: 19 September 2023
Revised: 6 October 2023
Accepted: 11 October 2023
Published: 13 October 2023

1. Introduction

Ferroelectric materials based on lead zirconate titanate $Pb(Zr_{1-x}Ti_x)O_3$ (PZT) have found extensive applications within the modern electronics industry, namely actuators, sensors, and various energy storage and harvesting devices [1–3]. PZT exhibits superior dielectric and piezoelectric properties. However, such a chemical composition contains more than 60 wt.% lead (Pb). As a result of the high volatility and toxicity of Pb, it raises much concern about environmental contamination during the production and disposal of PZT materials. In recent years, there has been renewed interest in developing lead-free ferroelectric materials as restrictions on the use of Pb-based materials become more stringent globally [4–6]. This has led to a new push in materials research, with major focus on $BaTiO_3$-based ferroelectrics [7]. Limited experimental results have shown that $BaTiO_3$ (BTO)-based ceramics and thin films are very promising alternatives for Pb-based systems in electromechanical device applications. Numerous Ca-, Sr-, and Zr-modified BTO materials have been developed, including the $Ba(Ti_{0.8}Zr_{0.2})$-$x(Ba_{0.7}Ca_{0.3})TiO_3$ (BZT-xBCT) crystal system [8]. Liu et al. found that a morphotropic phase boundary (MPB) exists at $x = 0.5$ for the BZT-xBCT system, which gives rise to a large piezoelectric response (d_{33} = ~620 pCN^{-1}) [8]. At present, to increase the ferroelectric, dielectric, and piezoelectric properties of BZT-xBCT, most of the research has focused on composition tuning around or near the MPB to enhance properties similar to that of Pb-based systems [9–11]. Alternatively, through the introduction of an $LaNiO_3$ buffer layer between the BZT-BCT thin film and the $Pt/Ti/SiO_2/Si$ substrate, Li et al. showed that the buffer layer could lead to better crystallinity and piezoelectric response due to the promotion of nucleation sites during growth [12]. Furthermore, it is widely acknowledged that orientation-engineered ferroelectric films exhibit significant variations in performance owing to their inherent anisotropy. This inherent anisotropy holds substantial importance in the context of materials design for microelectronic devices. Numerous reports have explored the connection between ferroelectric, dielectric, and piezoelectric characteristics, and orientation in various types of ferroelectric thin films, including those based on $PbZr_{1-x}Ti_xO_3$, $K_{0.5}N_{0.5}NbO_3$, $BiFeO_3$, and $Ba_{1-x}Sr_xTiO_3$ [13–16].

To date, limited research exists on the systematic investigation of the effects of different crystallographic orientations on the physical properties of BZT-*x*BCT. At the MPB, the anisotropic nature of BZT-*x*BCT arises from the crystal structure and arrangement of its constituent atoms. In relaxor ferroelectrics, the atomic arrangement is not perfectly ordered, resulting in microstructural frustrations within the crystal lattice. These localized regions encompass varying degrees of polarization direction with respect to the normal direction of the film, in particular, epitaxial films. These regions, known as polar nanoregions (PNRs), can contribute to the distinctive relaxor behavior largely in part due to the disordered state and the local structural fluctuations, where the polarization direction varies randomly on the microscopic scale [17]. The presence of PNRs in the nanoscale regions can lead to enhanced ferroelectric, dielectric, and piezoelectric properties. The disordered distribution of dipoles and ease of the dipoles to reorient once under an applied electric field make this relaxor-like behavior advantageous for fast-switching applications and energy storage devices as compared to nominal ferroelectrics that exhibit well-defined and highly organized ferroelectric domains possessing a high remanent polarization and large coercivity.

Here we report the structural, ferroelectric, and dielectric properties of epitaxial 0.5BZT-0.5BCT (or BZT-BCT for simplicity of future discussion) thin films grown on single crystal (001)-, (110)-, and (111)-oriented $SrTiO_3$ substrates via pulsed laser deposition, where epitaxial $SrRuO_3$ is used as a bottom electrode. The microstructures, ferroelectric, and frequency-dependent dielectric properties of the BZT-BCT thin films were investigated in relation to the respective crystallographic orientations. The coexistence of orthorhombic-tetragonal phases was identified in the (001)-oriented thin films, whereas the (110)- and (111)-oriented BZT-BCT films exhibited a single tetragonal phase. A large anisotropy exists in the frequency-dependent dielectric and tunability characteristics, suggesting the improved functional properties of the (001)-oriented film originate from fast domain switching and polarization rotation arising from the diminished energy barrier. The observation of the anisotropic properties of BZT-BCT thin films introduced by the crystallographic orientations of the substrates may open other possibilities for designing high-performance energy storage devices.

2. Results

2.1. Structural Analysis

Figure 1a–c show X-ray diffraction (XRD) 2θ-ω scan patterns of the BZT-BCT films on $SrRuO_3$ (SRO)-buffered $SrTiO_3$ (STO) substrates with different crystallographic orientations of the substrates. Well-defined reflections corresponding to the (001)-, (110)-, and (111)-oriented films, matching that of the corresponding substrate orientation, suggest the growth of highly oriented BZT-BCT films (see the section on Materials and Methods). The formation of pure BZT-BCT films on different oriented substrates can be attributed to the small lattice mismatch between the film and the underlying SRO/STO substrate. It is noted that SRO serves two purposes. First, it acts as a buffer layer to control the lattice-induced strain for epitaxial growth of BZT-BCT (lattice parameter ~4.002 Å). This is made evident by considering SRO as pseudo-cubic with a lattice constant ranging between 3.941 Å–3.956 Å, in comparison with STO with a lattice parameter of 3.905 Å. This method has been demonstrated by Park et al. [18], whereby inserting a very thin interlayer between the substrate and the main layer of epitaxial $Ba_{0.6}Sr_{0.4}TiO_3$, the strain state could be systematically controlled. It has been also reported that such a buffer layer with a lattice parameter between the film and the substrate can play a critical role in film quality [19]. Secondly, SRO has high thermal stability and conductivity. It can be used as a quality bottom electrode for the capacitors [20]. The full-width at half-maximum (FWHM) of (222), (220), and (002) is 0.93°, 1.29°, and 0.05° for the BZT-BCT films on (111), (110), and (001) STO, respectively (inset Figure 1a–c). The extremely narrow rocking curve of the (001)-oriented film indicates a much-improved crystalline quality of the BZT-BCT film grown under our optimized processing conditions. The rocking curve FWHM difference between different orientations could be due to the different growth dynamics. To further explore the in-plane alignment of the BZT-BCT thin films with respect to the SRO/STO, in-plane ϕ-scan measurements are performed. Figure 1d–f show the XRD in-plane ϕ-scans around the (220), (200), and (222) reflections of the BZT-BCT films on (111), (110), and (001) STO substrates, respectively. Characteristics of three-fold (120° apart), two-fold (180° apart), and four-fold (90° apart) symmetry of the film on STO substrates with different substrate orientations are understandable by considering the shape of the basal-plane of the STO substrates with different crystallographic orientations.

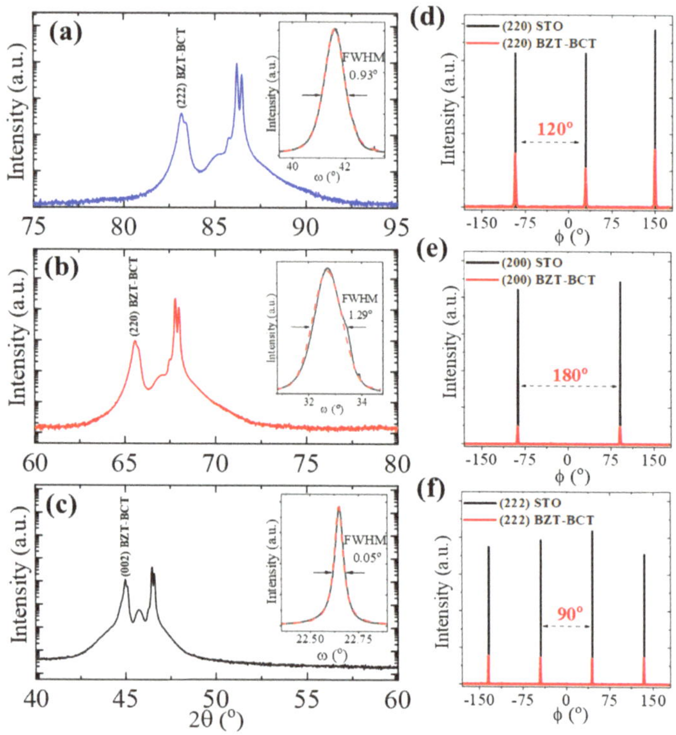

Figure 1. X-ray diffraction 2θ-scans of the BZT-BCT films on SRO-buffered (**a**) (111), (**b**) (110), (**c**) (001) STO substrates, where the unlabeled diffraction peaks are from the SRO and STO. Insets of (**a**–**c**) present the rocking curve of (222), (220), and (002) diffractions of the BZT-BCT films on their corresponding substrates. (**d**) In-plane ϕ-scan of the (220) reflection of the BZT-BCT thin film on (111) STO. (**e**) In-plane ϕ-scan of the (200) reflection of the BZT-BCT thin film on (110) STO. (**f**) In-plane ϕ-scan of the (222) reflection of the BZT-BCT thin film on (001) STO.

Based on the XRD 2θ-ω and ϕ-scans, the epitaxial relationship between the film and the substrate can be described as $(111)_{BZT\text{-}BCT}$ | | $(111)_{STO}$ (and $<220>_{BZT\text{-}BCT}$ | | $<220>_{STO}$), $(110)_{BZT\text{-}BCT}$ | | $(110)_{STO}$ (and $<002>_{BZT\text{-}BCT}$ | | $<002>_{STO}$), and $(001)_{BZT\text{-}BCT}$ | | $(001)_{STO}$

(and <222>$_{BZT-BCT}$ | | <222>$_{STO}$) for the BZT-BCT films on (111)-, (110)-, and (001)-oriented STO substrates, respectively.

It has been reported that BZT-BCT shows distinctive peak splitting along the (200) and (220), transforming into (002)/(200) and (202)/(220) doublets. Such an observation has been attributed to the phase shift to a tetragonal (T) phase structure [21,22]. To further investigate the phase structure of the epitaxial BZT-BCT films, Figure 2a–c shows the analysis of the corresponding deconvolution and peak fitting around the (002) at 2θ = ~45.0°, (220) at 2θ = ~65.6°, and (111) at 2θ = ~38.8° of the epitaxial BZT-BCT films on differently oriented substrates. The asymmetric nature of the (002) and (220) peaks for the BZT-BCT films on the (001) and (110) substrates, respectively, and the symmetrical nature of the (111) peak for the BZT-BCT film on the (111) STO substrate suggest the formation of tetragonal (T) phase films. Interestingly, on the (001)-oriented BZT-BCT film exhibits a superposition of the doublet T phase along with a single intense peak that can be defined as an orthorhombic phase. The formation of mixed phases of tetragonal and rhombohedral crystal structures has been reported for polycrystalline bulk-like ceramic samples [23,24].

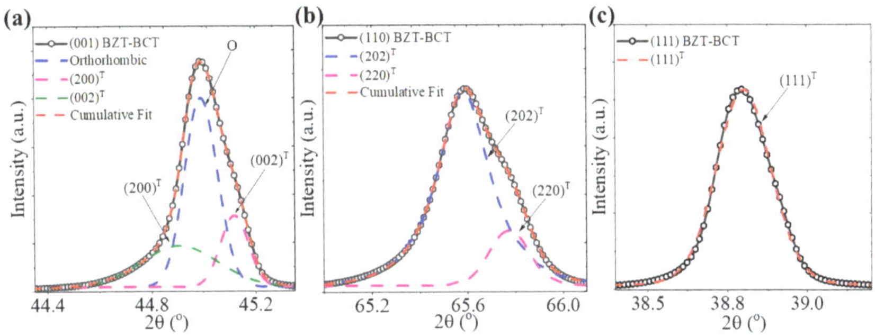

Figure 2. Enlarged view of the diffraction peaks and the resulting deconvolution and fitting of the (**a**) (002), (**b**) (220), and (**c**) (111) reflections.

It is noted that the coexistence of multiple phases in the BZT-BCT is still under intense study. For instance, some studies suggested BZT-BCT phase transitions from a low-temperature rhombohedral (R) phase to a room-temperature T phase at the MPB [8,11]. Recent work on the BZT-BCT phase diagram also indicated that an intermediate phase exists between the R and T phase transition [25]. High-resolution synchrotron X-ray characterization and more detailed studies of BZT-BCT have revealed that an MPB appears at the orthorhombic (O) to T phase transition [26,27]. At the MPB, the coexistence of the O + T phases results in enhanced ferroelectric and piezoelectric properties when stable and metastable phases exist, enabling direct domain and easy polarization switching. This is directly correlated to the minimum energy barrier in the mixed phase [28–30]. Our previous work has explored the microscopic mechanism of the relaxor-type behavior in BZT-BCT thin films. Our results have shown that the existence of high density nanodomains in BZT-BCT thin films and Sn doping further breaks down the domain size and reduces the polarization anisotropy [31].

2.2. *Ferroelectric and Energy Storage Properties*

Figure 3 shows the room temperature polarization vs. electric field (*P-E*)-hysteresis loops for the (001)-, (110)-, and (111)-oriented BZT-BCT films, at a fixed frequency of 1 kHz. Well-defined slim *P-E* loops are present for all films, confirming typical ferroelectric behavior of the materials. Interestingly, we have noticed that there is a clear trend for the remanent polarization (P_r) and coercive field (E_c) when the BZT-BCT films are deposited on different oriented substrates, i.e., $P_r^{(001)} < P_r^{(111)} < P_r^{(110)}$ and $E_c^{(001)} < E_c^{(111)} < E_c^{(110)}$. The values of some important parameters related to the ferroelectric materials are listed in Table 1. The absolute values of P_r and E_c for our (001)-, (110)-, and (111)-oriented BZT-BCT films can be compared to the ferroelectric properties reported by two separate studies conducted by Lou et al., one being an orientation dependent study where the reported P_r and E_c for their (001)-, (110)-, and (111)-oriented BZT-BCT epitaxial thin films on $La_{0.7}Sr_{0.3}MnO_3$-buffered STO substrates are 4.14 μC cm^{-2}, 2.39 μC cm^{-2}, and 1.67 μC cm^{-2} and 59.7 kV cm^{-1}, 39.6 kV cm^{-1}, and 60.2 kV cm^{-1}, respectively [32]. Furthermore, in an independent study of (110)-oriented BZT-BCT on SRO-buffered STO, Lou et al. reported a P_r of 2.01 μC cm^{-2} and an E_c of 187 kV cm^{-1} [33]. The variation could be attributed to the difference in growth techniques, growth conditions, film thickness, substrate quality, and the bottom electrode materials.

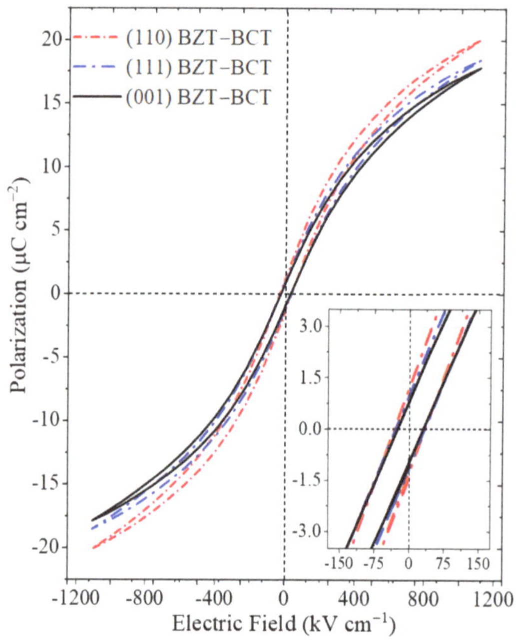

Figure 3. Polarization vs. electric field hysteresis loops for (001)-, (110)-, and (111)-oriented BZT-BCT films. The inset shows the magnification around the remanent polarization (P_r) and coercive field (E_c).

Table 1. Values of maximum polarization (P_{max}), remanent polarization (P_r), and coercive field (E_c) for the preferentially epitaxial BZT-BCT films on (001)-, (110)-, and (111)-oriented STO substrates.

Parameters	(001)	(110)	(111)
P_{max} (μC cm^{-2})	17.88	20.08	18.52
P_r (μC cm^{-2})	~0.84	~1.18	~0.96
E_c (kV cm^{-1})	~31.43	~33.88	~33. 69

The ferroelectric properties of the (001)-oriented BZT-BCT thin film differs from previous results based on the anisotropic properties of BTO-based films [34–36], as the energy barrier for polarization rotation is at the free energy minima from the phase coexistence seen in the (001)-oriented film. This may come from the reduction in the anisotropy energy in the BZT-BCT film with the coexistence of the O + T phase, which leads to the minimization of the orientation preferences of the PNRs, resulting in greater isotropic responses to the electric field as compared to the (110)- and (111)-oriented BZT-BCT films that exhibit a single T phase only.

To evaluate the effect of film orientation on the energy storage of the epitaxial BZT-BCT thin films with different crystallographic orientations, we calculated the energy density and efficiency (η) from the unipolar *P-E* loop based on the following equations:

$$W_{\text{rec}} = \int_{P_r}^{P_{max}} EdP \tag{1}$$

$$\eta\ (\%) = \int_{P_r}^{P_{max}} EdP / \int_{0}^{P_{max}} EdP \tag{2}$$

where W_{rec} is the recoverable energy storage density, P_{max} is the maximum polarization, P_r is the remanent polarization, and E is the electric field, respectively [37,38]. From the *P-E* loops shown in Figure 3, it is clear that the (001)-, (110)-, and (111)-oriented BZT-BCT films possess different P_{max} and P_r at an electric field of 1150 kV cm^{-1}, which is lower than their breakdown field. Figure 4 summarizes the energy density and efficiency of the different orientated BZT-BCT films. The (111)-oriented BZT-BCT thin film exhibits an η of 90.00, while the (110) and (001) films show slightly lower values of 89.82 and 87.62%, respectively. The energy density and efficiency of recent reports on BZT-*x*BCT are listed in Table 2.

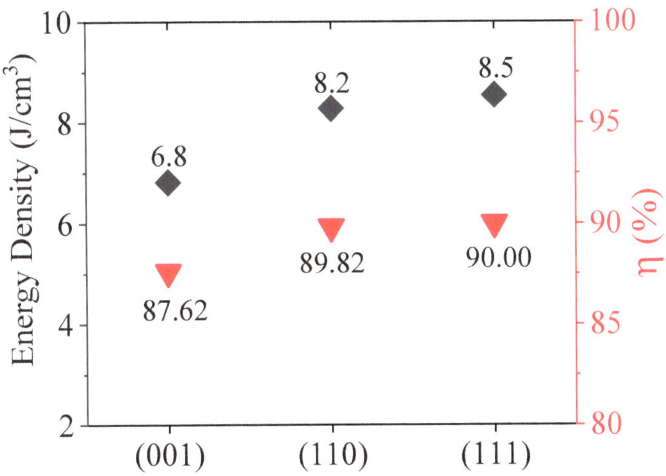

Figure 4. Energy storage density and efficiency of the (001)-, (110)-, and (111)-oriented BZT BCT films.

Table 2. Comparison of energy density and efficiency of different BZT-*x*BCT ceramics. * This work.

Material	Energy Efficiency (η %)	Energy Density (J/cm^3)	Electric Field (kV/cm)	Ref.
$[(BaZr_{0.2}Ti_{0.8})O_3]_{0.5}[Ba_{0.7}Ca_{0.3}Ti)O_3]_{0.5}$	52	0.10	40	[9]
$[(BaZr_{0.2}Ti_{0.8})O_3]_{0.5}[Ba_{0.7}Ca_{0.3}Ti)O_3]_{0.5}$	58	78 mJ	25	[11]
$[(BaZr_{0.1}Ti_{0.9})O_3]_{0.5}[Ba_{0.7}Ca_{0.3}Ti)O_3]_{0.5}$	74	164 mJ	-	[39]
$[(BaZr_{0.2}Ti_{0.8})O_3]_{0.6}[Ba_{0.7}Ca_{0.3}Ti)O_3]_{0.4}/ZnO$	74	2.61	282	[40]
$0.85[(BaZr_{0.2}Ti_{0.8})O_3]_{0.6}[Ba_{0.7}Ca_{0.3}Ti)O_3]_{0.4}$-$0.15SrTiO_3$	84	0.98	40	[41]
$[(BaZr_{0.2}Ti_{0.8})O_3]_{0.5}[Ba_{0.7}Ca_{0.3}Ti)O_3]_{0.5}$ (001)	87.62	6.8	1150	*
$[(BaZr_{0.2}Ti_{0.8})O_3]_{0.5}[Ba_{0.7}Ca_{0.3}Ti)O_3]_{0.5}$ (110)	89.82	8.2	1150	*
$[(BaZr_{0.2}Ti_{0.8})O_3]_{0.5}[Ba_{0.7}Ca_{0.3}Ti)O_3]_{0.5}$ (111)	90	8.5	1150	*

2.3. Dielectric and Tunability Properties

Figure 5a–c depict the dielectric constant (ε_r) vs. the electric field (E) characteristics at different frequencies, where the DC field was swept from positive to negative and back to positive. To evaluate the dielectric constant, ε_r, of the BZT-BCT films with different orientations, capacitance vs. voltage (*C-V*) measurements are taken at room temperature under a DC bias (330 kV cm^{-1}) at frequencies of 1 kHz, 10 kHz, and 100 kHz. It is well known that ferroelectric materials typically show distinctive butterfly-shaped ε_r vs. E loops, where the ε_r is at its maximum at near zero field. The value of the dielectric constant decreases with increasing the field. The scaling of the dielectric response with the applied field agrees well with the domain process of ferroelectrics [42]. On the other hand, it is very interesting to notice that the maximum ε_r of the BZT-BCT films is greatly affected by the orientation of the films. In other words, the BZT-BCT films show a large anisotropy in ε_r with the relationship of $\varepsilon_r^{(001)} > \varepsilon_r^{(110)} > \varepsilon_r^{(111)}$. The absolute value of the ε_r is 702, 546, and 459 at a frequency of 1 kHz for the (001)-, (110)-, and (111)-oriented BZT-BCT thin films, respectively. It should be emphasized that the ε_r of the epitaxial BZT-BCT films is significantly lower than their bulk counterparts (ε_r > 2400), potentially due to the effect of smaller grain size and substrate clamping [43,44]. Much smaller ε_r of epitaxial films such as $Ba_{1-x}Sr_xTiO_3$ in comparison with their bulk counterpart has also been reported, where the crystal structure distortion such as the lattice strain can also play an important role in the absolute value of ε_r for the given material [45,46]. We would also like to point out that the field at the maximum ε_r shifts slightly to the negative field direction for the (110)- and (111)-oriented films. This may be a result of the interface effects between the film and electrodes used in device fabrication.

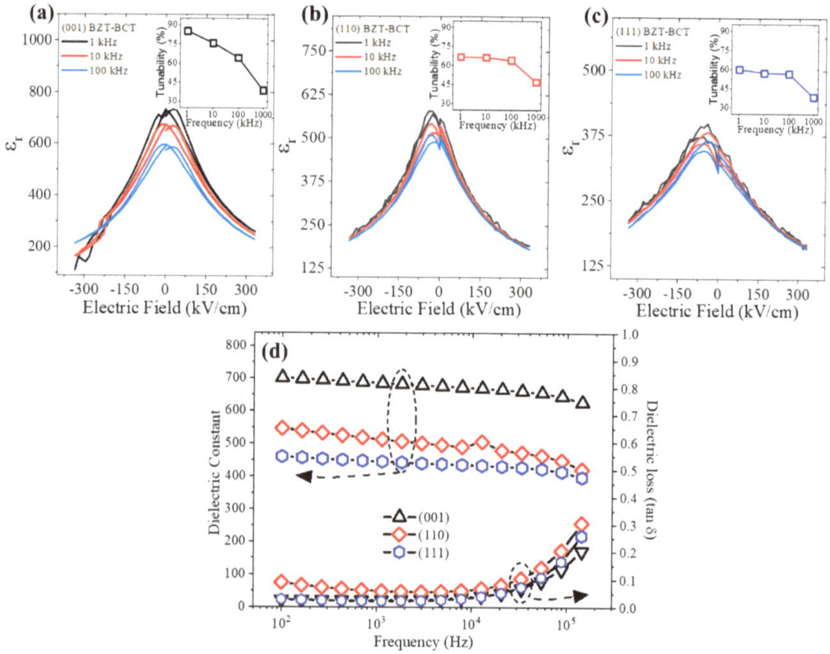

Figure 5. Dielectric constant vs. electric field at different frequencies of the (**a**) (001)-, (**b**) (110)-, and (**c**) (111)-oriented BZT-BCT thin films. The insets show the tunability (%) at a frequency of 1, 10, 100, and 1000 kHz. (**d**) Dielectric constant and dielectric loss as a function of frequency.

The dielectric tunability (DT) was extracted from the C-V data and calculated using:

$$DT\ (\%) = [\varepsilon_r(0) - \varepsilon_r(E)]/\varepsilon_r(0) \times 100 \tag{3}$$

where $\varepsilon_r(0)$ and $\varepsilon_r(E)$ are the dielectric constant at zero field and a given applied field, respectively. The inset of Figure 5a–c summarizes the tunability as a function of frequency for the three oriented BZT-BCT films. As can be seen from Figure 5, the (001)-oriented BZT-BCT thin film exhibits a moderately larger tunability with a relationship of

$DT^{(001)} > DT^{(110)} > DT^{(111)}$ at a relatively lower frequency (<100 kHz). The tunability of the BZT-BCT films at 1 kHz is 85.09, 66.67, and 59.89% for the (001)-, (110)-, and (111)-oriented films, respectively. It is noted that the tunability of the BZT-BCT films with different orientations decreases with increasing frequency. At relatively higher frequencies (>100 kHz), the tunability seems to become similar, regardless of the orientation of the films. The frequency-dependent ε_r and dielectric loss (tanδ) of the BZT-BCT films with different orientations are shown in Figure 5d. Throughout the entire frequency range, the ε_r demonstrated a descending order of $\varepsilon_r^{(001)} > \varepsilon_r^{(110)} > \varepsilon_r^{(111)}$. The dielectric loss of the BZT-BCT films, on the other hand, is in the same range with no large difference, although the (110)-oriented BZT-BCT film shows a slightly larger dielectric loss. Similar behavior was also reported in other ferroelectric films, which has been attributed to the space charging effects, interfacial diffusion, and contact resistance [47–49]. It is also important to note that all the BZT-BCT films show a small frequency dispersion of ε_r at frequencies from 100 Hz–100 kHz, which can be ascribed to the dynamics of the PNRs in the epitaxial BZT-BCT films as well as the permanent dipole moment retention [49].

3. Discussion

The polarization vector angle with respect to the surface normal changes with the crystallographic orientation direction for ferroelectric systems with tetragonal structure such as $BaTiO_3$. In the majority of studies of ferroelectric materials, anisotropic properties originate in the <110> and <111> directions [50–52]. For instance, in a study of the orientation dependence of $BaTiO_3$ thin films, Zhang et al. suggested that the further the tilting angle of the polar axis away from the crystallographic orientation, the greater the dielectric response becomes [34]. However, in the present study, we show that the superior dielectric response lies along the <001> direction. This is likely a result of phase coexistence (O + T) and tetragonal phase instability as described below.

Our work shows that (001)-oriented BZT-BCT films exhibit the coexistence of O + T phases, whereas the (110)- and/or (111)-oriented epitaxial BZT-BCT films do not. The enhancement of dielectric properties along certain crystallographic orientations could, for this reason, be explained by the local PNRs directly related to the microstructures, where the coexistence of different phases at room temperature and the specific composition of $x = 0.5$ can lead to a generally flattened energy barrier [53–57]. Under applied electric fields, the ease of domain switching and polarization rotation can come from the instability of the polarization direction at the MPB characterized by an ultra-low-energy barrier, leading to heightened ferroelectric and piezoelectric responses [58]. This is in contrast with the (110)- and (111) oriented epitaxial films where the XRD analysis presents a single-phase T structure. Furthermore, the single T phase at room temperature results in smaller fluctuations of polarization as the T phase is stable. It is therefore that an increase in the barrier height in the free energy profile and anisotropy energy may result in different properties [28].

We note that although both (110)- and (111)-orientated epitaxial films show relatively a lower dielectric constant as a result of the formation of stable single tetragonal phase, the nonequilibrium process of pulsed laser deposition may also play a role in the dielectric properties of the films. As all BZT-BCT thin films were grown under the same processing conditions, the ultimate stoichiometry and microstructure of the thin films can change with respect to the crystallographic orientation, which may result in defects (i.e., dislocations, impurities, vacancies, etc.) and the formation of defect-dipole complexes [59]. Furthermore, the residual stress and lattice strain vary by orientation, where the polarization displacement of the heterovalent cations (Ba^{2+}, Ca^{2+}, Ti^{4+}, Zr^{4+}) with respect to the oxygen octahedra of perovskite films affect the preferentially oriented BZT-BCT films that are largely seen in the dielectric response.

4. Materials and Methods

Epitaxial BZT-BCT thin films with a thickness of ~270 nm were synthesized on 50 nm thick conducting $SrRuO_3$ (SRO)-buffered $SrTiO_3$ substrates with orientations of (001), (110), and (111) via pulsed laser deposition (KrF excimer laser, λ = 248 nm). The thickness of the epitaxial BZT-BCT films were calculated via the deposition rate and cross-sectional transmission electron microscopy. The processing conditions were optimized to achieve high-crystallinity films. Briefly, the substrate temperature and oxygen partial pressure were maintained at 687 °C and 50 mTorr, respectively, during film growth. A laser energy of 1.5 J cm^{-2} at 5 Hz was used to ablate the BZT-BCT target. In order to obtain a uniform laser energy density on the target, a rectangular-shaped laser spot, defined by the image beam technique [60], was used in our experimental setup. After deposition, the films were annealed (in situ) at 500 °C in 760 Torr oxygen for one hour before being cooled down to room temperature at a ramping rate of 5 °C min^{-1}. To conduct dielectric property measurements, circular Au top electrodes (~100 nm thick) with a diameter of 350 μm defined by a shadow mask were then deposited by magnetron sputtering at room temperature. More detailed growth and optimization were reported elsewhere [31,61].

Capacitors with a configuration of Au/BZT-BCT/SRO/STO were characterized by measuring the polarization vs. electric field (*P-E*) hysteresis loops, and the capacitance vs. voltage (*C-V*) characteristics using a ferroelectric test system (Precision Premier II; Radiant Technologies, Inc., Albuquerque, NM, USA). The dielectric properties of the BZT-BCT films were further measured using a precision LCR meter (E4980, Keysight Agilent, Santa Rosa, CA, USA) at room temperature with a frequency range of 1 kHz–1 MHz. The crystallographic properties and microstructure of the films were analyzed via X-ray diffraction (XRD, Empyrean, Malvern PANalytical, Westborough, MA, USA).

5. Conclusions

In summary, epitaxial BZT-BCT thin films have been successfully grown on (001)-, (110)-, and (111)-oriented SRO-buffered STO substrates via pulsed laser deposition. XRD analysis reveals the coexistence of the orthorhombic and tetragonal phases at room temperature for the (001)-oriented film as compared to a single tetragonal phase for the (110)- and (111)-oriented films. The ferroelectric properties show typical ferroelectric behavior with all orientations exhibiting superior electrical energy efficiencies (η > 87%), suggesting such materials as quality candidates for potential energy storage applications. The dielectric constant and tunability extracted from *C-V* curves displays the existence of a large anisotropy between the (001)-, (110)-, and (111)-oriented epitaxial BZT-BCT films with a maximal ε_r and tunability response along the (001) BZT-BCT film. The present results suggest the properties of the (001)-oriented film is enhanced owing to the coexistence of orthorhombic and tetragonal phases, which can be attributed to the easing of direct domain switching and polarization rotation due to the low-energy barrier in its distorted state.

Author Contributions: Conceptualization: A.C. and Q.J.; methodology: N.C., A.C. and Q.J.; software: N.C. and H.Y.F.; validation: A.R.M., P.R., S.K., D.Z., A.C. and Q.J.; formal analysis: N.C., H.Y.F. and K.A.; investigation: N.C., H.Y.F. and K.A.; resources: A.C. and Q.J.; data curation: N.C.; writing—original draft preparation: N.C. and Q.J.; writing—review and editing: A.R.M., P.R., S.K., D.Z., A.C. and Q.J.; supervision: A.C. and Q.J.; funding acquisition: Q.J. All authors have read and agreed to the published version of the manuscript.

Funding: This research at the University at Buffalo (UB) was partially supported by the U.S. National Science Foundation (ECCS-1902623). N.C. and Q.J. acknowledge the CINT Users Program. The research at Los Alamos National Laboratory was supported by the NNSA's Laboratory Directed Research and Development Program and was performed, in part, at the Center for Integrated Nanotechnologies (CINT), an Office of Science User Facility operated for the U.S. Department of Energy Office of Science. Los Alamos National Laboratory, an affirmative action equal opportunity employer, is managed by Triad National Security, LLC, for the U.S. Department of Energy's NNSA, under contract 89233218CNA000001. N.C., A.C. and Q.J. acknowledge the support from DOE Office of Science Graduate Student Research Program (SCGSR). The SCGSR program is administered by the

Oak Ridge Institute for Science and Education for the DOE under contract number DE-SC0014664. P.R. and A.C. acknowledge the support from LANL Seaborg Institute.

Institutional Review Board Statement: Not applicable.

Informed Consent Statement: Not applicable.

Data Availability Statement: The data that support the findings of this study are available from the corresponding authors upon reasonable request.

Conflicts of Interest: The authors declare no conflict of interest. The funders had no role in the design of the study; in the collection, analyses, or interpretation of data; in the writing of the manuscript; or in the decision to publish the results.

Sample Availability: Not applicable.

References

1. Muralt, P.; Kholkin, A.; Kohli, M.; Maeder, T. Piezoelectric Actuation of PZT Thin-Film Diaphragms at Static and Resonant Conditions. *Sens. Actuators A Phys.* **1996**, *53*, 398–404. [CrossRef]
2. Gao, X.; Yang, J.; Wu, J.; Xin, X.; Li, Z.; Yuan, X.; Shen, X.; Dong, S. Piezoelectric Actuators and Motors: Materials, Designs, and Applications. *Adv. Mater. Technol.* **2019**, *5*, 1900716. [CrossRef]
3. Kang, M.-G.; Jung, W.-S.; Kang, C.-Y.; Yoon, S.-J. Recent Progress on PZT Based Piezoelectric Energy Harvesting Technologies. *Actuators* **2016**, *5*, 5. [CrossRef]
4. Tinte, S.; Burton, B.P.; Cockayne, E.; Waghmare, U.V. Origin of the Relaxor State inPb($B_X B_{1-X'}$)O_3. *Phys. Rev. Lett.* **2006**, *97*, 137601. [CrossRef]
5. Nguyen, M.D. Tuning the Energy Storage Performance, Piezoelectric Strain and Strain Hysteresis of Relaxor PLZT Thin Films through Controlled Microstructure by Changing the Ablation Rate. *J. Eur. Ceram. Soc.* **2019**, *39*, 2076–2081. [CrossRef]
6. Yin, Z.-W.; Luo, H.-S.; Wang, P.-C.; Xu, G.-S. Growth, Characterization and Properties of Relaxor Ferroelectric PMN-PT Single Crystals. *Ferroelectrics* **1999**, *229*, 207–216. [CrossRef]
7. Acosta, M.; Novak, N.; Rojas, V.; Patel, S.; Vaish, R.; Koruza, J.; Rossetti, G.A.; Rödel, J. $BaTiO_3$-Based Piezoelectrics: Fundamentals, Current Status, and Perspectives. *Appl. Phys. Rev.* **2017**, *4*, 041305. [CrossRef]
8. Liu, W.; Ren, X. Large Piezoelectric Effect in Pb-Free Ceramics. *Phys. Rev. Lett.* **2009**, *103*, 257602. [CrossRef]
9. Syal, R.; Goel, R.; De, A.; Kumar Singh, A.; Sharma, G.; Thakur, O.P.; Kumar, S. Flattening of Free Energy Profile and Enhancement of Energy Storage Efficiency near Morphotropic Phase Boundary in Lead-Free BZT-XBCT. *J. Alloys Compd.* **2021**, *873*, 159824. [CrossRef]
10. Ehmke, M.C.; Ehrlich, S.N.; Blendell, J.E.; Bowman, K.J. Phase Coexistence and Ferroelastic Texture in High Strain (1−X)Ba($Zr_{0.2}Ti_{0.8}$)O_3–X($Ba_{0.7}Ca_{0.3}$)TiO_3 Piezoceramics. *J. Appl. Phys.* **2012**, *111*, 124110. [CrossRef]
11. Jayakrishnan, A.R.; Alex, K.V.; Thomas, A.; Silva, J.P.B.; Kamakshi, K.; Dabra, N.; Sekhar, K.C.; Agostinho Moreira, J.; Gomes, M.J.M. Composition-Dependent XBa($Zr_{0.2}Ti_{0.8}$)O_3-(1-x)($Ba_{0.7}Ca_{0.3}$)TiO_3 Bulk Ceramics for High Energy Storage Applications. *Ceram. Int.* **2019**, *45*, 5808–5818. [CrossRef]
12. Oikawa, T.; Aratani, M.; Funakubo, H.; Saito, K.; Mizuhira, M. Composition and Orientation Dependence of Electrical Properties of Epitaxial Pb(Zr_xTi_{1-x})O_3 Thin Films Grown Using Metalorganic Chemical Vapor Deposition. *J. Appl. Phys.* **2004**, *95*, 3111–3115. [CrossRef]
13. Chen, W.; Wang, L.; Ren, W.; Niu, G.; Zhao, J.; Zhang, N.; Liu, M.; Tian, Y.; Dong, M. Crystalline Phase and Electrical Properties of Lead-Free Piezoelectric KNN-Based Films with Different Orientations. *J. Am. Ceram. Soc.* **2017**, *100*, 2965–2971. [CrossRef]
14. Sone, K.; Naganuma, H.; Miyazaki, T.; Nakajima, T.; Okamura, S. Crystal Structures and Electrical Properties of Epitaxial $BiFeO_3$ Thin Films with (001), (110), and (111) Orientations. *Jpn. J. Appl. Phys.* **2010**, *49*, 09MB03. [CrossRef]
15. Moon, S.E.; Kim, E.-K.; Kwak, M.-H.; Ryu, H.-C.; Kim, Y.-T.; Kang, K.-Y.; Lee, S.-J.; Kim, W.-J. Orientation dependent microwave dielectric properties of ferroelectric $Ba_{1-x}Sr_xTiO_3$ thin films. *Appl. Phys. Lett.* **2003**, *83*, 2166–2168. [CrossRef]
16. Li, W.L.; Zhang, T.D.; Hou, Y.F.; Zhao, Y.; Xu, D.; Cao, W.P.; Fei, W.D. Giant Piezoelectric Properties of BZT–0.5BCT Thin Films Induced by Nanodomain Structure. *RSC Adv.* **2014**, *4*, 56933–56937. [CrossRef]
17. Sharma, Y.; Lee, M.-C.; Pitike, K.C.; Mishra, K.K.; Zheng, Q.; Gao, X.; Musico, B.L.; Mazza, A.R.; Katiyar, R.S.; Keppens, V.; et al. High Entropy Oxide Relaxor Ferroelectrics. *ACS Appl. Mater. Interfaces* **2022**, *14*, 11962–11970. [CrossRef] [PubMed]
18. Park, B.H.; Peterson, E.J.; Jia, Q.X.; Lee, J.; Zeng, X.; Si, W.; Xi, X.X. Effects of very thin strain layers on dielectric properties of epitaxial $Ba_{0.6}Sr_{0.4}TiO_3$ films. *Appl. Phys. Lett.* **2001**, *78*, 533–535. [CrossRef]
19. Kang, K.T.; Corey, Z.J.; Hwang, J.; Sharma, Y.; Paudel, B.; Roy, P.; Collins, L.; Wang, X.; Lee, J.W.; Oh, Y.S.; et al. Heterogeneous Integration of Freestanding Bilayer Oxide Membrane for Multiferroicity. *Adv. Sci.* **2023**, *10*, 2207481. [CrossRef]
20. Rubi, D.; Vlooswijk, A.H.G.; Noheda, B. Growth of Flat $SrRuO_3$ (111) Thin Films Suitable as Bottom Electrodes in Heterostructures. *Thin Solid Film.* **2009**, *517*, 1904–1907. [CrossRef]
21. Puli, V.S.; Pradhan, D.K.; Pérez, W.; Katiyar, R. Structure, dielectric tunability, thermal stability and diffuse phase transition behavior of lead free BZT–BCT ceramic capacitors. *J. Phys. Chem. Solids* **2013**, *74*, 466–475. [CrossRef]

22. Li, B.; Ehmke, M.C.; Blendell, J.E.; Bowman, K.J. Optimizing Electrical Poling for Tetragonal, Lead-Free BZT–BCT Piezoceramic Alloys. *J. Eur. Ceram. Soc.* **2013**, *33*, 3037–3044. [CrossRef]
23. Hayati, R.; Fayazi, M.; Diargar, H.; Kaveh, M.; Tayebi, L. Electrical and mechanical properties of BZT − *x*BCT lead-free piezoceramics. *Int. J. Appl. Ceram. Technol.* **2020**, *17*, 1891–1898. [CrossRef]
24. Otta, S.; Kand, L.; Das, R.K.; Mohanta, V.R.; Roul, B.K.; Dash, B.B.; Kisan, B. Piezoelectric, Structural, Vibration and Optical Properties of Lead Free Based $0.5Ba(Ti_{0.8}Zr_{0.2})O_3$-$0.5(Ba_{0.7}Ca_{0.3})TiO_3$ Ceramic Sample. *Mater. Today Commun.* **2023**, *35*, 106191. [CrossRef]
25. Zhang, L.; Zhang, M.; Wang, L.; Zhou, C.; Zhang, Z.; Yao, Y.; Zhang, L.; Xue, D.; Lou, X.; Ren, X. Phase Transitions and the Piezoelectricity around Morphotropic Phase Boundary in $Ba(Zr_{0.2}Ti_{0.8})O_3$-x$(Ba_{0.7}Ca_{0.3})TiO_3$ Lead-Free Solid Solution. *Appl. Phys. Lett.* **2014**, *105*, 162908. [CrossRef]
26. Keeble, D.S.; Benabdallah, F.; Thomas, P.A.; Maglione, M.; Kreisel, J. Revised Structural Phase Diagram of $(Ba_{0.7}Ca_{0.3}TiO_3)$-$(BaZr_{0.2}Ti_{0.8}O_3)$. *Appl. Phys. Lett.* **2013**, *102*, 092903. [CrossRef]
27. Gao, J.; Dai, Y.; Hu, X.; Ke, X.; Li, S.; Zhang, L.; Wang, Y.; Wang, D.; Wang, Y.; Liu, Y.; et al. Phase Transition Behaviours near the Triple Point for Pb-Free $(1-x)Ba(Zr_{0.2}Ti_{0.8})O_3$-x$(Ba_{0.7}Ca_{0.3})TiO_3$ Piezoceramics. *Europhys. Lett.* **2016**, *115*. [CrossRef]
28. Acosta, M.; Khakpash, N.; Someya, T.; Novak, N.; Jo, W.; Nagata, H.; Rossetti, G.A.; Rödel, J. Origin of the Large Piezoelectric Activity in $(1-x)Ba(Zr_{0.2}Ti_{0.8})O_3-x(Ba_{0.7}Ca_{0.3})TiO_3$ ceramics. *Phys. Rev. B* **2015**, *91*, 104108. [CrossRef]
29. Zhukov, S.; Genenko, Y.A.; Acosta, M.; Humburg, H.; Jo, W.; Rödel, J.; von Seggern, H. Polarization Dynamics across the Morphotropic Phase Boundary in $Ba(Zr_{0.2}Ti_{0.8})O_3$-x$(Ba_{0.7}ca_{0.3})TiO_3$ Ferroelectrics. *Appl. Phys. Lett.* **2013**, *103*, 152904. [CrossRef]
30. Manjón-Sanz, A.; Culbertson, C.M.; Hou, D.; Jones, J.L.; Dolgos, M.R. Total Scattering and Diffraction Studies of Lead-Free Piezoelectric (1-x)$Ba(Zr_{0.2}Ti_{0.8})O_3$-$(Ba_{0.7}Ca_{0.3})TiO_3$ Deconvolute Intrinsic and Extrinsic Contributions to Electromechanical Strain. *Acta Mater.* **2019**, *171*, 79–91. [CrossRef]
31. Yuan, R.; Kumar, A.; Zhuang, S.; Cucciniello, N.; Lu, T.; Xue, D.; Penn, A.; Mazza, A.R.; Jia, Q.X.; Liu, Y.; et al. Machine Learning-Enabled Superior Energy Storage in Ferroelectric Films with a Slush-like Polar State. *Nano Lett.* **2023**, *23*, 4807–4814. [CrossRef] [PubMed]
32. Luo, B.; Wang, D.Y.; Duan, M.; Li, S. Orientation-Dependent Piezoelectric Properties in Lead-Free Epitaxial $0.5BaZr_{0.2}Ti_{0.8}O_3$-$0.5Ba_{0.7}Ca_{0.3}TiO_3$ Thin Films. *Appl. Phys. Lett.* **2013**, *103*, 122903. [CrossRef]
33. Luo, B.; Xu, Y.; Zhang, F.; Wang, T.; Yao, Y. Dielectric Tunability Properties in (110)-Oriented Epitaxial $0.5Ba(Ti_{0.8}Zr_{0.2})O_3$-$0.5(Ba_{0.7}Ca_{0.3})TiO_3$ Thin Films Prepared by PLD Method. *Materials* **2020**, *13*, 4771. [CrossRef]
34. Zhang, W.; Cheng, H.; Yang, Q.; Hu, F.; Ouyang, J. Crystallographic Orientation Dependent Dielectric Properties of Epitaxial $BaTiO_3$ Thin Films. *Ceram. Int.* **2016**, *42*, 4400–4405. [CrossRef]
35. Raeder, T.M.; Holstad, T.S.; Nylund, I.-E.; Einarsrud, M.-A.; Glaum, J.; Meier, D.; Grande, T. Anisotropic in-plane dielectric and ferroelectric properties of tensile-strained BaTiO3 films with three different crystallographic orientations. *AIP Adv.* **2021**, *11*, 025016. [CrossRef]
36. Yu, Z.; Guo, R.; Bhalla, A.S. Dielectric Behavior of $Ba(Ti_{1-x}Zr_x)O_3$ Single Crystals. *J. Appl. Phys.* **2000**, *88*, 410–415. [CrossRef]
37. Zhang, L.; Jing, R.; Huang, Y.; Hu, Q.; Alikin, D.; Shur, V.Y.; Gao, J.; Wei, X.; Zhang, L.; Liu, G.; et al. Enhanced antiferroelectric-like relaxor ferroelectric characteristic boosting energy storage performance of $(Bi_{0.5}Na_{0.5})TiO_3$-based ceramics via defect engineering. *J. Mater.* **2022**, *8*, 527–536. [CrossRef]
38. Kim, J.; Saremi, S.; Acharya, M.; Velarde, G.; Parsonnet, E.; Donahue, P.; Qualls, A.; Garcia, D.; Martin, L.W. Ultrahigh Capacitive Energy Density in Ion-Bombarded Relaxor Ferroelectric Films. *Science* **2020**, *369*, 81–84. [CrossRef] [PubMed]
39. Swain, A.B.; Subramanian, V.; Murugavel, P. The Role of Precursors on Piezoelectric and Ferroelectric Characteristics of 0.5BCT-0.5BZT Ceramic. *Ceram. Int.* **2018**, *44*, 6861–6865. [CrossRef]
40. Jayakrishnan, A.R.; Silva, J.P.B.; Kamakshi, K.; Annapureddy, V.; Mercioniu, I.F.; Sekhar, K.C. Semiconductor/Relaxor 0–3 Type Composites: A Novel Strategy for Energy Storage Capacitors. *J. Sci. Adv. Mater. Devices* **2021**, *6*, 19–26. [CrossRef]
41. Jayakrishnan, A.R.; Alex, K.V.; Kamakshi, K.; Silva, J.; Sekhar, K.C.; Gomes, M. Enhancing the Dielectric Relaxor Behavior and Energy Storage Properties of $0.6Ba(Z_{r0.2}Ti_{0.8})O_3$–$0.4(Ba_{0.7}Ca_{0.3})TiO_3$ Ceramics through the Incorporation of Paraelectric $SrTiO_3$. *J. Mater. Sci. Mater. Electron.* **2019**, *30*, 19374–19382. [CrossRef]
42. Thomas, R.; Varadan, V.K.; Komarneni, S.; Dube, D.C. Diffuse Phase Transitions, Electrical Conduction, and Low Temperature Dielectric Properties of Sol–Gel Derived Ferroelectric Barium Titanate Thin Films. *J. Appl. Phys.* **2001**, *90*, 1480–1488. [CrossRef]
43. Yan, X.; Zheng, M.; Gao, X.; Zhu, M.; Hou, Y. High-Performance Lead-Free Ferroelectric BZT–BCT and Its Application in Energy Fields. *J. Mater. Chem. C* **2020**, *8*, 13530–13556. [CrossRef]
44. Boota, M.; Houwman, E.P.; Dekkers, M.; Nguyen, M.D.; Vergeer, K.H.; Lanzara, G.; Koster, G.; Rijnders, G. Properties of epitaxial, (001)- and (110)-oriented $(PbMg_{1/3}Nb_{2/3}O_3)_{2/3}$-$(PbTiO_3)_{1/3}$ films on silicon described by polarization rotation. *Sci. Technol. Adv. Mater.* **2016**, *17*, 45–57. [CrossRef]
45. Park, B.H.; Gim, Y.; Yang, F.; Jia, Q.X.; Lu, P. High Nonlinearity of $Ba_{0.6}Sr_{0.4}TiO_3$Films Heteroepitaxially Grown on MgO Substrates. *Appl. Phys. Lett.* **2000**, *77*, 2587–2589. [CrossRef]
46. Gim, Y.; Hudson, T.D.; Yang, F.; Kwon, C.; Findikoglu, A.T.; Gibbons, B.J.; Park, B.H.; Jia, Q.X. Microstructure and Dielectric Properties of $Ba_{1-x}Sr_xTiO_3$ Films Grown on $LaAlO_3$ Substrates. *Appl. Phys. Lett.* **2000**, *77*, 1200–1202. [CrossRef]
47. Song, S.; Zhai, J.; Gao, L.; Yao, X. Orientation-Dependent Dielectric Properties of $Ba(Sn_{0.15}Ti_{0.85})O_3$ Thin Films Prepared by Sol–Gel Method. *J. Phys. Chem. Solids* **2009**, *70*, 1213–1217. [CrossRef]

48. Das, R.R.; Bhattacharya, P.; Katiyar, R.S. Enhanced Ferroelectric Properties in Laser-Ablated $SrBi_2Nb_2O_9$ Thin Films on Platinized Silicon Substrate. *Appl. Phys. Lett.* **2002**, *81*, 1672–1674. [CrossRef]
49. Pontes, F.M.; Pontes, L.; Leite, E.R.; Longo, E.; Santos, E.; Mergulhão, S.; Chiquito, A.J.; Pizani, P.S.; Lanciotti, F., Jr.; Boschi, T.M.; et al. Influence of ca Concentration on the Electric, Morphological, and Structural Properties of (Pb,Ca)TiO_3 Thin Films. *J. Appl. Phys.* **2002**, *91*, 6650–6655. [CrossRef]
50. Imai, H.; Kanno, I.; Yokokawa, R.; Wasa, K.; Kotera, H. Orientation Dependence of Transverse Piezoelectric Properties of Epitaxial $BaTiO_3$ Films. *Jpn. J. Appl. Phys.* **2010**, *49*, 09MA09. [CrossRef]
51. Du, X.; Wang, Q.-M.; Belegundu, U.; Bhalla, A.; Uchino, K. Crystal Orientation Dependence of Piezoelectric Properties of Single Crystal Barium Titanate. *Mater. Lett.* **1999**, *40*, 109–113. [CrossRef]
52. Sando, D. Strain and Orientation Engineering in ABO_3 Perovskite Oxide Thin Films. *J. Phys. Condens. Matter* **2022**, *34*, 153001. [CrossRef]
53. Acosta, M.; Novak, N.; Jo, W.; Rödel, J. Relationship between Electromechanical Properties and Phase Diagram in the $Ba(Zr_{0.2}Ti_{0.8})O_3$–$x(Ba_{0.7}Ca_{0.3})TiO_3$ Lead-Free Piezoceramic. *Acta Mater.* **2014**, *80*, 48–55. [CrossRef]
54. Du, B.; Ma, W. Phenomenological Modeling of Phase Transitions and Electrocaloric Effect in $Ba(Zr_{0.2}Ti_{0.8})O_3$–$(Ba_{0.7}Ca_{0.3})TiO_3$. *J. Am. Ceram. Soc.* **2018**, *102*, 2604–2610. [CrossRef]
55. Du, B.; Ma, W. Phase boundary and electric field induced polarization rotation in lead-free $Ba(Zr_{0.2}Ti_{0.8})O_3$-(Ba0.7$Ca_{0.3}$)TiO_3. *Phase Transit.* **2020**, *93*, 666–677. [CrossRef]
56. Nahas, Y.; Akbarzadeh, A.; Prokhorenko, S.; Prosandeev, S.; Walter, R.; Kornev, I.; Íñiguez, J.; Bellaiche, L. Microscopic Origins of the Large Piezoelectricity of Leadfree (Ba,Ca)(Zr,Ti)O_3. *Nat. Commun.* **2017**, *8*, 15944. [CrossRef] [PubMed]
57. Yang, T.; Ke, X.; Wang, Y. Mechanisms Responsible for the Large Piezoelectricity at the Tetragonal-Orthorhombic Phase Boundary of (1-x)$BaZr_{0.2}Ti_{0.8}O_3$-x$Ba_{0.7}Ca_{0.3}TiO_3$ System. *Sci. Rep.* **2016**, *6*, 33392. [CrossRef]
58. Ahluwalia, R.; Lookman, T.; Saxena, A.; Cao, W. Domain-Size Dependence of Piezoelectric Properties of Ferroelectrics. *Phys. Rev. B* **2005**, *72*, 014112. [CrossRef]
59. Sahoo, K.K.; Katoch, R.; Brajesh, K.; Garg, A.; Gupta, R. Improved ferroelectric response of pulsed laser deposited $BiFeO_3$-$PbTiO_3$ thin films around morphotropic phase boundary with interfacial $PbTiO_3$ buffer layer. *J. Appl. Phys.* **2020**, *127*, 064101. [CrossRef]
60. Dowden, P.C.; Bi, Z.; Jia, Q.X. Method for Controlling Energy Density for Reliable Pulsed Laser Deposition of Thin Films. *Rev. Sci. Instrum.* **2014**, *85*, 025111. [CrossRef]
61. Chen, A.; Khatkhatay, F.; Zhang, W.; Clement, J.H.; Liang, J.-T.; Wang, H. Strong Oxygen Pressure Dependence of Ferroelectricity in $BaTiO_3$/$SrRuO_3$/$SrTiO_3$ Epitaxial Heterostructures. *J. Appl. Phys.* **2013**, *114*, 124101. [CrossRef]

materials MDPI

Article

Influence of Lattice Mismatch on Structural and Functional Properties of Epitaxial $Ba_{0.7}Sr_{0.3}TiO_3$ Thin Films

Jonas Wawra [1,2], Kornelius Nielsch [1,2] and Ruben Hühne [1,*]

1 Leibniz Institute for Solid State and Materials Research, Helmholtzstrasse 20, D-01069 Dresden, Germany; j.wawra@ifw-dresden.de (J.W.); k.nielsch@ifw-dresden.de (K.N.)
2 Institute for Applied Physics, TUD Dresden University of Technology, D-01062 Dresden, Germany
* Correspondence: r.huehne@ifw-dresden.de

Abstract: Substrate-induced strains can significantly influence the structural properties of epitaxial thin films. In ferroelectrics, this might lead to significant changes in the functional properties due to the strong electromechanical coupling in those materials. To study this in more detail, epitaxial $Ba_{0.7}Sr_{0.3}TiO_3$ films, which have a perovskite structure and a structural phase transition close to room temperature, were grown with different thicknesses on *RE*ScO_3 (*RE*–rare earth element) substrates having a smaller lattice mismatch compared to $SrTiO_3$. A fully strained $SrRuO_3$ bottom electrode and Pt top contacts were used to achieve a capacitor-like architecture. Different X-ray diffraction techniques were applied to study the microstructure of the films. Epitaxial films with a higher crystalline quality were obtained on scandates in comparison to $SrTiO_3$, whereas the strain state of the functional layer was strongly dependent on the chosen substrate and the thickness. Differences in permittivity and a non-linear polarization behavior were observed at higher temperatures, suggesting that ferroelectricity is supressed under tensile strain conditions in contrast to compressive strain for our measurement configuration, while a similar reentrant relaxor-like behavior was found in all studied layers below 0 °C.

Keywords: ferroelectrics; phase transition; relaxor; epitaxy; barium strontium titanate; thin films; pulsed laser deposition

Citation: Wawra, J.; Nielsch, K.; Hühne, R. Influence of Lattice Mismatch on Structural and Functional Properties of Epitaxial $Ba_{0.7}Sr_{0.3}TiO_3$ Thin Films. *Materials* **2023**, *16*, 6036. https://doi.org/10.3390/ma16176036

Academic Editor: Dana Popescu

Received: 7 August 2023
Revised: 28 August 2023
Accepted: 31 August 2023
Published: 2 September 2023

1. Introduction

Whereas ferroelectric oxide thin films are already used in many modern devices, research has been focusing over the last few years on exploiting their properties for novel applications [1]. For example, their opto-electronic properties are tested in high-frequency modulators [2–4], while the excellent dielectric and switching characteristics are promising for energy storage purposes [5,6]. Additionally, the pyroelectric properties might be beneficial for energy-harvesting devices as well as for the use in eco-friendly and efficient solid-state cooling [7,8]. The materials showing the highest potential are lead-containing relaxor solid solutions, which reveal extraordinary electromechanical responses. The relaxor-like behavior is described by the appearance of nanoscale regions, resulting in a spontaneous polarization even at high temperatures [9]. This feature is often explained by the chemical disorder in the system [10], but unfortunately, the underlying mechanism leading to the relaxor behavior remains unclear [11,12]. Additionally, the high toxicity of lead is detrimental; therefore, major efforts are made to develop lead-free alternatives [13]. One of those is the well-known perovskite ferroelectric barium titanate $BaTiO_3$. While its properties are inferior for many applications in comparison to its lead-containing counterparts, the substitution of cations is an effective way to create solid solutions in order to enhance the ferroelectric characteristics [14–17]. Among them is $Ba_{1-x}Sr_xTiO_3$, which is commonly used in varactors [18,19] due to its high dielectric constant and tunability [20]. The replacement of Ba^{2+} with smaller Sr^{2+} results in a stabilization of the higher symmetric perovskite phases and therefore reduces the phase transition temperatures with increasing substitution x [21] compared to pure $BaTiO_3$. Additionally, many studies report that the type of phase transition in this material shifts from a classic ferroelectric transition to a diffuse or relaxor behavior with increasing Sr content [22,23]. This is consistent with our own studies on epitaxial $Ba_{1-x}Sr_xTiO_3$ films, where a relaxor-like transition was observed in the frequency-dependent permittivity on samples with a composition of $Ba_{0.7}Sr_{0.3}TiO_3$ (BST) [24]. Since the strain induced by the substrate can alter the properties of epitaxial thin films as demonstrated by theoretical calculations [25,26] as well as experimental investigations [27,28], the choice of substrate material offers an easy way to modify ferroelectric films. However, the precise control of the strain in an epitaxial thin film is usually not straightforward, in particular for samples exceeding the critical thickness, which can be as low as a few nm only in some systems [29]. If the film thickness increases further in such a case, the formation of dislocations is observed, resulting in a relaxation during growth. The mechanical strain in these film is then not only a function of the difference between the thermal expansion coefficient of film and substrate but also of the deposition temperature, thickness and the dislocation density [25], where some factors can be hard to quantify. To study the effect of strain on the prototypical lead-free solid solution BST, epitaxial films on REScO$_3$ (RE is a rare earth element) substrates were prepared, which have a much smaller lattice mismatch in comparison to the previously used $SrTiO_3$ (STO). This allows growing films with a higher critical thickness as well as preparing samples with equal thicknesses but different strain states, which will be used to compare their electrical properties.

2. Materials and Methods

$SrRuO_3$ (SRO) and BST layers were grown on different single crystal substrates (CrysTec GmbH, Berlin, Germany) by a standard on axis pulsed laser deposition (PLD) process, using a KrF laser with a wavelength of $\lambda = 248\,\text{nm}$. The laser was focused as a rectangular spot with a size of $A = 3\,\text{mm}^2$ on stoichiometric polycrystalline targets, giving an energy density of $w = 1\,\text{J}\,\text{cm}^{-2}$. The SRO target was homemade from polycrystalline powder, whereas the BST target was prepared at Fraunhofer IKTS Dresden by a conventional solid-state reaction method. During the deposition, the oxygen background pressure was kept at $p_{O_2} = 0.01\,\text{mbar}$, whereas the distance between the target and substrate was set to 50 mm. First, an SRO buffer layer with a thickness of 20 nm was deposited at a substrate temperature $T_S = 750\,°\text{C}$ and a laser frequency of $f = 3\,\text{Hz}$. Next, the temperature was increased to $T_S = 900\,°\text{C}$ to deposit the BST layer with a frequency of $f = 5\,\text{Hz}$. Subsequently, the sample was cooled back down to room temperature and removed from the vacuum chamber.

Since the deposition rate of both layers was already known from previous studies [24], the number of pulses could be adjusted to reach the desired thickness. A Bruker DektakXT stylus profilometer (Bruker, Billerica, MA, USA) was used to check the thickness after the growth and no significant variation in dependence of the used substrate was found. Surface morphology measurements were carried out using a Bruker Icon atomic force microscope (AFM) in tapping mode.

Afterwards, a photolithographically prepared shadow mask was put on the sample surface. Through this mask, 5 nm Ti as well as 100 nm Pt were sputtered on the surface of the samples at room temperature to create round electrodes with a radius of $r = 100\,\mu\text{m}$. Later, the shadow mask was removed by a standard lift-off process.

The microstructure of the samples was characterized by X-ray diffraction (XRD) utilizing a Bruker D8 Advance with Co-K_α radiation for the 2θ-θ scans. Additionally, a Philips X'Pert four-cycle goniometer (Malvern Panalytical, Almelo, The Netherlands) equipped with a Cu tube and a primary Ge monochromator was used to perform reciprocal space map (RSM) and rocking curve measurements.

The electrical characterization of the chosen samples was performed by contacting two probes to the Pt electrodes and applying a field along the out-of-plane direction through the whole film thickness, using the SRO buffer layer as a bottom electrode. In this setup, the capacitance C could be measured by an HP 4284 precision LCR Meter (Keysight, Santa Rosa, CA, USA) using an alternating voltage signal with a frequency between 0.1 and 10 kHz with 50 mV amplitude. Afterwards, the relative permittivity was calculated by assuming two equal plate capacitors in series using $\varepsilon_r = C \cdot 2d/(\varepsilon_0 \cdot \pi r^2)$, with the vacuum permittivity ε_0, the film thickness d and the area of the Pt contact πr^2. Polarization measurements were carried out with a Radiant Technologies Multiferroic tester (Radiant Technologies, Inc., Albuquerque, NM, USA) in the same contact configuration. For the electric field-dependent loops, a triangular profile with a frequency of 1 kHz was used. First, a positive pulse was applied to generate a defined polarization state in the film before a whole loop was measured anti-clockwise. All electrical characterizations were performed in a Linkam LTSE420-P probe stage (Linkam Scientific Instruments Ltd., Redhill, UK) that allowed for temperature-dependent measurements while cooling down the sample from 250 °C to −150 °C.

3. Results

The mostly used parameter to quantify the strain in hetero-epitaxial layer architectures is the lattice mismatch $\alpha = (a_s - a_f)/a_s$ between the in-plane lattice parameter of the substrate a_s and the film a_f. STO is a cubic perovskite material with the space group $Pm\bar{3}m$ (no. 221). Its lattice parameter is $a_{STO} = 3.905$ Å, which leads to a mismatch of −1.9% for bulk BST, that crystallizes in the same space group with a lattice parameter of $a_{BST} = 3.980$ Å at room temperature. In bulk, a structural phase transition to the ferroelectric tetragonal structure with the space group *P4mm* (no. 99) is observed slightly below room temperature. As in pure $BaTiO_3$, the ferroelectricity in this phase originates from the spontaneous polarization caused by the shift of position between the positively charged Ti^{4+} cation and the negatively charged O^{2-} oxygen octahedra. Since there are no other cubic oxide single crystals with a smaller mismatch to BST available, orthorhombic $\textit{RE}ScO_3$ single crystals were chosen as an alternative, which crystallize in the space group *Pbnm* (no. 62). If cut along the (1 1 0)-plane, a rectangular surface structure is obtained. The lattice parameters of the almost quadratic surface unit cell with $a_{1\bar{1}0}$ parallel to the [1 $\bar{1}$ 0] direction and a_{001} parallel to the [0 0 1] can be calculated from the orthorhombic lattice parameters a, b and c with the relations $a_{1\bar{1}0} = 0.5\sqrt{a^2 + b^2}$ and $a_{001} = 0.5c$ respectively. For our studies, $DyScO_3$ (DSO), $GdScO_3$ (GSO) and $NdScO_3$ (NSO) were selected to achieve a broad range of mismatches, ranging from compressive to tensile strain conditions for BST layers. The lattice constants as well as the size of the surface unit cell of the used $\textit{RE}ScO_3$ crystals are summarized in Table 1.

Table 1. Lattice parameters a, b and c of the orthorhombic unit cell and the calculated values of the surface unit cell of the (1 1 0) surface for the used $\textit{RE}ScO_3$ substrates. Data taken from [30].

Crystal	a/Å	b/Å	c/Å	$a_{1\bar{1}0}$/Å	a_{001}/Å
DSO	5.440	5.704	7.903	3.946	3.952
GSO	5.480	5.746	7.932	3.970	3.966
NSO	5.575	5.776	8.003	4.014	4.002

From those lattice parameters, the mismatches along both in-plane directions to the bulk (1 0 0) BST lattice parameters are calculated that will lead to a biaxial strain state. The results are summarized for all substrates in Table 2.

Table 2. Calculated lattice mismatch α in % at room temperature between both in-plane directions of the substrates and the cubic BST.

	STO	DSO		GSO		NSO	
	a	$a_{1\bar{1}0}$	a_{001}	$a_{1\bar{1}0}$	a_{001}	$a_{1\bar{1}0}$	a_{001}
BST	−1.92	−0.86	−0.71	−0.25	−0.35	0.85	0.55

3.1. Structural Characterization

Before the top contacts were prepared, AFM scans of the BST/SRO heterostructure surface were taken; the results are shown in Figure 1. In all cases, a dense and smooth surface without any notable features is observed; i.e., no significant influence of the substrate could be identified. The root-mean-square roughness was between 130 and 170 pm for all films studied, which is comparable to the results of previous studies [24].

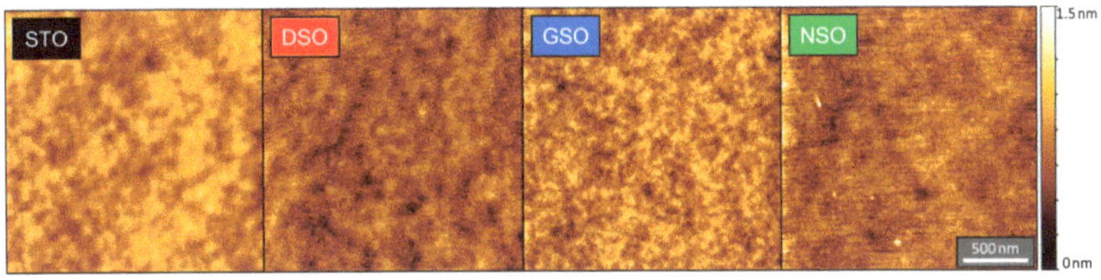

Figure 1. AFM scans of 250 nm BST layers on different SRO-buffered substrates. From left to right: STO, DSO, GSO and NSO.

Figure 2a shows 2θ-θ scans for the complete Pt/BST/SRO architectures on the different substrates. All BST films had a thickness of roughly 250 nm. In all samples, sharp peaks with high intensity are visible next to the out-of-plane reflexions of the substrates, which can be indexed with the (0 0 l) planes of the BST structure, confirming the successful c-axis-oriented growth. This peak is slightly shifted toward lower angles compared to the corresponding bulk peak for the films on STO, DSO and GSO, which corresponds to a larger out-of-plane lattice parameter indicating a tetragonal distortion of those layers. Due to the tensile strain condition on NSO, this shift is not observed in the corresponding scan. The structure on NSO also shows the pseudo-cubic (0 0 2) reflection of the SRO buffer layer, which is not visible for the other spectra, due to the small difference between the pseudo-cubic lattice parameter of SRO ($a_{SRO(pc)}$ = 3.929 Å) and the substrate values as well as the small layer thickness. The peak observed in all scans at $2\theta = 46.7°$ results from the (1 1 1) planes of the Pt electrodes.

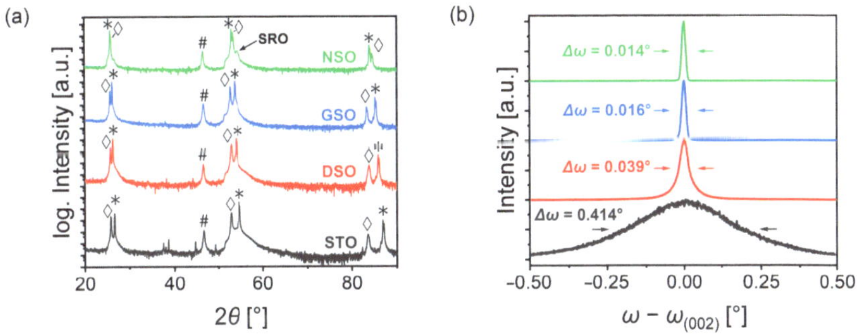

Figure 2. (**a**) 2θ-θ XRD scans of 250 nm BST layer architectures on different substrates. Corresponding substrate peaks are marked with $*$, (0 0 l) BST peaks are marked with $\Diamond$ and Pt peaks are marked with #. (**b**) Rocking curves for the same samples.

To obtain more detailed information on the crystalline quality of the BST-films, (0 0 2) rocking curve scans were measured for these samples. The results are shown in Figure 2b. It is clearly visible that the films on the *RE*ScO$_3$ substrates show significant smaller full width at half maximum (FWHM) $\Delta\omega$ values compared to the film on STO. *RE*ScO$_3$ crystals are grown with a higher quality using the Czochralski method in comparison to STO crystals, which are grown via the Verneuil method [30,31]. However, rocking curve scans of the substrates (Figure A1a in Appendix A) revealed a much smaller difference in the

FWHM values (0.006° for the REScO$_3$ substrates vs. 0.011° for STO). We assume that the higher quality of the REScO$_3$ substrates in combination with the smaller lattice mismatch significantly reduced the mosaicity on the films compared to the sample on STO (see RSM measurements below).

XRD texture measurements (Figure A1b in Appendix A) confirmed a pure epitaxial growth on all substrates. Detailed RSM measurements were performed to acquire additional insights on the quality of our samples and to calculate the in- and out-of-plane lattice parameters of the different layers. Exemplarily, the scans around the (1 0 3) BST peak are shown for each substrate in Figure 3a. The highest intensities peaks originate from the corresponding substrates. As expected, those peaks shift to lower Q_x and Q_z values due to the increasing unit cell parameters from STO to NSO.

The weak reflections elongated along the Q_z direction can be assigned to the (1 0 3) pseudo-cubic planes of the thin SRO buffer layer. No broadening or shift along the Q_x direction is observed, demonstrating that the in-plane lattice parameters of the substrate are transferred to the SRO layer. The center of the SRO peaks was used to calculate the out-of-plane lattice parameter c_{SRO} in dependence of the average lattice mismatch to the corresponding substrates, using the literature pseudo-cubic bulk value of 3.929 Å. The results are shown in Figure 3b. A decrease in the out-of-plane lattice parameter with increasing mismatch is observed due to of a fully strained growth. A small increase in the pseudo-cubic unit cell of roughly 2% is observed. This is usually associated with small losses of the volatile RuO_2 in the SRO layer due to the high deposition temperatures and the low oxygen background pressure during PLD growth, as reported by other groups [32,33]. However, the RuO_2 losses seem to be relatively small as higher deficiencies would lead to the formation of the Ruddlesden–Popper phase $Sr_3Ru_2O_7$ [34], which would lead to an even larger increase in the out-of-plane lattice parameter or an additional diffraction peak, which was not observed in any of our samples.

In all four RSM maps, the BST peak is clearly visible; however, the shape and position show notable differences. Similar to the rocking curve measurements, the film on STO shows a broad peak, which is elongated in the Q_x direction At the same time, the maximum is at a lower Q_x value compared to the substrate reflex. This indicates that dislocations are formed early during layer growth, which results in a relaxed strain state with a high mosaicity resulting in the broad peak. A more narrow spot along Q_x is observed on DSO; instead, the peak is elongated along the Q_z direction. Accordingly, dislocation might be introduced at a later point during growth, which indicates a coherent growth of the first layers on SRO before strain relaxation starts. In comparison, the film peaks on GSO and NSO show nearly no broadening; i.e., a fully strained growth case is observed in both cases. This is consistent with the sharpest rocking curve for these samples (compare Figure 2b). A strong shift to a higher Q_z value is found for the film on NSO due to the much smaller out-of-plane lattice parameter, originating from the tensile strain state due to the large in-plane lattice parameter of the substrate.

To study the influence of the BST layer thickness, additional films were grown with different pulse numbers and studied with RSM measurements. The position of the maximum intensity from the (1 0 3), (0 1 3), (0 $\bar{1}$ 3) and ($\bar{1}$ 0 3) BST reflexes were used to determine the lattice parameters [35], which are summarized in Figure 3c. For DSO substrates, a strong dependence on thickness was found for thinner samples. In particular, the thinnest film shows equal in-plane lattice parameters as the substrate, pointing to a fully strained state of our BST layers up to a thickness of 100 nm on DSO. If the thickness increases further, the in-plane lattice parameters become larger, whereas the out-of-plane lattice parameter decreases due to relaxation processes during growth. For about 500 nm thick layers, almost no difference was found for the lattice parameters in comparison to the 250 nm thick film. For films on GSO, a similar behavior was observed; however, the relaxation process starts at a higher thickness compared to DSO, which is expected due to the smaller lattice mismatch. Therefore, only the 500 nm thick film shows larger in-plane lattice constants than the substrate as well as a slight decrease in the out-of-plane parameter. Interestingly, all films on NSO show roughly the same lattice parameters, and no indication for relaxation was found up to a thickness of 500 nm despite the fact that the lattice mismatch is larger than for films on GSO. In this case, even an orthorhombic structure can be assumed, as all lattice parameters are different from each other.

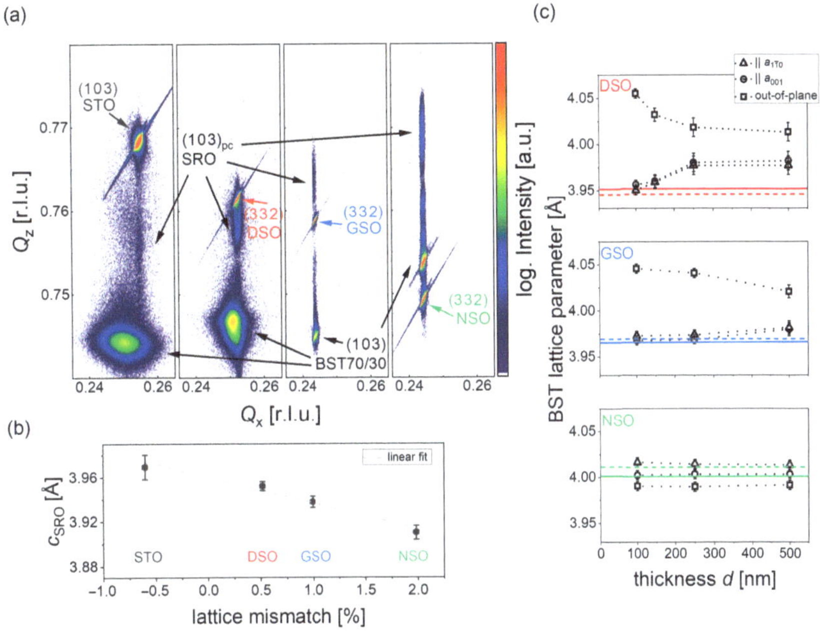

Figure 3. (**a**) RSMs around the (1 0 3) BST (250 nm) peak on the different substrates. From left to right: STO, DSO, GSO, NSO. (**b**) Out-of-plane lattice parameter of the SRO bottom electrode in dependence of the lattice mismatch induced by the different substrates. (**c**) Thickness dependence of the BST lattice parameters on the different $REScO_3$ substrates. From top to bottom: DSO, GSO, NSO. The colored lines represent the literature value for the pseudo-cubic lattice parameter of the corresponding crystal along the [0 0 1] direction and the dashed colored line along the [1 $\bar{1}$ 0] direction.

3.2. Electrical Characterization

To study the ferroelectric behavior of the grown BST layer architectures, the dependence of the polarization P on the electric field E was measured at different temperatures. Selected results are shown for the 250 nm films in Figure 4a–c.

With decreasing temperature, the remnant and saturation polarization P_r and P_s, as well as the coercive field E_c increase for all samples, and S-shaped hysteresis loops are observed at low temperatures. On DSO, the increase in P_r and P_s is continuous when cooling the sample down from 150 to −50 °C, while no major change is observed when cooling further down to −150 °C. The structure on GSO shows a small change of the hysteresis loop between 150 and 50 °C, whereas a significantly higher P_r value is observed during cooling toward −50 °C. At lower temperatures, an additional increase in P_r and E_c was found for this sample. Overall, the film architecture grown on GSO shows a higher P_s compared to the sample on DSO, which might be a result of the higher tetragonal distortion due to the fully strained growth on GSO, as discussed in the previous section.

The sample on NSO, which showed lower out-of-plane than in-plane parameters, reveals lower polarizations values as the films on the other substrates. Even at −150 °C, this film architecture results in a much slimmer polarization loop compared to the ones on DSO and GSO. While the S-shape of the loop is still identifiable at high temperatures for the samples grown on DSO and GSO, an almost linear $P(E)$ dependence is measured for the film on NSO above 50 °C. This behavior is even better visible, when the derivative of the polarization $\partial P/\partial V$ (which is equal to the capacitance per unit area) is plotted versus the applied field as shown in Figure 4d–f. Here, the non-linear behavior with two peaks around the coercive field demonstrates the ferroelectric switching of our capacitors. In case of films on NSO, the $\partial P/\partial V$ curves become nearly field independent at higher temperatures, indicating that the ferroelectricity is vanishing. Instead, the curve at 50 °C shows a hysteresis at higher positive fields, which is a typical sign when leakage has a notable contribution to the corresponding $P(E)$ loop. This feature increases at 150 °C, suggesting that leakage increase with temperature.

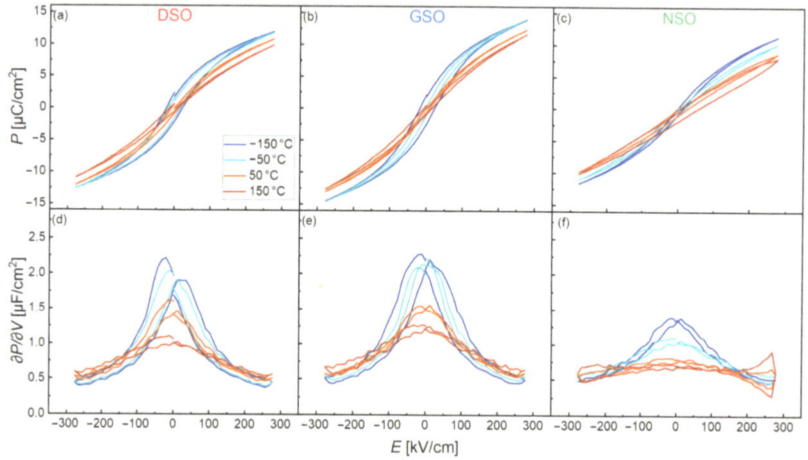

Figure 4. $P(E)$ loops for Pt/BST/SRO layer architectures with a dielectric layer thickness of 250 nm on (**a**) DSO, (**b**) GSO and (**c**) NSO. The corresponding derivative $\partial P/\partial V$ are shown in (**d–f**), respectively.

Since a strong dependence of P on the out-of-plane lattice parameter was found for the about 250 nm thick samples on the different $RE\mathrm{ScO}_3$ substrates and a comparable decrease was found for the out-of plane lattice parameter when increasing the BST layer thickness to 500 nm on GSO, the $P(E)$ loops of those two samples were compared. The results are shown for the lowest temperature in Figure 5a as an example. While the coercive field shows almost no change, the polarization values increase despite the smaller out-of-plane lattice parameter of the thicker layer. This behavior might be explained with a reduced clamping to the substrate, which allows larger parts of the film to align along the applied field [36,37], which in turn leads to a higher measured polarization. Again, the derivative of the polarization loops was calculated and compared as shown in Figure 5b. The main difference in those curves is an additional offset along the $\partial P/\partial V$-axis, indicating a different linear contribution for the capacitance. Here, the offset is roughly half the value as the thickness of the BST layer doubles. In ferroelectric capacitors, the linear component is often associated with interface layers that show significantly reduced functional properties, often referred to as "dead layers" [38,39], and therefore contribute as a linear dielectric offset to the ferroelectric non-linear capacitance. This linear behavior does not change in dependence of the substrate, as shown in Figure 4d–f for the samples with a layer thickness of about 250 nm. The thicker the ferroelectric film becomes, the lower the relative contribution of the interfaces and therefore the linear contribution, which might offer an explanation for the observed behavior.

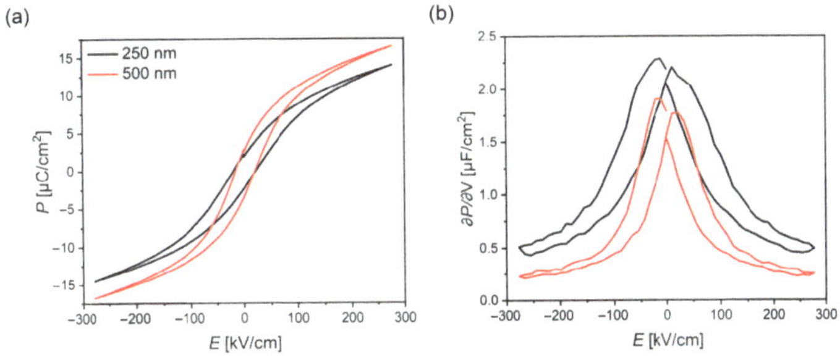

Figure 5. Comparison of the (**a**) $P(E)$ loops and (**b**) its derivative $\partial P/\partial V$ at $-150\,°C$ of a film architecture with a 250 nm or 500 nm BST layer, respectively, grown on GSO.

To acquire a more comprehensive picture of the temperature-dependent dielectric behavior of our layer architectures, the low field permittivity was measured at different frequencies. The results are shown in Figure 6. In the following, the results will be discussed first for the about 250 nm thick films (i.e., Figure 6a–c). The observed dependencies can be divided roughly into different temperature regimes. At high temperatures (i.e., above 100 °C), the measured permittivity is frequency dependent for all samples, which is unexpected for a conventional ferroelectric material. We assume that this frequency dependence is a result of the high leakage in the architectures grown on NSO, as already discussed for the hysteresis loops (compare Figure 4). As the leakage is typically reduced with decreasing temperatures, the permittivity becomes less frequency dependent at medium temperatures between 125 and 0 °C. Since no notable contributions from leakage were observed in the hysteresis loops even at 150 °C for the films on DSO and GSO and the frequency dependence is still observed for the medium-temperature range, it might have another origin in these samples. It should be noted at this place that a similar behavior is often observed in measurement configurations with a poorly conducting electrode [40]. While SRO is usually considered as a good conducting oxide, its resistivity is significantly higher compared to a pure metal like Pt and might be responsible for the observed behavior. It was shown that tensile strain leads to an increase in the ferromagnetic Curie temperature in SRO [41], which results in a higher conductivity and explains why this effect is reduced in the structure on NSO where the highest tensile strain is induced into the oxide electrode layer. A low conductivity of the bottom electrode might also lead to a reduction in permittivity [42]; therefore, our results should be interpreted with caution. Nonetheless, the ε_r values are considerably higher on DSO and GSO when compared to NSO, where the conduction of the oxide layer seems to have less impact, indicating that compressive strain increase the permittivity in comparison to the tensile strain. Since compressive strain typically favors the dielectric dipoles in perovskite ferroelectrics to order along the out-of-plane direction [25], higher depolarization fields are expected, which lead to a higher ε_r. Additionally, the layers on DSO and GSO show, in contrast to NSO, a broad peak for the maximum permittivity at $T = 50\,°C$ and $T = 40\,°C$, respectively.

A frequency-dependent peak is typically observed in relaxor ferroelectrics due to an increasing size or ordering of polar regions inside the material with decreasing temperature up to a certain point T_m. The absence of such a peak for films on NSO might originate from the tensile strain leading to a preferred polarization along the in-plane direction [43]. Therefore, the increased polar order is barely visible in measurements along the out-of-plane direction, as in our case. Nonetheless, a frequency-dependent decrease in permittivity is observed in those samples at low temperatures between −20 and −80 °C. At the same time, peaks are observed for the dielectric loss tan δ at the same temperatures. Such a phase shift between the small amplitude signal and the response of the dielectric film at a point, where the permittivity decreases, is often observed in materials, which show a reentrant relaxor behavior. It is associated with a reduced mobility or freezing of the growing polar regions. At a certain point, those materials go into a state similar to a spin glass, where the dielectric dipoles are randomly aligned, and no reorientation is possible with small fields [44]. Even though the loss peaks are slightly smaller, less frequency dependent and the reduction in permittivity is less significant due to the broad peak above room temperature, a similar behavior is observed on DSO and GSO in the same temperature range. As demonstrated by the polarization measurements, the hysteresis loops reveal an increase in the coercive field and the polarization values, indicating a transition from the more relaxor-like behavior to a classic ferroelectric.

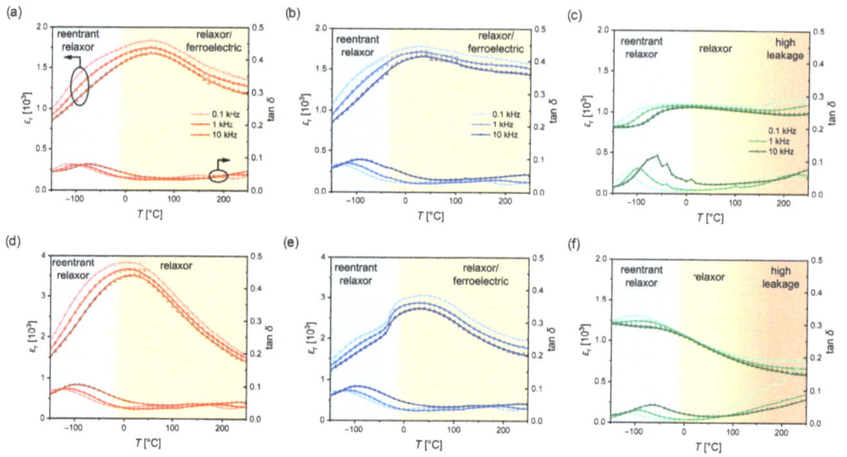

Figure 6. Real part of the relative permittivity ε_r and dielectric loss tan δ as a function of temperature T at different frequencies of film architectures with a 250 nm and 500 nm BST layer on (**a**,**d**) DSO, (**b**,**e**) GSO and (**c**,**f**) NSO.

Similar permittivity measurements were performed for the layer architectures with a BST thickness of about 500 nm. The results are summarized in Figure 6d–f. A significant increase in permittivity is found for the films on DSO and GSO (note the different scale). Again, the mechanical boundary conditions resulting from clamping on the substrate surface might explain this observation. With increased layer thickness, these effects are less prominent due to relaxation effects. Since the critical thickness of the BST layer is significantly lower on DSO compared to GSO, such relaxation effects might have a larger impact on the films on DSO, resulting in a higher permittivity increase. In contrast, no significant increase is observed for thicker BST layers on NSO, which is consistent with the fact that no relaxation was observed even for the 500 nm thick film. It should be also noticed that the layers on DSO show a more relaxor-like behavior in comparison to all other films, as T_m slightly shifts to lower temperatures with decreasing frequency. While this is not clearly visible for the films on GSO, a more pronounced change in slope of ε_r is observed in the low-temperature range. For all three layer architectures, the dielectric loss shows a similar behavior as for the thinner BST layers, indicating similar reentrant-like features at low temperatures.

4. Conclusions

In summary, the use of *RE*ScO$_3$ substrates improved the crystalline quality of the epitaxial layer architectures in comparison to the previously studied STO and results in a coherent growth of the BST-SRO heterostructures up to thicknesses of over 100 nm. Polarization measurements revealed a ferroelectric behavior even at higher temperatures (up to 150 °C) under compressive strain, whereas a suppression of those properties was found under tensile strain. Overall, the slim polarization loops as well as the permittivity measurements indicate a relaxor-like behavior at temperatures above 50 °C, whereas a reentrant-like transition below 0 °C is observed under tensile and compressive strain.

Additionally, the electrode material, interfaces and clamping effects might have a significant influence on the measurements, as indicated by the comparison of structures with a thicker dielectric layer. A temperature-dependent structural characterization by X-ray diffraction might help to further understand how the strain influence the properties of the prepared dielectric layers.

Author Contributions: Conceptualization, R.H.; investigation, J.W. and R.H.; resources, K.N. and R.H.; writing—original draft preparation, J.W.; writing—review and editing, R.H.; supervision, K.N. and R.H.; project administration, R.H.; funding acquisition, R.H. All authors have read and agreed to the published version of the manuscript.

Funding: This work is partially funded by the DFG under grant no. HU1726/8.

Institutional Review Board Statement: Not applicable

Informed Consent Statement: Not applicable

Data Availability Statement: All data in this work are available on request by contact with the corresponding author.

Acknowledgments: The authors thank M. Kühnel for technical support as well as C. Molin and S. E. Gebhardt (Fraunhofer IKTS Dresden) for the provision of the BST target.

Conflicts of Interest: The authors declare no conflict of interest.

Appendix A

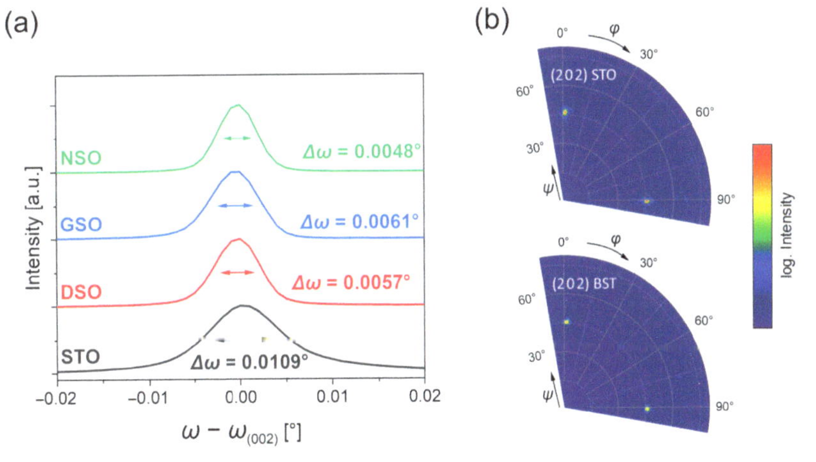

Figure A1. (**a**) Rocking curve scans of the four different substrates. (**b**) Texture measurement of the (2 0 2) planes of a STO single crystal substrate (**top**) and the corresponding BST layer (**bottom**).

References

1. Martin, L.W.; Rappe, A.M. Thin-film ferroelectric materials and their applications. *Nat. Rev. Mater.* **2016**, *2*, 16087. [CrossRef]
2. He, C.; Wen, Y.; Wen, Y.; Zhou, K.; Deng, C.; Li, Q.; Lu, Y. Optical Properties of PT-Based Relaxor Ferroelectric Crystals. *Cryst. Res. Technol.* **2023**, *58*, 2200197. [CrossRef]
3. Liu, J.; Shen, Z.; Yao, W.; Zhao, Y.; Mukherjee, A.K. Visible and infrared transparency in lead-free bulk $BaTiO_3$ and $SrTiO_3$ nanoceramics. *Nanotechnology* **2010**, *21*, 075706. [CrossRef] [PubMed]
4. Dey, S.K.; Kooriyattil, S.; Pavunny, S.P.; Katiyar, R.S.; Subramanyam, G. Analyses of Substrate-Dependent Broadband Microwave (1–40 GHz) Dielectric Properties of Pulsed Laser Deposited $Ba_{0.5}Sr_{0.5}TiO_3$ Films. *Crystals* **2021**, *11*, 852. [CrossRef]
5. Sharma, A.P.; Pradhan, D.K.; Pradhan, S.K.; Bahoura, M. Large energy storage density performance of epitaxial BCT/BZT heterostructures via interface engineering. *Sci. Rep.* **2019**, *9*, 16809. [CrossRef] [PubMed]
6. Zhao, P.; Wang, H.; Wu, L.; Chen, L.; Cai, Z.; Li, L.; Wang, X. High-performance relaxor ferroelectric materials for energy storage applications. *Adv. Energy Mater.* **2019**, *9*, 1803048. [CrossRef]

7. Veerapandiyan, V.; Benes, F.; Gindel, T.; Deluca, M. Strategies to improve the energy storage properties of perovskite lead-free relaxor ferroelectrics: A review. *Materials* **2020**, *13*, 5742. [CrossRef] [PubMed]
8. Scott, J. Electrocaloric materials. *Annu. Rev. Mater. Res.* **2011**, *41*, 229–240. . [CrossRef]
9. Cross, L.E. Relaxor ferroelectrics. *Ferroelectrics* **1987**, *76*, 241–267. [CrossRef]
10. Samara, G.A. The relaxational properties of compositionally disordered ABO_3 perovskites. *J. Phys. Condens. Matter* **2003**, *15*, R367. [CrossRef]
11. Cohen, R. Relaxors go critical. *Nature* **2006**, *441*, 941–942. [CrossRef]
12. Takenaka, H.; Grinberg, I.; Liu, S.; Rappe, A.M. Slush-like polar structures in single-crystal relaxors. *Nature* **2017**, *546*, 391–395. [CrossRef]
13. Panda, P. Environmental friendly lead-free piezoelectric materials. *J. Mater. Sci.* **2009**, *44*, 5049–5062. [CrossRef]
14. Sharma, A.P.; Behera, M.K.; Pradhan, D.K.; Pradhan, S.K.; Bonner, C.E.; Bahoura, M. Lead-free relaxor-ferroelectric thin films for energy harvesting from low-grade waste-heat. *Sci. Rep.* **2021**, *11*, 111. [CrossRef] [PubMed]
15. Maiti, T.; Guo, R.; Bhalla, A. Structure-property phase diagram of $BaZr_xTi_{1-x}O_3$ system. *J. Am. Ceram. Soc.* **2008**, *91*, 1769–1780. [CrossRef]
16. Ren, P.; Gehringer, M.; Huang, B.; Hoang, A.P.; Steiner, S.; Klein, A.; Frömling, T. High field electroformation of sodium bismuth titanate and its solid solutions with barium titanate. *J. Mater. Chem. C* **2021**, *9*, 3334–3342. [CrossRef]
17. Bell, A.J. Ferroelectrics: The role of ceramic science and engineering. *J. Eur. Ceram. Soc.* **2008**, *28*, 1307–1317. [CrossRef]
18. Entesari, K.; Rebeiz, G.M. RF MEMS, BST, and GaAs varactor system-level response in complex modulation systems. *Int. J. RF Microw. Comput.-Aided Eng.* **2008**, *18*, 86–98. [CrossRef]
19. Chun, Y.H.; Hong, J.S.; Bao, P.; Jackson, T.J.; Lancaster, M.J. BST-Varactor Tunable Dual-Mode Filter Using Variable Z_C Transmission Line. *IEEE Microw. Wirel. Components Lett.* **2008**, *18*, 167–169. [CrossRef]
20. Wei, X.; Yao, X. Nonlinear dielectric properties of barium strontium titanate ceramics. *Mater. Sci. Eng. B* **2003**, *99*, 74–78. [CrossRef]
21. Acosta, M.; Novak, N.; Rojas, V.; Patel, S.; Vaish, R.; Koruza, J.; Rossetti, G., Jr.; Rödel, J. $BaTiO_3$-based piezoelectrics: Fundamentals, current status, and perspectives. *Appl. Phys. Rev.* **2017**, *4*, 041305. [CrossRef]
22. Kathait, G.S.; Rohilla, V.; Thapliyal, P.; Biswas, D.; Singh, S. Effect of different strontium content on dielectric properties of barium strontium titanate ceramic. *Int. J. Latest Technol. Eng. Manag. Appl. Sci.* **2017**, *6*, 75.
23. Tyunina, M.; Levoska, J. Coexistence of ferroelectric and relaxor properties in epitaxial films of $Ba_{1-x}Sr_xTiO_3$. *Phys. Rev. B* **2004**, *70*, 132105. [CrossRef]
24. Magalhaes, B.; Engelhardt, S.; Molin, C.; Gebhardt, S.; Nielsch, K.; Hühne, R. Electrocaloric temperature changes in epitaxial $Ba_{1-x}Sr_xTiO_3$ films. *J. Alloys Compd.* **2022**, *891*, 162041. [CrossRef]
25. Ban, Z.G.; Alpay, S. Phase diagrams and dielectric response of epitaxial barium strontium titanate films: A theoretical analysis. *J. Appl. Phys.* **2002**, *91*, 9288–9296. [CrossRef]
26. Koukhar, V.; Pertsev, N.; Waser, R. Thermodynamic theory of epitaxial ferroelectric thin films with dense domain structures. *Phys. Rev. B* **2001**, *64*, 214103. [CrossRef]
27. Shaw, T.; Suo, Z.; Huang, M.; Liniger, E.; Laibowitz, R.; Baniecki, J. The effect of stress on the dielectric properties of barium strontium titanate thin films. *Appl. Phys. Lett.* **1999**, *75*, 2129–2131. [CrossRef]
28. Choi, K.J.; Biegalski, M.; Li, Y.; Sharan, A.; Schubert, J.; Uecker, R.; Reiche, P.; Chen, Y.; Pan, X.; Gopalan, V.; et al. Enhancement of ferroelectricity in strained $BaTiO_3$ thin films. *Science* **2004**, *306*, 1005–1009. [CrossRef]
29. Kawai, M.; Kan, D.; Isojima, S.; Kurata, H.; Isoda, S.; Shimakawa, Y.; Kimura, S.; Sakata, O. Critical thickness control by deposition rate for epitaxial $BaTiO_3$ thin films grown on $SrTiO_3$ (001). *J. Appl. Phys.* **2007**, *102*, 114311. [CrossRef]
30. Uecker, R.; Velickov, B.; Klimm, D.; Bertram, R.; Bernhagen, M.; Rabe, M.; Albrecht, M.; Fornari, R.; Schlom, D. Properties of rare-earth scandate single crystals (Re = Nd − Dy). *J. Cryst. Growth* **2008**, *310*, 2649–2658. . [CrossRef]
31. Guguschev, C.; Galazka, Z.; Kok, D.J.; Juda, U.; Kwasniewski, A.; Uecker, R. Growth of $SrTiO_3$ bulk single crystals using edge-defined film-fed growth and the Czochralski methods. *CrystEngComm* **2015**, *17*, 4662–4668. . [CrossRef]
32. Schwarzkopf, J.; Dirsyte, R.; Devi, A.; Schmidbauer, M.; Wagner, G.; Fornari, R. Depositions of $SrRuO_3$ thin films on oxide substrates with liquid-delivery spin MOCVD. *Thin Solid Films* **2010**, *518*, 4675–4679. [CrossRef]
33. Schraknepper, H.; Bäumer, C.; Gunkel, F.; Dittmann, R.; De Souza, R. Pulsed laser deposition of $SrRuO_3$ thin-films: The role of the pulse repetition rate. *APL Mater.* **2016**, *4*, 126109. [CrossRef]
34. Ohnishi, T.; Takada, K. Epitaxial thin-film growth of $SrRuO_3$, $Sr_3Ru_2O_7$, and Sr_2RuO_4 from a $SrRuO_3$ target by pulsed laser deposition. *Appl. Phys. Express* **2011**, *4*, 025501. [CrossRef]
35. Bauer, G.; Li, J.; Holy, V. High resolution X-ray reciprocal space mapping. *Acta Phys. Pol. A* **1996**, *89*, 115–127. [CrossRef]
36. Schwarzkopf, J.; Fornari, R. Epitaxial growth of ferroelectric oxide films. *Prog. Cryst. Growth Charact. Mater.* **2006**, *52*, 159–212. [CrossRef]
37. Pesquera, D.; Parsonnet, E.; Qualls, A.; Xu, R.; Gubser, A.J.; Kim, J.; Jiang, Y.; Velarde, G.; Huang, Y.L.; Hwang, H.Y.; et al. Beyond substrates: Strain engineering of ferroelectric membranes. *Adv. Mater.* **2020**, *32*, 2003780. . [CrossRef]
38. Stengel, M.; Vanderbilt, D.; Spaldin, N.A. Enhancement of ferroelectricity at metal–oxide interfaces. *Nat. Mater.* **2009**, *8*, 392–397. [CrossRef] [PubMed]

39. Wang, Y.; Niranjan, M.K.; Janicka, K.; Velev, J.P.; Zhuravlev, M.Y.; Jaswal, S.; Tsymbal, E.Y. Ferroelectric dead layer driven by a polar interface. *Phys. Rev. B* **2010**, *82*, 094114. [CrossRef]
40. Tyunina, M. Size effects and dielectric behaviour in ferroelectric heterostructures. *J. Phys. Condens. Matter* **2006**, *18*, 5725. [CrossRef]
41. Dirsyte, R.; Schwarzkopf, J.; Schmidbauer, M.; Wagner, G.; Irmscher, K.; Anooz, S.B.; Fornari, R. Impact of epitaxial strain on the ferromagnetic transition temperature of $SrRuO_3$ thin films. *Thin Solid Films* **2011**, *519*, 6264–6268. [CrossRef]
42. Tyunina, M.; Levoska, J. Dielectric anomalies in epitaxial films of relaxor ferroelectric $(PbMg_{1/3}Nb_{2/3}O_3)_{0.68} - (PbTiO_3)_{0.32}$. *Phys. Rev. B* **2001**, *63*, 224102. [CrossRef]
43. Everhardt, A.S.; Matzen, S.; Domingo, N.; Catalan, G.; Noheda, B. Ferroelectric domain structures in low-strain $BaTiO_3$. *Adv. Electron. Mater.* **2016**, *2*, 1500214. [CrossRef]
44. Bokov, A.A.; Ye, Z.G. Reentrant phenomena in relaxors. In *Nanoscale Ferroelectrics and Multiferroics: Key Processing and Characterization Issues, and Nanoscale Effects*; Wiley: Hoboken, NJ, USA, 2016; pp. 729–764. . [CrossRef]

materials

MDPI

Article

Structural, Magnetic, and Magneto-Optical Properties of Thin Films of BaM Hexaferrite Grown by Laser Molecular Beam Epitaxy

Boris Krichevtsov [1], Alexander Korovin [1], Sergey Suturin [1], Aleksandr A. Levin [1,*], Ivan Lobov [2], Andrey Telegin [2], Andrey Badalyan [1], Vladimir Sakharov [1], Igor Serenkov [1], Maxim Dorogov [3] and Nikolai Sokolov [1]

[1] Ioffe Institute, Politechnicheskaya 26, 194021 St. Petersburg, Russia; boris@mail.ioffe.ru (B.K.); amkorovin@mail.ioffe.ru (A.K.); suturin@mail.ioffe.ru (S.S.); andrey.badalyan@mail.ioffe.ru (A.B.); vlad.sakh@rambler.ru (V.S.); r.ilin@mail.ioffe.ru (I.S.); nsokolov@fl.ioffe.ru (N.S.)

[2] M.N. Mikheev Institute of Metal Physics, 18 S. Kovalevskaya Str., 620108 Ekaterinburg, Russia; i_lobov@imp.uran.ru (I.L.); telegin@imp.uran.ru (A.T.)

[3] Institute of Advanced Data Transfer Systems, ITMO University, Kronverksky Pr. 49, 197101 St. Petersburg, Russia; mvdorogov@itmo.ru

* Correspondence: aleksandr.a.levin@mail.ioffe.ru

Abstract: Thin films of BaM hexaferrite ($BaFe_{12}O_{19}$) were grown on α-Al_2O_3(0001) substrates by laser molecular beam epitaxy. Structural, magnetic, and magneto-optical properties were studied using medium-energy ion scattering, energy dispersive X-ray spectroscopy, atomic force microscopy, X-ray diffraction, magneto-optical spectroscopy, and magnetometric techniques, and the dynamics of magnetization by ferromagnetic resonance method. It was shown that even a short time annealing drastically changes the structural and magnetic properties of films. Only annealed films demonstrate magnetic hysteresis loops in PMOKE and VSM experiments. The shape of hysteresis loops depends on thickness of films showing practically rectangular loops and high value of remnant magnetization (M_r/M_s~99%) for thin films (50 nm) and much broader and sloped loops in thick (350–500 nm) films. The magnitude of magnetization $4\pi M_s \approx 4.3$ kG in thin films corresponds to that in bulk BaM hexaferrite. Photon energy and sign of bands in magneto-optical spectra of thin films correspond to ones observed earlier in bulk samples and films of BaM hexaferrite. FMR spectra of 50 nm films at 50 GHz consist of a number of narrow lines. The width of main line ΔH~20 Oe is lower than has been reported up to now.

Keywords: hexaferrite; molecular beam epitaxy; crystal structure; hysteresis loops; XRD; magneto-optical Kerr effect; VSM; FMR

Citation: Krichevtsov, B.; Korovin, A.; Suturin, S.; Levin, A.A.; Lobov, I.; Telegin, A.; Badalyan, A.; Sakharov, V.; Serenkov, I.; Dorogov, M.; et al. Structural, Magnetic, and Magneto-Optical Properties of Thin Films of BaM Hexaferrite Grown by Laser Molecular Beam Epitaxy. *Materials* **2023**, *16*, 4417. https://doi.org/10.3390/ma16124417

Academic Editor: Dana Popescu

Received: 28 April 2023
Revised: 8 June 2023
Accepted: 13 June 2023
Published: 15 June 2023

1. Introduction

One of the fundamental limitations hindering the increase in the speed and volume of information transmission and processing is the release of Joule heat during the transport of charge carriers. A possible way to solve this problem is by the utilization of magnonic devices based on the use of spin waves packets propagating in magnetic nanoheterostructures [1–3]. In this regard, the problem of creating thin film materials in which it is possible to excite, control, and record weakly damping spin waves arises. For these purposes, intensive studies of nanostructures based on garnet ferrites [4–9], spinel ferrites [10–14], and orthoferrites [15–17] were carried out.

Desirable parameters for the practical application of magnonic structures are out-of-plane orientation of the magnetization without an external magnetic field, a small value of the ferromagnetic resonance (FMR) line width, and a high value of the magnetic moment, which means a large value of the film thickness [18]. As previous studies show, it is difficult to fulfill all these conditions. Thin films of ferrite garnets, in particular, yttrium iron garnet, which have record-breaking minimum FMR line widths, are characterized by low values of the magnetic anisotropy field and demonstrate a planar orientation of the magnetization in the absence of a magnetic field [5–8]. Ferrimagnetic spinels, in particular Ni ferrite (NFO), have higher values of magnetization and magnetic anisotropy field but also exhibit in-plane orientation of magnetization and higher values of FMR line widths [10–12].

Thin films of ferrimagnetic hexaferrites and, in particular, Ba-hexaferrite of M-type (BaM, $BaFe_{12}O_{19}$), are currently attracting much attention due to the unique magnetic parameters of this material. A review of the synthesis, properties, and applications of hexaferrites can be found in ref. [19]. Compared with the above-mentioned magnetic garnets, spinels and orthoferrites, hexaferrites have a number of advantages. The magnetization of BaM hexaferrite at room temperature (RT) $4\pi M_s$~4 kG is higher than in yttrium iron garnet (YIG) $4\pi M_s$~1.7 kG and nickel ferrite $4\pi M_s$~3.3 kG. The uniaxial magnetic anisotropy field H_a in BaM hexsaferrite is $H_a \cong 1.75$ T, that is two orders of magnitude higher than in YIG and one order higher than that of NFO. Due to the high value of H_a, devices based on hexaferrites can operate at frequencies up to $f = 60$ GHz. Moreover, textured polycrystalline hexaferrites can be created with a significant remnant magnetization, which in some cases makes it possible to avoid the use of external magnets.

For these reasons, a number of works have recently been carried out on the fabrication of thin films of BaM hexaferrite by various methods and the study of their magnetic properties. The results of these works can be found in review articles [18–20]. Thin films of hexaferrites were fabricated by pulsed laser deposition (PLD) [21–25], liquid phase epitaxy (LPE) [26–28], direct current magnetron sputtering [29], screen printing (SP) [30], and metallo-organic decomposition (MO) [31] methods on different substrates (sapphire, Al_2O_3, MgO, GdGa-garnet (GGG), 6H–SiC). Comparison of films prepared by different methods, presented in [18], shows that the films of high crystal quality, prepared by PLD or LPE, show an out-of-plane orientation of *c*-axis and small values of FMR lines widths (~30–60 Oe) but also small values of remanence M_r for out-of-plane hysteresis loops. For this reason, for the application of these films in microwave devices an external magnetic field is needed. Contrary to that, the films of lower crystal quality, fabricated by SP [30], have high hysteresis loop squareness (M_r/M_s~0.9), showing realization of "self-bias" effect, but large values of FMR line width. Thus, obtaining hexaferrite films with high crystal quality and high self-bias values is very desirable. Note that epitaxial films of BaM hexsaferrite with in-plane orientation of easy axis and "self bias" effect were synthesized by direct current magnetron sputtering on α-plane (11–20) of single-crystal sapphire substrates [29].

Intensive studies of films and nanostructures based on hexaferrites continue to the present. Recent papers on this topic, in particular, present the results of studying strontium hexaferrite ($SrFe_{12}O_{19}$) films obtained by pulsed laser deposition [32], the effect of La substitution on the structural and chemical properties of barium hexaferrite [33], the magnetic properties of nanostructured $BaFe_{12}O_{19}$ synthesized by sol–gel spontaneous combustion [34], mechanical regulation magnetic properties of uniaxial anisotropic thin films of hexaferrite [35] etc. The main objectives of presented article were fabricating epitaxial thin (with thickness h = 50–500 nm) films of hexaferrite BaM ($BaFe_{12}O_{19}$) grown by laser molecular beam epitaxy on α-Al_2O_3(0001) substrates, studying the chemical composition, surface morphology, crystal structure, static and dynamic magnetic properties, and magneto-optical spectroscopy. Much attention is paid to the comparison of composition, structural, and magnetic properties of as-grown and annealed structures and dependence of magnetic properties on film thickness.

2. Materials and Methods

Thin films (thickness h~50–500 nm) were grown using the laser molecular beam epitaxy (LMBE) method on α-Al_2O_3 (0001) substrates with an installation produced by Surface GmbH (Hückelhoven, Germany). The films were grown in an oxygen atmosphere at a pressure of p = 0.04–0.06 mBar at growth temperatures of T_{gr} = 750 °C–1000 °C. The flow of matter onto the substrate was created by ablation of a stoichiometric $BaFe_{12}O_{19}$ (BaM) target by a KrF excimer laser with a wavelength of 248 nm. The target was fabricated according to the standard technology for the manufacture of ferrites from a charge containing oxides $BaCO_3$ and Fe_2O_3. To determine the best growth parameters, more than 25 structures were grown in various modes, i.e., at various growth temperatures, oxygen pressures, pulse repetition rates, and various annealing modes. For our study, we used both the as-grown structures and structures after annealing procedure. For post-growth annealing, the samples after preparation were removed from the growth chamber and annealed up to 4 h (240 min) in air at the temperature of 1000 °C. Table 1 shows growth and annealing parameters of the samples (single-layer structures) presented in this paper.

Table 1. Number of structure, film thickness h, growth temperature $T_{gr,}$ oxygen pressure p, annealing time t_{ann}, and annealing temperature T_{ann}.

# Sample Number	Film Thickness h, nm	Growth Temperature T_{gr}, °C	Oxygen Pressure p, mBar	Annealing Time t_{ann} (min)/Annealing Temperature T_{ann}, °C
8790	50	850	0.06	no
8939	50	850	0.04	no
8940A	50	850	0.004	no
8940B	50	850	0.004	120/1000
8941A	50	850	0.06	no
8941B	50	850	0.06	30/1000
8944	50	900	0.06	no
8947	50	1000	0.06	no
8948A	50	750	0.06	no
8948B	50	750	0.06	10/1000
8948B$^+$	50	750	0.06	20/1000
8948C	50	750	0.06	60/1000
8948C$^+$	50	750	0.06	120/1000
8949	50	925	0.06	no
8952 **	50 + 1	800 + 900	0.004	no
8954A	300	750	0.06	no
8954B	300	750	0.06	60/1000
8960A	500	700	0.06	no
8960B	500	700	0.06	180/1000
8960C	500	700	0.06	120/1000
8963A	250	550	0.06	no
8963B	250	550	0.06	60/1000
8963C	250	550	0.06	30/1000
8963D	250	550	0.06	10/1000
9001A	500	700	0.06	60/1000
9001B	500	700	0.06	240/1000
9001C	500	700	0.06	no

** #8952 two-stage sample BFO 50 nm 0.004 mbar O_2 800 °C/BFO 0.8 nm 0.004 mbar O2 900 °C/Al_2O_3(0001).

The chemical composition of the BaM films was studied using energy dispersive X-ray spectroscopy (EDX) (Ultim MAX 100 with AZtec software, Oxford Instruments, Oxford,

UK) on a TESCAN scanning electron microscope (SEM) (TESCAN ORSAY HOLDING, a.s.; Brno-Kohoutovice, Czech Republic) with a spectrum analyzer attachment (Oxford Instruments plc, Abingdon, Oxford, UK). The analysis was carried out at the 20 keV energy range and 800,000 count limit. The measurements were carried out on annealed and unannealed samples of 50 nm thick. On each sample, the EDX spectra were measured at four points.

The chemical composition of the films was also studied by medium-energy ion scattering (MEIS) method, which makes it possible to study nanometer-thick films of various compositions and, in particular, to determine the film thickness and its inhomogeneity, elemental composition, and its depth distribution in a non-destructive way. The method includes recording the energy spectra of backscattered ions (BSI) and determining the parameters of the film based on a comparison of the measured spectra with the calculated or standard ones obtained in the study of films of a known composition. The films were probed with H^+ and He^+ ions with an energy of 227 keV. BSI spectra were recorded by electrostatic analyzer, providing the depth resolution in the near-surface region of 0.5–1 nm.

To characterize the surface morphology and crystal quality of grown films, the Al_2O_3(0001) substrates and films grown on them were monitored using an NT-MDT atomic force microscope (NT-MDT LLC, Zelenograd, Russia) in the semicontact mode at RT. Atomic force microscopy (AFM) measurements were carried out both for the as-grown and annealed single-layer structures, as well as for films obtained under various growth conditions.

The growth of a hexaferrite layer on the substrate surface was controlled by reflection high-energy electron diffraction (RHEED). The crystal structure of the grown layers was studied in situ using three-dimensional (3D) mapping of the diffraction intensity distribution. To do this, a series of RHEED patterns were taken during rotation of the sample about the normal to the surface. As a result of φ scanning with a step of 0.5° in φ, a series of 900 images was measured, from which a 3D map of the reciprocal space was obtained using the software. This made it possible to obtain a 3D distribution of the diffraction intensity in the reciprocal lattice and the projection of the intensity distribution onto the chosen reciprocal lattice plane. To analyze the RHEED patterns, the method of 3D mapping of diffraction patterns was applied.

X-ray diffraction (XRD) measurements were carried out using desktop powder X-ray diffractometer D2 Phaser (Bruker AXS, Karlsruhe, Germany) constructed in vertical Bragg–Brentano (reflection) θ-θ geometry and supplied with linear semiconductor position-sensitive detector LYNXEYE and Cu-K_α radiation (wavelength λ = 1.5418 Å) of X-ray tube with copper anode filtered with Ni-foil filter. For the XRD measurements, the film samples were placed at low-background single-crystal Si(119) sample holder. To reduce the influence of effects of preferential orientation, during the measurements, the samples were rotated around the axis coinciding with the axis of the goniometer of the diffractometer. Corrections for zero shift $\Delta 2\theta_{zero}$ and displacement $\Delta 2\theta_{displ}$ to obtain corrected Bragg angle values $2\theta_B$ [36] were determined based on additional XRD measurements of samples immersed in NaCl powder calibrated using XRD standard powder Si640f (NIST, Gaithersburg, MD, USA), so that the NaCl powder and sample surfaces coincided with and were illuminated by the X-ray beam. Due to the design features of the desktop X-ray diffractometer used during measurements, the temperature in the sample chamber was kept at 313 ± 1 K. All XRD patterns were recorded in the range of diffraction angles 2θ from 6° to 141° with an angle step $\Delta 2\theta_{step}$ = 0.02° using the symmetric scanning mode θ-2θ.

The analysis of XRD patterns was carried out using the same methods as described earlier, for example, in [36]. These methods include use the program *EVA* [37] for determination of the XRD reflection parameters and for X-ray phase analysis using the Powder Diffraction File-2 database (PDF-2) [38], calculation of unit cell parameters from the corrected Bragg angle values $2\theta_B$ and Miller indices *hkl* of the reflections using program *Celsiz* [39], and determination of the microstructural parameters utilizing WHP [40] and SSP [41] techniques encoded in the program *SizeCr* [42], which takes into account the pseudo-Voigt (pV) type [43] of XRD reflections in the calculations. The microstructural parameters obtained from the XRD data are mean sizes D of areas of coherent X-ray scattering (crystallites) and absolute mean values of microstrains ε_s in them, and mean sizes D_0 of crystallites in model of zero-microstrain ($\varepsilon_s = 0$). When calculating the WHP and SSP points, the coefficients $K_{\text{strain}} = 4$ and $K_{\text{Scherrer}} = 0.94$ of the Wilson–Stokes [44] and Scherrer [45] equations were used, respectively, which connect the corresponding contributions to FWHM with the values of ε_s and D. Some other details of the analysis are given in the Supporting Materials.

In addition to the XRD technique operating in Bragg–Brentano geometry, XRD studies were performed using reciprocal space mapping, similar to the technique used to analyze high-energy electron diffraction data. For this, a Super Nova diffractometer (Agilent Technologies, Inc.; Santa Clara, CF, USA) operating in kappa geometry with a two-dimensional (2D) detector (Atlas S2 CCD) and an X-ray gun with a copper cathode (λ = 1.5418 Å) was used. Mapping (as in the case of RHEED) consisted in measuring a series of XRD patterns depending on the angle of rotation around the normal to the sample. When the samples were rotated around the normal with a step of 0.5°, a series of images (360 images in a series) were measured, from which 3D maps of the reciprocal space were obtained using the software.

The study of the static magnetic properties of $BaFe_{12}O_{19}/Al_2O_3(0001)$ single-layer structures was carried out using magnetometric and magneto-optical methods. The magnetization curves were measured by means of vibrating sample magnetometry (VSM) method using a vibrating magnetometer (Lake Shore Cryotronics, Westerville, OH, USA) with the magnetic field oriented both along the normal and in the plane of the structure. The magnetic field H varied in the range from +20 to −20 kOe. Measurements were carried out in the temperature range T = 100–300 K. Magnetization values were obtained using the measured magnetic moments, film thickness, and area. The saturation magnetization M_s and coercive fields H_c were calculated from hysteresis loops. To do this, dependences linear in the magnetic field were extracted from the experimental hysteresis loops, which are observed after saturation and are due to the magnetization induced by the magnetic field in the substrate. These loops were used to define the saturation magnetization as the magnetization independent of the magnetic field in strong fields, and the coercive field as the magnetic field corresponding to zero magnetization.

The magnetic field dependences of the polar magneto-optical Kerr effect (PMOKE) were measured on a polarimetric setup at wavelength λ = 405 nm at almost normal incidence of linearly polarized light (angle of incidence~1°). The magnitude of the magnetic field varied in the range from +25 to −25 kOe. The orientation of the magnetic field H was normal to the surface. During slow scanning of the magnetic field, the rotation of the plane of polarization of the reflected light was measured. To increase the sensitivity of polarization plane rotation measurements, the polarization of the incident (or reflected) light was modulated at a frequency f~400 Hz by the amplitude $\alpha_{\max}$~1° using a Faraday cell. The sensitivity of rotation measurements was $\delta\alpha$~1″.

The spectral and field dependencies of the PMOKE were measured in the photon energy range E_{ph} = 1.5–4 eV at RT in magnetic fields up to $H = \pm 15$ kOe. A compact shielded electromagnet was used to magnetize the sample at a frequency of ~2 Hz. During measurements, the magnetic field was oriented normally to the sample surface. The linearly polarized light incident from the monochromator was S-polarized. The analyzer was oriented at 45° to the plane of polarization of the incident light. The PMOKE value was calculated from the reflected light intensities for positive and negative magnetic fields H. The angle of incidence of light on the structure was θ = 52°. For getting spectral dependence of PMOKE at constant magnetic field H the intensity of reflected light was measured for $+H$ and $-H$ magnetic fields at each wavelength. For getting magnetic field dependence of PMOKE, the measurements were carried out for different magnitudes of magnetic field at constant wavelength λ. In addition, the spectral dependencies of the reflection coefficient of the samples were measured.

Spectral dependencies of transversal magneto-optical Kerr effect (TKE) were measured for in-plane orientation of magnetic field H oriented perpendicular to the light incidence plane. The magnitude of TKE = $\Delta I/I_0$ is a ratio between difference in reflected light intensities for positive $+H$ and negative $-H$ magnetic field $\Delta I = I(+H) - I(-H)$, and reflected light intensity I_0 in demagnetized state.

Magnetorefractive effect (MRE) MRE^S was measured in geometry of TKE but for S-polarization of incident light. The magnitude of $MRE^S(H) = \Delta I^{MRE}/I_0$ where $\Delta I^{MRE} = I(0) - I(H)$. FMR was studied in thin (thickness h = 50 nm) films of hexaferrite $BaFe_{12}O_{19}$. For measurements, a JEOL-PE3 electron paramagnetic resonance radiospectrometer (JEOL Ltd., Tokyo, Japan) with a magnet providing a magnetic field H up to 1.3 T was used. Microwaves were generated using a Г4-141 backward wave lamp generator operating in the frequency range of f = 38–53 GHz. FMR studies were carried out at a frequency f = 50 GHz. The recording system of the radiospectrometer was used to register the FMR signal. Magnetic field modulation with amplitude δH = 5 Oe and frequency f = 100 kHz were used to increase the sensitivity of measurements.

3. Results and Discussion

3.1. Chemical Composition

EDX spectra were measured from both as-grown and annealed films. Typical EDX spectrum measured by means of SEM is presented in Figure 1. Table 2 shows the relative concentrations of Fe and Ba atoms in the as-grown and annealed films calculated from measured EDX spectra. In both cases, the Fe:Ba atomic ratio averaged over four measured points is close to the ideal composition (12:1), in particular (11.7:1) in the as-grown structure and (10.4:1) in the annealed one.

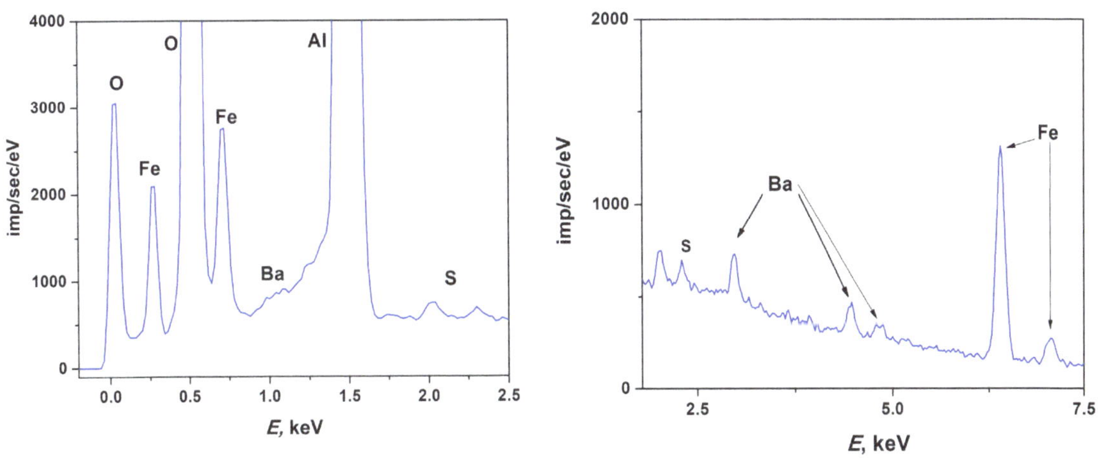

Figure 1. Typical EDX spectra of BaM structures measured by SEM for energy regions (**a**) 0–2.5 keV and (**b**) 2.5–7.5 keV.

More accurate results were obtained using the MEIS technique. Figure 2 shows the experimental MEIS spectra measured with use of He^+ ions and model spectra of the unannealed (#8948A) and annealed (#8948C) samples.

Table 2. Fe:Ba concentration ratio for films with thickness h = 50 nm grown at 750 °C, before (film #8948A) and after annealing at 1000 °C for 10 min (film #8948B).

	As-Grown Sample #8948A		Annealed Sample #8948B	
Point Number	**Fe:Ba Mass Ratio**	**Fe:Ba Atomic Ratio**	**Fe:Ba Mass Ratio**	**Fe:Ba Atomic Ratio**
1	5.0	12.1	4.1	9.925
2	4. 7	11.2	4.2	10.07
3	5	12	4.0	9.511
4	4.8	11.4	4.8	11.45

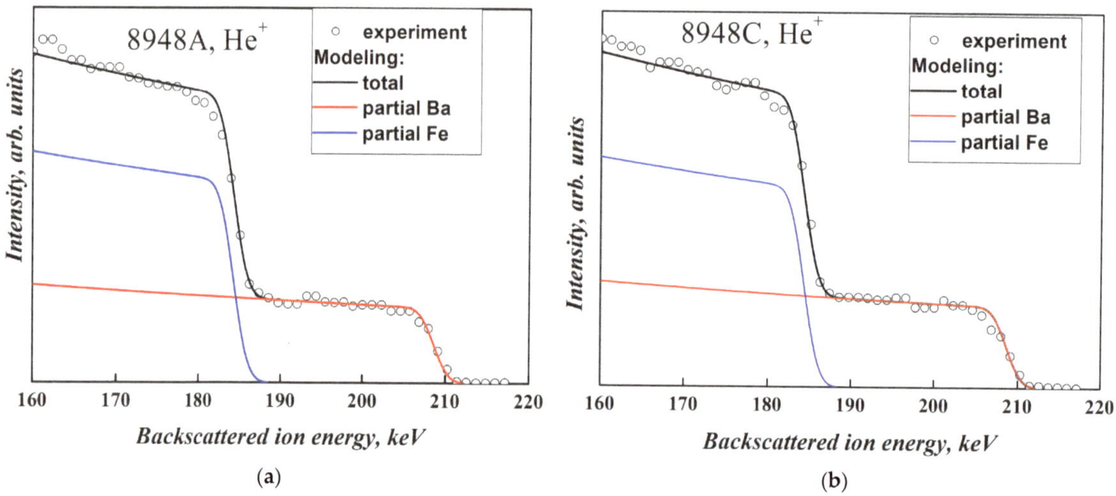

Figure 2. Experimental (circles) and model (solid lines) MEIS spectra measured using He^+ ions for films with thickness h = 50 nm grown at 750 °C, (**a**) before (film #8948A) and (**b**) after annealing at 1000 °C for 60 min (film #8948C). Red and blue lines show contributions of Ba to Fe ions correspondingly in total signal, black line shows the modeling of total signal.

Model MEIS spectra were calculated using our original code, utilizing the basic principles of Rutherford Backscattering Spectrometry. The stopping cross sections for protons and helium ions were taken from [46–48]. The simulation showed that the Fe:Ba ratio is equal to 11.0 and 10.2 for the as-grown (#8948A) and annealed (#8948C) sample, respectively, which correlates with the data obtained using SEM. As can be seen from Figure 2, the decrease in the Fe:Ba ratio after annealing is mainly associated with a decrease in the Fe concentration, and the concentration of Ba ions changes slightly.

To measure the film thickness and its variations over area, as well as to estimate the oxygen content, MEIS experiments with H^+ ions were carried out (Figure 3).

Simulation of the spectra presented in Figure 3 shows that the lateral density N of $Ba_xFe_yO_z$ molecules with $x + y + z = 1$ (i.e., the total number of Ba, Fe, O atoms per 1 cm^2 of the film) is $N = 470 \cdot 10^{15}$ molecules/cm^2 and $N = 580 \cdot 10^{15}$ molecules/cm^2 in the as-grown and annealed films, respectively. An estimate of the film thickness h based on the assumption that the mass density of the films corresponds to the density of $BaFe_{12}O_{19}$ crystals (i.e., $\rho = 5.01$ g/cm^3) gives values h = 54 nm and 67 nm for the samples #8948A and #8948C, correspondingly, which is close to the RHEED results (h = 50 nm).

It is important to note that, in the as-grown film, the value of N does not depend on the in-plane coordinates, i.e., the film is laterally homogeneous. In contrast, in the annealed film, this parameter is characterized by a remarkable dispersion of $80 \cdot 10^{15}$ atoms/cm^2. In fact, the shape of the low-energy Ba + Fe signal front is rather sharp for as-grown sample (180–195 keV, Figure 3a) and flat in the annealed film (170–195 keV, Figure 3b).

An indirect estimate of the oxygen content was obtained by comparing the intensity of signal from the film (right sides in Figure 3a,b) and the substrate signal (left side). Such an estimate showed the atomic ratio O:Ba~(14:1) and ~(19:1) for the as-grown and annealed 8948 film, respectively.

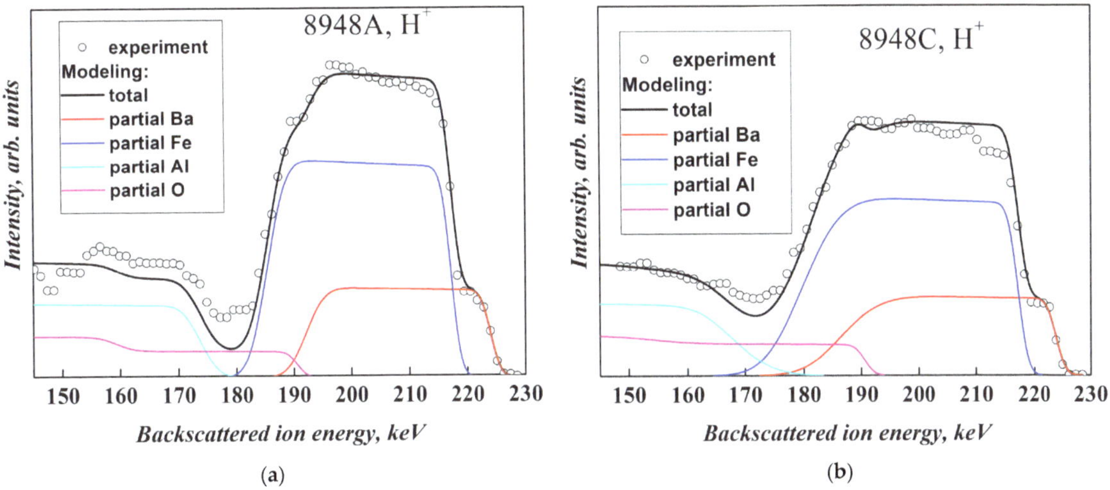

Figure 3. Experimental (circles) and model (solid lines) MEIS spectra measured using H^+ ions for films with thickness h = 50 nm grown at 750 °C, (**a**) before (film #8948A) and (**b**) after annealing at 1000 °C for 60 min (film #8948C). Red, blue, cyan, and magenta lines in (**a**,**b**) show contributions to total signal of Ba, Fe, Al, and O ions, correspondingly, and the black line represents total calculated spectrum.

Thus, both approaches, EDX and MEIS, show that the atomic ratios in the as-grown and annealed films are close to those expected in $BaFe_{12}O_{19}$. This indicates that the number of ions on the substrate surface after deposition is sufficient for the nucleation and crystallization of BaM hexaferrite. The Fe:Ba atomic ratio in the annealed sample is slightly lower than in the as-grown one, which indicates that the film composition is depleted of iron upon annealing at high temperatures. After annealing, both the layer thickness and their dispersion increase, which may be due to $BaFe_{12}O_{19}$ crystallization during annealing. At the same time, as a result of annealing, the O:Ba atomic ratio approaches the ideal one.

3.2. Surface Morphology

Figure 4 shows the AFM images of structures #8954A (as-grown, h = 300 nm) and #8954C (annealed in duration of 60 min, h = 300 nm). The unannealed film #8954A demonstrates a rather smooth surface with no evidence of the presence of nanocrystallites (Figure 4a). The average roughness of such a surface is a root mean square (RMS) value of ~2 nm over an area of 2 × 2 μm^2. In contrast, the surface of the annealed sample #8954C (Figure 4b) consists of a set of nanocrystallites with a pronounced faceting, which is typical for crystals with a 6th-order axis directed approximately normal to the substrate. The structure contains nanocrystals in the form of regular hexagons, hexagons with different side lengths, and triangles. Note that the orientations of the sides of the nanocrystallite faces located at different edges of the pattern are practically parallel to each other, which indicates a correlation of crystallographic directions in them.

Thus, studies of the surface morphology using AFM showed that the surface of the annealed films is composed of closely packed nanocrystallites in which a 6th-order axis is oriented approximately normal to the surface, which confirms the presence of BaM hexaferrite. In contrast, the presence of nanocrystallites is not observed in the AFM images of unannealed structures.

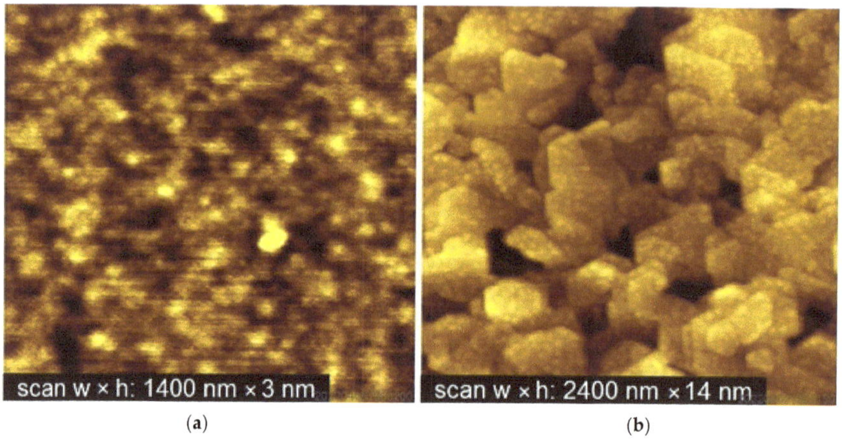

Figure 4. AFM images of the surface of BaM hexaferrite layers (h = 300 nm) grown at 750 °C (**a**) before (film #8954A) and (**b**) after annealing at 1000 °C for 60 min (film #8954C).

3.3. RHEED Study of Crystal Structure

Figure 5 presents reciprocal space images obtained by 3D mapping of RHEED patterns in films #8954A (T_{gr} = 750 °C, as-grown), 8954B (T_{gr} = 750 °C, annealing at T_{ann} = 1000 °C for 60 min), and #8947 (T_{gr} = 1000 °C, as-grown).

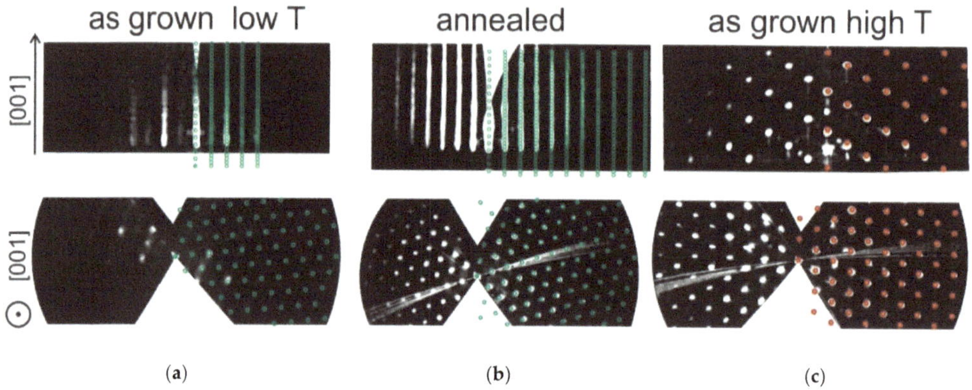

Figure 5. Three-dimensional reconstruction of RHEED data for (**a**) film #8954A grown at 750 °C (h = 300 nm) without annealing, (**b**) #8954B (h = 300 nm) grown at 750 °C and annealed for 60 min at 1000 °C, and (**c**) #8947 grown at 1000 °C without annealing (h = 50 nm). The green and red circles represent the nodes of the model lattice calculated using the parameters of the BaM hexaferrite bulk crystal.

The RHEED patterns of hexaferrite layers grown at temperatures of 700–850 °C (not annealed, see Table 1) more or less correspond to the bulk structure of BaM hexaferrite (Figure 5a). However, the increased width of the reflections and the uneven distribution of the intensity of the reflections in these patterns indicate a relatively low crystalline quality of the grown films. It is also seen that every second rod has a significantly higher intensity, which indicates a strong violation of the long-range order. In contrast, films grown at 700–850 °C and then annealed in air at 1000 °C, show RHEED images ideally modeled by the BaM bulk lattice (Figure 5b). It can be seen that the number of reflections and the signal-to-background ratio are much better compared to the unannealed samples. All rods have the same intensity, which indicates the presence of long-range order. Films grown at temperatures T_{gr} = 900–1000 °C differ significantly from samples grown at lower temperatures. The RHEED patterns of such samples are well modeled by the α-Fe_2O_3 (hematite) lattice (Figure 5c). The formation of α-Fe_2O_3 is apparently associated with repeated sputtering (re-evaporation caused by elevated temperature) of Ba ions at high temperatures. Accounting for the re-sputtering process can explain the effect of annealing at 1000 °C on the crystal quality of films grown at low temperatures. In films grown at T_{gr} = 700–850 °C, the re-sputtering is small and the Ba ions are distributed more or less uniformly inside the hexaferrite film. Annealing, on the other hand, leads to a redistribution of Ba positions and the occupation of Ba ions in places corresponding to the BaM crystal structure, which leads to the formation of a high-quality BaM layer with improved magnetic properties (see below). Film growth at 1000 °C is accompanied by the disappearance of Ba ions from the film and the formation of the α-Fe_2O_3 structure.

Thus, the results of studying the crystal structure using RHEED confirm the conclusions of the previous section devoted to the study of surface morphology using AFM. In the unannealed films, the RHEED patterns do not show the presence of a well-defined BaM hexaferrite structure with long-range order. They may contain uncorrelated small nuclei. A clear picture of RHEED, which fully corresponds to the presence of the hexaferrite structure, appears only in annealed films.

3.4. XRD Studies

The results of XRD measurements of the film samples are shown in Figure 6.

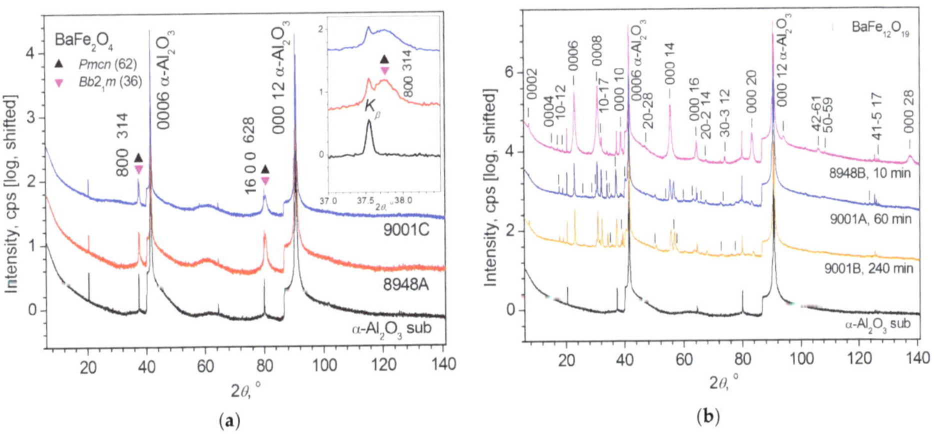

Figure 6. XRD patterns (θ-2θ scans) of the samples #9001 and #8948, and α-Al_2O_3(0001) substrate (**a**) before annealing, and (**b**) after annealing (see Table 1). The duration of annealing of each sample is indicated. On panel (**a**), different triangular symbols show the angular positions of the observed reflections of various possible modifications of $BaFe_2O_4$ (space group (sp. gr.) $Bb2_1m$ (36) and sp. gr. *Pmcn* (62)) according to the unit cell parameters from the PDF-2 database, entries 00-046-0113 and 01-077-2337, respectively. The inset in (**a**) shows part of the XRD patterns of the films before annealing in vicinity of the $BaFe_2O_4$ reflection with Miller indices *hkl* = 800 for modification with sp. gr. $Bb2_1m$ (36) (or *hkl* = 314 for phase with sp. gr. *Pmcn* (62)) on an enlarged scale. In panel (**b**), the observed reflections of $BaFe_{12}O_{19}$ (sp. gr. $P6_3/mmc$ (194) according to PDF-2 entry 00-027-1029) are marked with bar symbols at the positions of the Bragg angles $2\theta_B^{calc}$ obtained using the $BaFe_{12}O_{19}$ unit cell parameters calculated in this work. The Miller–Bravais *hkil* indices of some selected $BaFe_{12}O_{19}$ reflections are indicated. For better visualization, the XRD patterns in (**a**,**b**) are shifted vertically.

The obtained interplanar distances in unannealed samples and the calculated parameters of unit cells in annealed samples are given in Tables 3 and 4. The results of determining the parameters of the microstructure from the analysis of the observed pseudo-Voigt (pV) profiles ($0.636 < FWHM/B_{int} < 0.939$ [39]) of XRD reflections using the Williamson–Hall plot (WHP) and the Size–Strain plot (SSP) are presented in the same Tables 3 and 4. The WHP and SSP graphs are given in Figures S1 and S2 of Supplementary Materials.

Table 3. Interplanar spacings *d* and microstructure parameters (average sizes D_0 of crystallites in the model without microstrains ($\varepsilon_s = 0$), mean sizes *D* of crystallites and absolute mean values of microstrains ε_s in them) of the observed crystalline phase(s) $BaFe_2O_4$ in unannealed samples #8948A and #9001C. Mean profile type criteria $FWHM/B_{int}$ averaged over all observed XRD reflection and are shown.

Sample	Interplane Distance *d*, Å	D_0, nm $FWHM/B_{int}$	WHP *D*, nm ε_s, %	SSP *D*, nm ε_s, %
#8948A [a] (without annealing)	0.23813(6) [b] 0.11914(1) [c]	54(28) 0.673(1)	– [d] 0.159(8)	– [d] 0.159(8)
#9001C [a] (without annealing)	0.23736(6) [b] 0.11898(1) [c]	30(11) 0.6721(1)	80(12) 0.22(1)	80(41) 0.22(3)

[a] Only two reflections are observed, which can be attributed to different orders of one reflection of various modifications of $BaFe_2O_4$ (space group (sp. gr.) $Bb2_1m$ (36), *a* = 19.042(4) Å, *b* = 5.3838(7) Å, *c* = 8.4445(7) Å at ambient temperature 298 K, PDF-2 entry 00-046-0113 or sp. gr. *Pmcn* (62), *a* = 17.3469(39) Å, *b* = 9.3358(24) Å, *c* = 10.8818(19) Å at ambient temperature 298 K, PDF-2 entry 01-077-2337); in WHP and SSP graphs, the calculated value of R_{cod} = 100% has no physical meaning. [b] Interplanar spacing *d* corresponding to the reflection with Miller indices *hkl* = 800 of the $BaFe_2O_4$ modification with sp. gr. $Bb2_1m$ (36) or *hkl* = 314 of the $BaFe_2O_4$ phase with sp. gr. *Pmcn* (62). According to the unit cell parameters of the PDF-2 entries 00-046-0113 and 01-077-2337, *d* = 0.23800 Å and 0.23803 Å for phases with space groups $Bb2_1m$ (36) and *Pmcn* (62), respectively. [c] Interplanar spacing *d* corresponding to the reflection with Miller indices *hkl* = 16 0 0 of the $BaFe_2O_4$ modification with sp. gr. $Bb2_1m$ (36) or *hkl* = 628 of the $BaFe_2O_4$ phase with sp. gr. *Pmcn* (62). According to the unit cell parameters of the PDF-2 entries 00-046-0113 and 01-077-2337 *d* = 0.11901 Å and 0.11901 Å for both modifications. [d] The graphical line profile analysis technique (WHP or SSP, respectively) evidences the dominant contribution of microstrain to broadening of the observed reflections and no influence of size broadening ("*D* = ∞").

Table 4. Parameters *a* and *c* of the hexagonal unit cell and microstructure parameters (average sizes D_0 of crystallites in the model without microstrains ($\varepsilon_s = 0$), mean sizes *D* of crystallites and absolute mean values of microstrains ε_s in them) of the observed crystalline phase $BaFe_{12}O_{19}$ [a] in annealed samples #8948B, #9001a, and #9001B. Mean profile type criteria $FWHM/B_{int}$ averaged over all observed XRD reflection and are shown. Coefficients of determination R_{cod} of straight WHP (SSP) lines are presented.

Sample	*a*, Å *c*, Å	D_0, nm $FWHM/B_{int}$	WHP		SSP	
			R_{cod}, %	*D*, HM ε_s, %	R_{cod}, %	*D*, HM ε_s, %
#8948B (annealing 10 min)	5.8853(18) 23.120(9)	45(17) 0.69(3)	2.03	45(17) 0	95.22	38(2) 0.03(36)
9001A (annealing 60 min)	5.8618(15) 23.164(6)	53(25) 0.80(10)	4.39	50(8) 0.066(30)	71.95	145(18) 0.173(12)
#9001B (annealing 240 min)	5.8573(45) 23.075(20)	52(23) 0.81(9)	15.06	61(1) 0.117(25)	76.04	100(10) 0.176(17)

[a] According to PDF-2 00-027-1029 entry of PDF-2, *a* = 5.892 Å, *c* = 23.198 Å at 298 K for $BaFe_{12}O_{19}$ (sp. gr. $P6_3/mmc$ (194)).

Estimated standard deviations (e.s.d.s., see the note in the Supplementary Materials) of the structural and microstructural parameters are shown in Tables 3 and 4 in round brackets. As is seen, the *D* and ε_s values obtained by WHP and SSP techniques are close in the limits of e.s.d. for the samples without annealing (Table 3) and for the sample #8948D after annealing (Table 4). For the sample #9001C after different annealing durations, there are significant differences (Table 4). However, high values of the R_{cod} coefficient (R_{cod} = 71.95–95.22%) in the case of SSP analysis compared with significantly lower values of R_{cod} = 2.03–15.06% for WHP indicate more accurate values of microstructural parameters obtained using SSP, which were used for further analysis.

In the unannealed films, only two reflections from the film are present in XRD patterns. These reflections are well-identified as reflections of two possible orthorhombic modifications of $BaFe_2O_4$ (Figure 6a). Both reflections are reflections of a different order from the same type of parallel diffracting atomic planes. No reflections from a set of planes of another type are observed, which indicates a strong predominant orientation of the unannealed films. Judging by the Miller indices of the reflections, there is a preferential orientation along the [100] or [314] direction depending on the possible modification of $BaFe_2O_4$. Due to the small number of XRD reflections, measurements do not allow one to choose between these two modifications of $BaFe_2O_4$.

All observed non-substrate reflections of annealed films 8948B, 9001A, and 9001B grown at T_{gr} = 700–750 °C (Table 1) are attributed to the $BaFe_{12}O_{19}$ crystalline phase (sp. gr. $P6_3/mmc$ (194)). There are no reflections of other crystalline phases, including reflections of hematite (α-Fe_2O_3, sp. gr. $R\bar{3}c$ (167)), which are formed in films grown at higher T_{gr} = 900–1000 °C (see Section 3.3 and Figure S3 of Supporting Materials). Thus, annealing leads to crystallization in all studied films of the $BaFe_{12}O_{19}$ compound of hexagonal syngony with a predominant orientation along the [*hkil*] = [0001] direction (Figure 6b), as evidenced by the presence of a large set of different orders of the 000*l* reflection with increased intensities. However, the predominant orientation is not as strong as in unannealed films, since XRD patterns also contain not only 000*l* reflections, but also *hkil* reflections.

Thus, XRD studies showed that only the orthorhombic $BaFe_2O_4$ phase, which consists of highly stressed nanocrystallites (ε_s = 0.159(8)–0.22(3)%, Table 4), manifests itself in unannealed samples. After annealing, the $BaFe_2O_4$ phase disappears in all structures and the hexagonal $BaFe_{12}O_{19}$ phase is present. In a #8948B structure (film thickness h = 50 nm), a 10 min annealing virtually eliminates microstrains (ε_s = 0.03(36)% in comparison to ε_s = 0.159(8)% in the unannealed sample). In thick films 9001A and 9001B (h = 500 nm), annealing for 60 to 240 min does not result in peeling, but reduces the amount of microstrain from ε_s = 0.22(3)% to 0.17(1)–0.18(2)% (Table 4). The results of XRD studies of the films using reciprocal space mapping are shown in Figures 7 and 8.

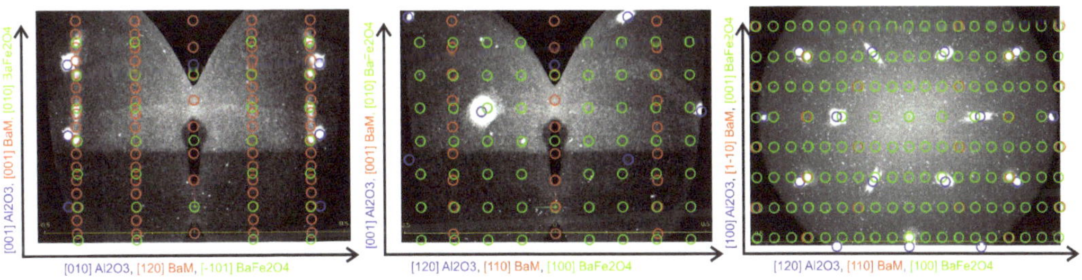

(**a**) (**b**) (**c**)

Figure 7. Cross sections of reciprocal space maps from an unannealed sample 300 nm thick in different planes. Blue, red, and green circles in (**a**–**c**) correspond to calculated pattern of Al_2O_3, BaM, and $BaFe_2O_4$, respectively.

Figure 7 shows that the substrate reflections are well modeled by the reciprocal lattice of a bulk sapphire crystal. In contrast, the film reflections in none of the azimuths correspond to the BaM model reciprocal lattice (red circles in Figure 7a,b). In the direction normal to the sample, the interplanar spacing coincides with good accuracy with the interplanar spacing in $BaFe_2O_4$ (green circles in Figure 7a,b). However, in the plane of the sample (Figure 7c), apparently, there is a large disorder.

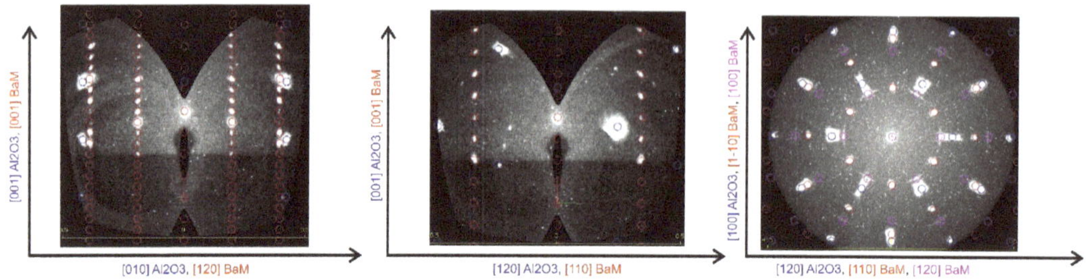

(**a**) (**b**) (**c**)

Figure 8. Cross sections of reciprocal space maps from an annealed sample 300 nm thick in various planes. Blue, red, and violet circles in (**a–c**) correspond to Al_2O_3, dominant phase of BaM, and textured phase of BaM, respectively.

For the annealed sample, XRD mapping data (Figure 8a–c) show good agreement between the model reciprocal lattice of BaM and the observed reflections. It can be concluded that the dominant lattice of the BaM film is rotated by 30° relative to the sapphire lattice in the sample plane. However, it can be seen that in the section constructed in the plane of the sample (Figure 8c), in addition to the dominant BaM lattice (red circles), there is one more phase with interplanar spacings corresponding to the reciprocal BaM lattice. This phase does not have a 30° turn and tends to be textured (has a fairly large, on the order of several degrees, spread in rotation angles around the normal to the sample surface), since the reflections from this phase have the shape of sphere sectors. According to the intensity ratios, it can be said that the volume of the textured phase is much smaller than the volume of the BaM dominant phase. Note that some of the reflections present in the data obtained when measuring in reflection geometry (θ-2θ scans, Figure 6b) using a linear (1D) detector were not reliably recorded in the course of 3D mapping; this may be due to the lower sensitivity of the 2D detector used in mapping (on the specular reflection curves, the data reflections have an intensity several orders of magnitude lower than the main BaM phase). Apparently, the XRD reflections found in the reflection geometry are related to the textured phase found using XRD mapping. It can be hoped that with further optimization of the annealing conditions, it will be possible to get rid of the textured BaM phase, which, in our opinion, should lead to an improvement in the magnetic characteristics of the BaM films.

In conclusion of this section, let us consider the results obtained from the point of view of understanding the processes occurring during growth. Film growth at the growth temperature T_{gr} = 1000 °C leads to the appearance on the substrate mainly of hematite α-Fe_2O_3 phase. Obviously, this is due to the strong re-evaporation of Ba ions from the film at such a high growth temperature, which excludes the appearance of hexaferrite. A decrease in the growth temperature to T_{gr} = 700–750 °C is no longer accompanied by a strong evaporation of Ba ions, and the amount of Ba, Fe, and O ions on the substrate required to obtain BaM hexaferrite turns out to be quite sufficient, both in annealed and unannealed structures. However, immediately after film growth, it is not hexaferrite that is formed on the substrate, but rather highly strained $BaFe_2O_4$ nanocrystallites with a spinel structure. This happens, on the one hand, because the growth process by the LMBE method is a very nonequilibrium process, in which a large amount of material evaporated from the target falls on the substrate in a very short time, which should eventually form a hexaferrite structure, and on the other hand, due to the complex structure of hexaferrite, the unit cell of which consists of a certain sequence of spinel and hexagonal blocks. It is the spinel block $BaFe_2O_4$ that is one of the "bricks" of the hexaferrite lattice. It seems that the formation of "bricks" at the first stage is required for the construction of hexaferrite, and their ordering into the structure of hexaferrite at the second stage. Perhaps, in the films obtained immediately after growth, we observe the result of only the first stage. The implementation of the second stage requires annealing at T_{ann} = 1000 °C, which is higher than the growth temperature. Despite the high annealing temperature, evaporation of Ba ions from the film, at least for a short annealing time of 10 min, does not occur, since they are already embedded in the Ba_2FeO_4 spinel blocks.

Note that annealing must be carried out in air, i.e., at high oxygen concentrations. Experiments have shown that annealing at T_{ann} = 1000 °C directly in the growth chamber, at a low oxygen pressure of up to 0.2 mbar, does not lead to the formation of a hexaferrite structure. Apparently, this indicates that, during annealing, a larger amount of oxygen must be introduced into the film from the outside.

As a result, the formed crystalline hexaferrite phase in thin 50 nm films (#8948B), annealed in a short time of 10 min, is characterized by the nanocrystallite sizes D~40 nm comparable with the film thickness, absence of microstrains in them and the unit cell parameters that are closest to the PDF-2 tabular values (Table 4).

Annealing in air at T_{ann} = 1000 °C of thick 500 nm-films at long durations of 60 min (#9001A) and 240 min (#9001B) leads to formation of larger crystallites with noticeable microstrains (according to SSP analysis, D are larger than ~100 nm, ε_s~0.1%). In addition, an increase in the annealing duration leads to a sequential decrease in the values of the unit cell parameters. This is probably due to the fact that during long-term annealing, nevertheless, some of the Ba atoms can still evaporate (with the concomitant evaporation of oxygen or a change in the Fe valence to maintain electrical neutrality), but with the preservation of the hexaferrite structure. As a result of a larger difference from the tabular values of the lattice parameters, microstrains appear in the emerging hexaferrite nanocrystallites.

3.5. Static Magnetic Properties

3.5.1. PMOKE Measurements

The study of magnetic hysteresis loops using PMOKE polarimetric technique showed a significant difference between unannealed and annealed structures. Most of the as-grown structures do not show the presence of any magnetic moment. Only in a few structures were observed weak and diffuse loops, the shape of which is far from rectangular (Figure 9a).

In contrast, in all as-grown films, even in those that did not show magnetic properties, annealing led to the appearance of pronounced and almost rectangular magnetic hysteresis loops (Figure 9b). The effect of annealing on the shape of hysteresis loops in films #8948 (h = 50 nm), #8963 (h = 250 nm), and #9001C (h = 500 nm) is shown in Figure 9c,e,f. The shape of the hysteresis loops of the annealed structures is rather close to rectangular (Figure 9b,c). It is important that the remanent PMOKE value is very close to the saturation one. This indicates that after saturation of the film in a strong field, the direction of the residual magnetization in zero field remains practically normal to the plane of the film. Figure 9d shows the value of PMOKE in saturation (proportional to M_s) and relative remanence M_r/M_s versus annealing time t_{ann} in structure #8948 (h = 50 nm). With an increase in t_{ann} to 120 min, the value of M_s first increases sharply, and in the range t_{ann}~40–120 min it changes very weakly. The relative residual magnetization M_r/M_s in the range of 10–120 min slightly increases from 0.9 to 1.

It is important to note that the shape of the hysteresis loops in the annealed structures essentially depends on the thickness h of the hexaferrite layer. Figure 9e,f show hysteresis loops in films #8963 (h = 250 nm) and #9001 (h = 500 nm) with different annealing time. The shape of the loops in these structures is far from being rectangular, due to the tightening of the branches at $|\mathbf{H}| > |\mathbf{H}_c|$ in sample #8963 and with a strong slope of the branches in sample #9001. Note that an increase in the annealing time in such a structure leads to "deterioration" of the loops, i.e., to an increase in H_c and an even greater difference in the shape of the loop from a rectangular one.

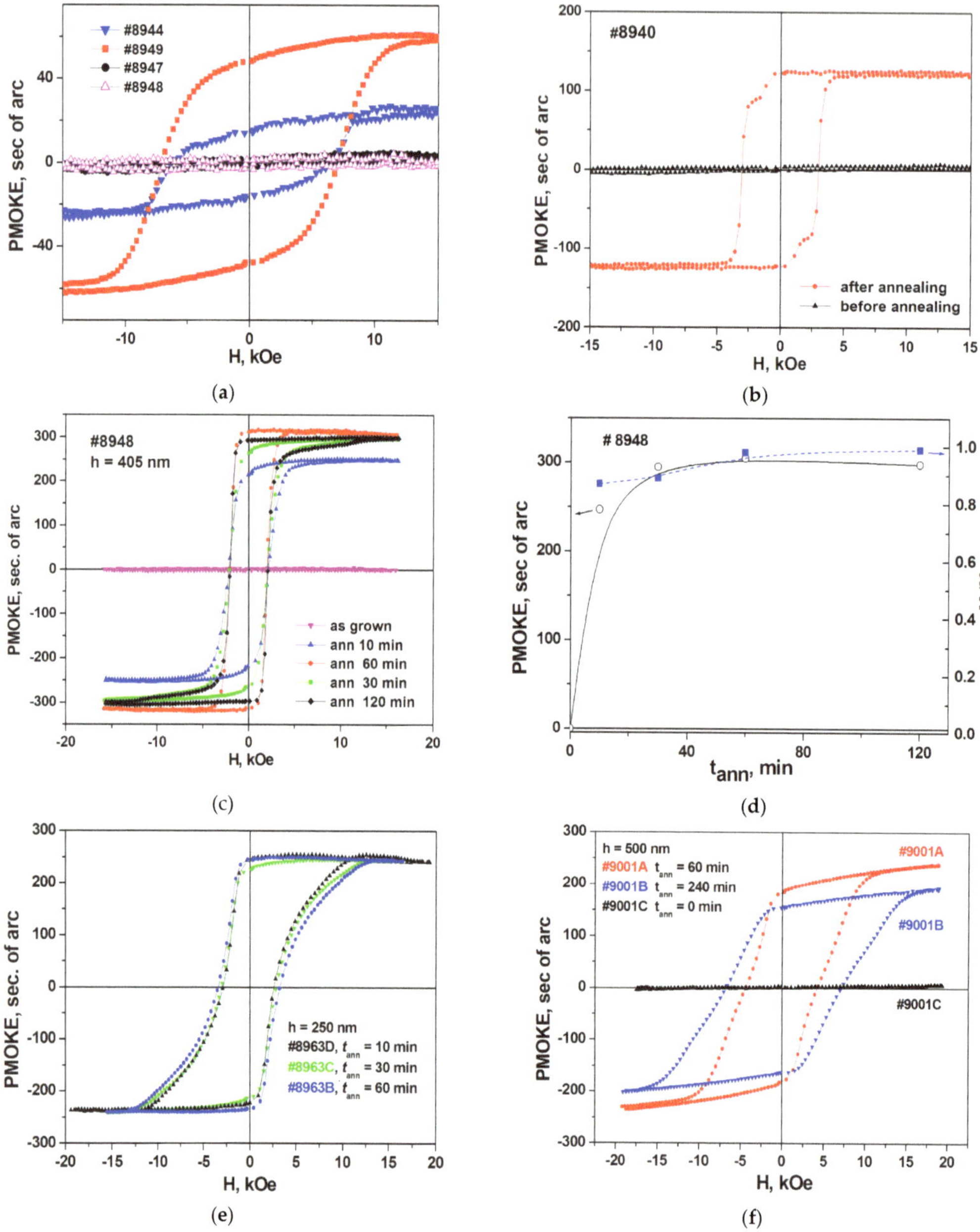

Figure 9. PMOKE magnetization curves of unannealed samples (**a**) #8944, #8949, #8947, #8948, and (**b**) sample #8940 before and after annealing, and (**c**) PMOKE magnetization curves of the sample #8948 (T_{gr} = 750 °C, film thickness h = 50 nm) versus annealing time (t_{ann} = 0 (i.e, as grown), 10 min, 30 min, 60 min, and 120 min) at temperature T_{ann} = 1000 °C. (**d**) Annealing time dependence of the PMOKE value in saturation (circles) and the ratio of remanence to saturation magnetization M_r/M_s (blue squares). PMOKE magnetization curves versus annealing time (**e**) of the sample #8963 (h = 250 nm, annealing time t_{ann} = 10 min, 30 min, and 60 min) and (**f**) of the sample #9001 (h = 500 nm, t_{ann} = 60 min and 240 min).

Thus, studies of hysteresis loops using PMOKE have shown that only annealed structures demonstrate magnetic properties. They exhibit loops with a large remanence magnetization M_r/M_s, which is required for microwave devices based on direct bulk spin waves. The narrowest and most rectangular loops appear in samples with thin (h = 50 nm) hexaferrite layers. In structures with layer thickness h = 250–500 nm, the loop rectangularity decreases with increasing h. An increase in the annealing time in such structures leads to an even greater deviation of the loop shape from a rectangular shape.

3.5.2. VSM Measurements

The magnetization curves of annealed samples #8948C (h = 50 nm, T_{ann} = 1000 °C for t_{ann} = 60 min) and #8960C (h = 500 nm, T_{ann} = 1000 °C, t_{ann} = 120 min) measured by VSM are obtained for a magnetic field oriented normally (Figure 10b,d) and in the plane of the BaM layer (b,d). In contrast to the loops observed in PMOKE, the magnetic loops manifest themselves against the background of linear dependencies $M(H)$ due to the substrate. This background does not manifest itself in PMOKE measurements, since light at a wavelength of 405 nm does not reach the substrate even at a film thickness h = 50 nm. Hysteresis loops of thin #8948C film for in-plane and out-of-plane orientations of the magnetic field are different, as expected. In sample #8960C with a film thickness h = 500 nm, a completely different picture is observed in comparison to the thin (h = 50 nm) sample #8948C (cf. Figure 10d,c).

For #8960, the hysteresis loop in the out-of-plane geometry (the same shape as the PMOKE loop) turns out to be comparable in size to the loop in the in-plane geometry (Figure 10d). This indicates that the film #8960 contains regions with a very wide distribution of the direction of the axis of easy magnetization, i.e., the film is very inhomogeneous. This correlates with the conclusions of the previous Section 3.5.1 devoted to PMOKE studies, according to which thin films (h = 50 nm) of BaM hexaferrite after annealing turn out to be much more uniform than thick films (h = 250–500 nm).

BaM hexaferrite films on Al_2O_3(0001) substrates should have uniaxial anisotropy with an easy magnetization axis normal to the surface, so the magnetization switching in out-of-plane geometry should occur by domain wall nucleation and motion, and in in-plane geometry by magnetization rotation. The appearance of a weak loop in the $M(H)$ dependence in the in-plane geometry in thin film #8948C (Figure 10c) indicates that the film also is not ideal due to presence in it of regions with the direction of the easy axis somewhat different from the normal one.

The appearance of irreversible behavior of the magnetization in in-plane geometry in BaM hexaferrite films was associated with the existence of an interdiffusion layer caused by mutual diffusion of substrate and film ions at the interface [49]. Such diffusion can lead to the appearance of regions with slightly different orientations of the easy axis and, as a result, magnetic hysteresis loops in such a geometry. One of the arguments in favor of this mechanism was a considerable increase in the hysteresis loop, observed in a film, grown on an Al_2O_3 substrate by laser deposition in in-plane geometry after the sample was annealed at a temperature of 900 °C for 60 min.

In our samples with thin layers of BaM hexaferrite (h = 50 nm), the in-plane magnetic hysteresis is small. In thick (h = 250–500 nm) hexaferrite films strong hysteresis loops are observed in this geometry. In general, this corresponds to the possible manifestation of an interdiffusion layer, since the growth time of thick films is longer than that of thin ones. However, judging by the behavior of hysteresis loops in out-of-plane geometry, annealing affects thin and thick films differently. In thin films, this leads to an increase in the M_r/M_s ratio with increasing annealing time (Figure 9d), i.e., to a greater rectangularity of the loop. In thick films, on the contrary, annealing leads to a deterioration in the shape of the loop and an increase in H_c. This indicates the manifestation of more complex loop formation mechanisms, at least in thin films.

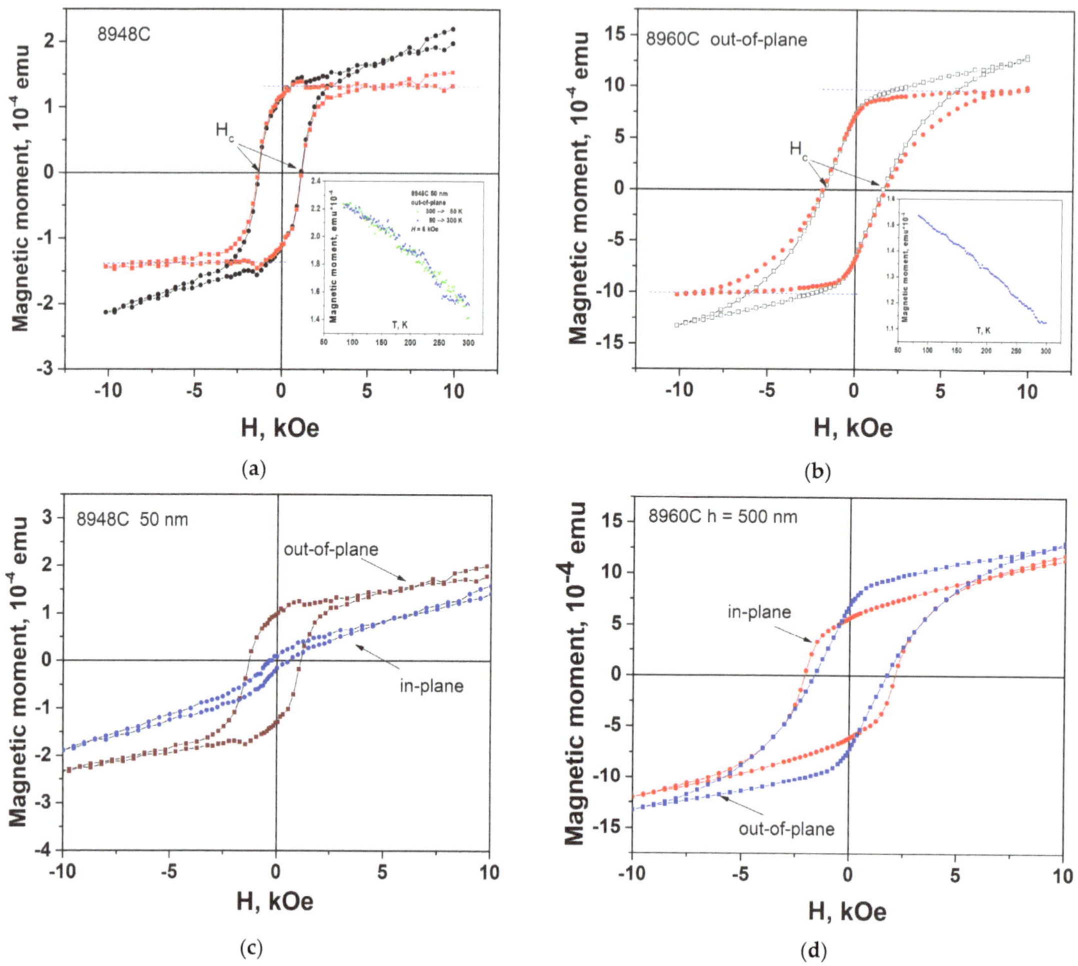

Figure 10. Magnetization curves measured with VSM of the samples (**a**,**c**) #8948C (h = 50 nm, t_{ann} = 60 min) and (**b**,**d**) #8960C (h = 500 nm, t_{ann} = 120 min). The VSM measurements for films (**a**) #8948C and (**b**) #8960C are carried out in magnetic field H normal to the film surface. The red dots in (**a**,**b**) are obtained after subtracting the linear dependence associated with the substrate. Coercive field H_c obtained after the subtraction of the linear dependence is indicated in (**a**,**b**). Insets in (**a**,**b**) show temperature dependence of magnetic moment. (**c**) Comparison of hysteresis loops for out-of-plane and in-plane orientation of the magnetic field in #8948C. (**d**) Comparison of hysteresis loops for out-of-plane and in-plane orientation of the magnetic field in structure #8960C.

The magnetization of the structures was calculated from the experimentally measured value of magnetic moment, using the known layer thickness h, and the sample area S. In film #8948C (h = 50 nm, t_{ann} = 60 min), the saturation magnetization is M_s = 340 emu/cm^3 ($4\pi M_s$ = 4.3 kG), which is in good agreement with the value for a bulk crystal. The magnetization value for thick film is much lower M_s = 230 emu/cm^3 ($4\pi M_s$ = 2.9 kG). A decrease in measurement temperature T leads to an increase in magnetization M_s in both thin and thick films (see insets in Figure 10a,b).

Thus, studies of the magnetic moment using VSM showed that the magnetization of a samples with a thin film of hexaferrite BaM (h = 50 nm) depends significantly on the orientation of the magnetic field relative to the plane of the structure, in accordance with what is observed in bulk samples. In contrast, in the sample with a thick hexaferrite layer (h = 500 nm), this difference turns out to be insignificant, which indicates a strong spread in the orientation of the easy magnetization axis in this structure. Also important is the difference in the magnitude of magnetization in structures with thin (h = 50 nm) and thick (h = 500 nm) BaM hexaferrite films. The low values of magnetization in a film with a thickness of h = 500 nm indicate an inhomogeneous structure of thick films.

3.6. Magnetooptical Spectroscopy

The spectral dependencies of PMOKE and transverse Kerr effect (TKE) measured for the #8948C (T_{gr} = 750 °C, h = 50 nm) sample are shown in Figure 11a. As is known [31,50,51], the measured rotation of the light polarization plane in BaM hexaferrite films in the region of photon energies $E_{ph} > 2.6$ eV is related to the polar Kerr effect only. At lower photon energies, the Faraday effect adds to the rotation as the film becomes more transparent. In this photon energy region, the position of the bands in the PMOKE spectrum and their intensity depend on the ratio between the light wavelength and the film thickness [49]. A characteristic feature of the PMOKE spectrum is the appearance of strong bands of different signs in the region of E_{ph}~3.18 eV (PMOKE ≈ -0.15 deg) and E_{ph}~4.6 eV (PMOKE~0.28 deg) [49]. Similar bands were also observed in BaM hexaferrite films, grown by the metallo-organic decomposition method [37], at E_{ph}~3.15 eV and 4.25 eV, as well as in sputtered $PbFe_{12}O_{19}$ [52] and $SrFe_{12}O_{19}$ [53]. The nature of these bands is associated with a charge transfer (from Fe^{3+} to O^{2-}) optical transitions for Fe^{3+} ions in octahedral and tetrahedral positions [49]. Figure 11 clearly shows manifestation of PMOKE band at E_{ph}~3.2 eV (PMOKE = -0.12 deg), a change of the PMOKE sign at E_{ph}~3.6 eV, and an increase of PMOKE positive values above E_{ph}~3.6 eV.

The TKE spectrum measured for H = 4 kOe (Figure 11b) also shows the band centered at E_{ph}~3.2 eV. Note that because magnitude of TKE is proportional to in-plane magnetization component, which saturates at $H = H_a$~15 kOe, the values of TKE measured for H = 4.9 kOe are smaller than when all magnetization is oriented in the plane of the film (see inset in Figure 11b).

The optical band at $E_{ph} \approx 3.2$ eV manifests itself also in the spectrum of magneto-reflection effect MRE^S, which is measured in geometry of TKE, but for the S-polarization of incident light (Figure 11c). MRE^S quadratically depends on magnetic field H (see inset in Figure 11c), in contrast to TKE, which reveals a linear dependence on the magnetic field H (see inset in Figure 11b). This shows that MRE^S is proportional to the square of in-plane magnetization component caused by applied in-plane magnetic field H, in contrast to TKE which is linear in this component. The reflectivity quadratic in magnetization components was discovered in 1969 [54] and named the orientational magneto-optical effect. Later, it was studied in Fe–Ni, Fe–Ti, Fe–V alloys [55], orthoferrites, orthochromites, and orthomanganides [56].

To conclude this subsection, we note that the studies of magneto-optical spectroscopy show that both the PMOKE and TKE spectra and the MRE spectra confirm the presence of BaM hexaferrite on the (0001) Al_2O_3 substrate in the studied samples.

3.7. Magnetization Dynamics

The FMR spectrum at a frequency F = 50 GHz with a magnetic field direction normal to the film plane for the annealed sample #8948C (h = 50 nm, T_{ann} = 1000 °C, t_{ann} = 60 min) is shown in Figure 12. The FMR band consists of a set of narrow FMR lines with resonant fields in the range H_{res} = 5.6 kOe–5.8 kOe. Resonant field and width of the main FMR line are H_{res} = 5.76 kOe and ΔH_{res} = 20 Oe, respectively. The presence of a set of lines at lower fields is apparently associated with the magnetic inhomogeneity of the layer due to the spread in the anisotropy field, the direction of the easy axis, and so on. We note that we observed such a splitting of FMR lines earlier in YIG/GGG(111) structure [6]. Using the values of the resonant field for the main line H_{res} = 5.76 kOe, the value of the magnetization obtained using the VSM, $4\pi M_s$ = 4.3 kG, the measurement frequency F = 50 GHz, we obtained an estimate of the anisotropy field H_a = 16.4 kOe, which is quite close to the values H_a = 17.0 kOe for $4\pi M_s$ = 4.3 kG in the $BaFe_{10.5}Mn_{1.5}O_{19}/Al_2O_3(0001)$ structure [57].

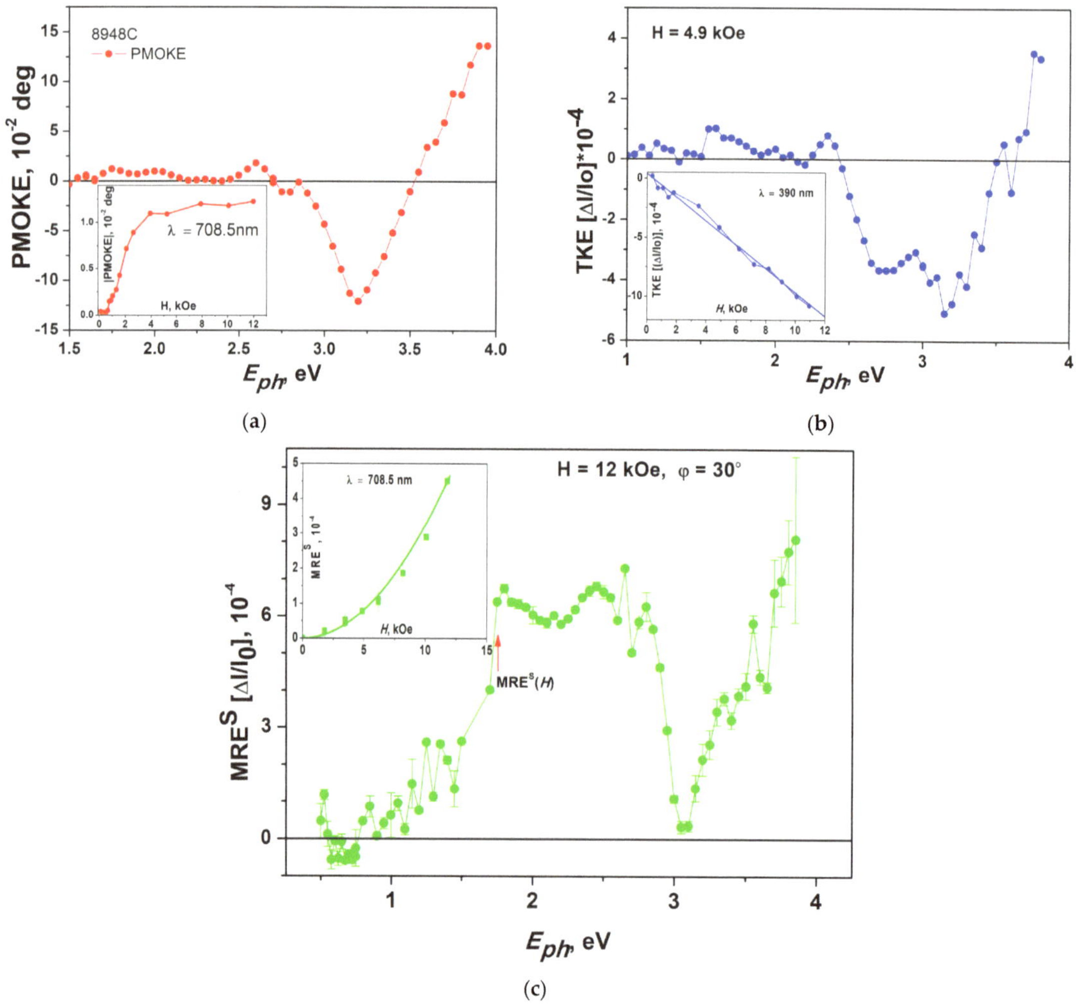

Figure 11. Magneto-optical spectra of #8948C sample. (**a**) PMOKE spectrum measured with out-of-plane magnetic field H = 4 kOe. Inset shows magnetic field dependence of PMOKE measured with spectroscopic setup. Note that the shape of magnetic field dependence is different from one measured with polarimetric setup (Figure 9) because in spectroscopic setup the magnitude of PMOKE(H) is obtained as a difference between signals measured for $+H$ and $-H$ magnetic fields. (**b**) TKE measured with in-plane magnetic field H = 4.9 kOe. Inset shows magnetic field dependence of TKE for λ = 390 nm. Red dashed line in the inset corresponds to fitting of the experimental points, linear with respect to H. (**c**) MRE^S spectrum. Inset shows magnetic field dependence of MRE^S for λ = 708 nm. Red dashed line in the inset corresponds to fitting of the experimental points, quadratic with respect to H.

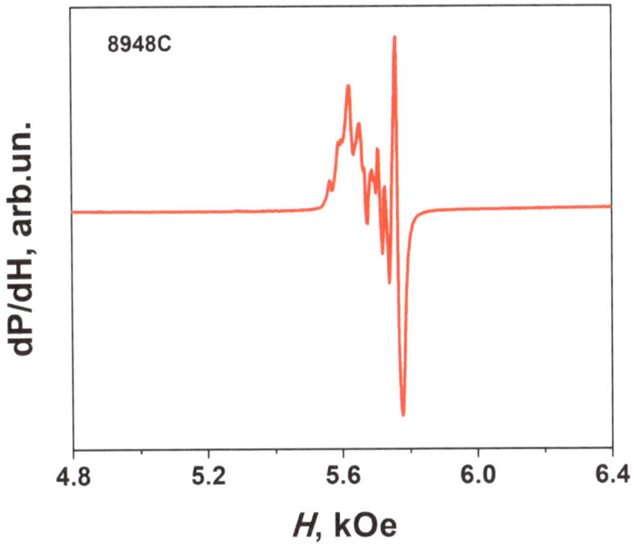

Figure 12. FMR spectrum in annealed #8948C sample (h = 50 nm, T_{ann} = 1000 °C, t_{ann} = 60 min).

4. Conclusions

The main result of these studies, carried out using various experimental techniques, consists in observation of almost rectangular hysteresis loops for out-of-plane orientation of magnetic field in heterostructures with a thin epitaxial layer of BaM hexaferrite grown by laser molecular beam epitaxy on Al_2O_3(0001) substrate. Such loops are observed in thin (h~50 nm) films grown at T_{gr} = 700 °C after a short (t_{ann}~5–10 min) annealing in air at T_{ann} = 1000 °C.

It is important to note that the key role in the formation of the epitaxial structure of hexaferrite is played by the evaporation of Ba ions from the substrate during growth process. If BaM films are grown at a temperature T_{gr} = 1000 °C, then hematite α-Fe_2O_3 is mainly formed on the substrate, because of a strong evaporation of Ba ions from the substrate at this temperature. During film growth at T_{gr} = 700 °C, the amount of Ba, Fe, and O ions on the substrate is sufficient for the formation of $BaFe_{12}O_{19}$. Nevertheless, at a stage of "as-grown", mainly $BaFe_2O_4$ nanocrystallites are formed on the Al_2O_3 surface. Upon subsequent annealing at T_{gr} = 1000 °C, $BaFe_2O_4$ recrystallizes into BaM hexaferrite. In this case, Ba ions do not evaporate from the film.

An important result is the fact that the structural and magnetic properties of the annealed films depend significantly on the film thickness. In thin (h~50 nm) annealed $BaFe_{12}O_{19}$ films, a hexaferrite crystal structure is realized with the direction of the hexagonal axis and the easy magnetization axis close to the normal to the plane of the structure. The magnitude of the magnetic moment M and the anisotropy magnetic field H_a in such films are close to those of bulk BaM samples, and relatively narrow FMR lines are observed in them. The magnitude of the magnetization and the anisotropy field is close to similar films known from the literature.

In contrast, thicker $BaFe_{12}O_{19}$ films (h = 250 nm–500 nm), also annealed at T_{ann} = 1000 °C, exhibit a structure with strong fluctuations in the direction of the easy magnetization axis, which leads to the appearance of hysteresis loops for in-plane the magnetic field comparable to the loops observed for out-of-plane one, as well as strong differences in the shape of the loops from the rectangular ones. In addition, such films are characterized by significantly lower values of spontaneous magnetization.

In our opinion, to obtain thick epitaxial BaM hexaferrite layers comparable in magnetic characteristics to thin ones, a more complex growth protocol can be used, in which the growth process consists of several stages, each of which includes the growth of a thin (h~50 nm) layer and its annealing in air at T_{ann} = 1000 °C.

Supplementary Materials: The following supporting information can be downloaded at: https://www.mdpi.com/article/10.3390/ma16124417/s1, Supplementary Materials S1: Powder XRD; Figure S1: (a,c) WHP and (b,d) SSP graphs for the samples (a,b) #8948A and (c,d) #8948B; Figure S2: (a,c,e) WHP and (b,d,f) SSP graphs for the samples (a,b) #9001C, (c,d) #9001A, and (e,f) #9001B. Figure S3: XRD patterns (θ-2θ scans) of the annealed samples #9001A,B and #8948, and α-Al_2O_3(0001) substrate. Red lines show positions of hematite α-Fe_2O_3 reflections according to PDF-2. References [36–45] are cited in the Supplementary Materials.

Author Contributions: Conceptualization, B.K.; formal analysis, B.K., A.K., S.S., A.A.L., I.L., A.T., A.B., V.S., I.S., M.D. and N.S.; investigation, B.K., A.K., S.S., A.A.L., I.L., A.T., A.B., V.S., I.S. and M.D.; writing—original draft preparation, B.K., A.K., A.A.L., A.T., V.S. and N.S.; writing—review and editing, B.K.; project administration; N.S.; funding acquisition, B.K. All authors have read and agreed to the published version of the manuscript.

Funding: The study was supported by the Russian Science Foundation grant No. 22-22-00768, https://rscf.ru/en/project/22-22-00768/ (accessed on 13 January 2022).

Institutional Review Board Statement: Not applicable.

Informed Consent Statement: Not applicable.

Data Availability Statement: Data are contained within the article or Supplementary Materials.

Acknowledgments: The authors thank A. Urbanovich, I. Pribytkova, and R. Burkovsky from Peter the Great St. Petersburg Polytechnic University (Russia) for X-ray diffraction measurements using a Super Nova diffractometer (Agilent) and S.V. Naumov (M.N. Mikheev Institute of Metal Physics, Ekaterinburg, Russia) for his assistance in the preparation of $BaFe_{12}O_{19}$ target samples. The powder XRD characterizations were performed using equipment from the Joint Research Center "Material science and characterization in advanced technology" (Ioffe Institute, St. Petersburg, Russia).

Conflicts of Interest: The authors declare no conflict of interest.

References

1. Kruglyak, V.V.; Demokritov, S.O.; Grundler, D. Magnonics. *J. Phys. D Appl. Phys.* **2010**, *43*, 264001. [CrossRef]
2. Nikitov, S.A.; Kalyabin, D.V.; Lisenkov, I.V.; Slavin, A.N.; Barabanenkov, Y.N.; Osokin, S.A.; Sadovnikov, A.V.; Beginin, E.N.; Morozova, M.A.; Sharaevsky, Y.P.; et al. Magnonics: A new research area in spintronics and spin wave electronic. *Phys. Uspekhi* **2015**, *58*, 1002–1028. [CrossRef]
3. Grundler, D. Nanomagnonocs. *J. Phys. D Appl. Phys.* **2016**, *49*, 391002. [CrossRef]
4. Serga, A.A.; Chumak, A.V.; Hillebrands, B. YIG magnonics. *J. Phys. D Appl. Phys.* **2010**, *43*, 264002. [CrossRef]
5. Howe, B.M.; Emori, S.; Jeon, H.; Oxholm, T.M.; Jones, J.G.; Mahalingam, K.; Zhuang, Y.; Sun, N.X.; Brown, G.J. Pseudomorphic Yttrium Iron Garnet Thin Films with Low Damping and Inhomogeneous Linewidth Broadening. *IEEE Magn. Lett.* **2015**, *6*, 2013–2016. [CrossRef]
6. Sokolov, N.S.; Fedorov, V.V.; Korovin, A.M.; Suturin, S.M.; Baranov, D.A.; Gastev, S.V.; Krichevtsov, B.B.; Maksimova, K.Y.; Grunin, A.I.; Bursian, V.E.; et al. Thin yttrium iron garnet films grown by pulsed laser deposition: Crystal structure, static, and dynamic magnetic properties. *J. Appl. Phys.* **2016**, *119*, 023903. [CrossRef]
7. Chang, H.; Li, P.; Zhang, W.; Liu, T.; Hoffmann, A.; Deng, L.; Wu, M. Thin yttrium iron garnet films grown by pulsed laser deposition: Crystal structure, static, and dynamic magnetic properties. *IEEE Magn. Lett.* **2014**, *5*, 10–13. [CrossRef]
8. Arsad, A.Z.; Zuhdi, A.W.M.; Ibrahim, N.B.; Hannan, M.A. Recent Advances in Yttrium Iron Garnet Films: Methodologies, Characterization, Properties, Applications, and Bibliometric Analysis for Future Research Directions. *Appl. Sci.* **2023**, *13*, 1218. [CrossRef]
9. Nosov, A.P.; Dubinin, S.S.; Starichenko, D.V.; Ivanov, D.V.; Kobelev, A.V.; Kravtsov, E.A.; Ryabukhina, M.V.; Antropov, N.O.; Bessonov, V.D.; Naumov, S.V.; et al. Specific features of the magnetic anisotropy of thin yttrium iron garnet films prepared by pulsed laser deposition. *Phys. Met. Metallogr.* **2018**, *119*, 1062–1067. [CrossRef]
10. Chinnasamy, C.N.; Yoon, S.D.; Yang, A.; Baraskar, A.; Vittoria, C.; Harris, V.G. Effect of growth temperature on the magnetic, microwave, and cation inversion properties on $NiFe_2O_4$ thin films deposited by pulsed laser ablation deposition. *J. Appl. Phys.* **2007**, *101*, 09M517. [CrossRef]
11. Singh, A.V.; Khodadadi, B.; Mohammadi, J.B.; Keshavarz, S.; Mewes, T.; Negi, D.S.; Ranjan, D.; Galazka, Z.; Uecker, R.; Gupta, A. Bulk Single Crystal-Like Structural and Magnetic Characteristics of Epitaxial Spinel Ferrite Thin Films with Elimination of Antiphase Boundaries. *Adv. Mater.* **2017**, *29*, 1701222. [CrossRef] [PubMed]

12. Krichevtsov, B.; Gastev, S.; Mashkov, K.; Kaveev, A.; Korovin, A.; Lutsev, L.; Suturin, S.; Lobov, I.; Telegin, A.; Lomov, A.; et al. Magnetization reversal in $NiFe_2O_4/SrTiO_3$ nanoheterostructures grown by laser molecular beam epitaxy. *J. Phys. Conf. Ser.* **2019**, *1389*, 012106. [CrossRef]
13. Bursian, V.E.; Kaveev, A.R.; Korovin, A.M.; Krichevtsov, B.B.; Lutsev, L.V.; Suturin, S.M.; Sawada, M.; Sokolov, N.S. Bulk-Like Dynamic Magnetic Properties of Nickel Ferrite Epitaxial Thin Films Grown on $SrTiO_3$(001) Substrates. *IEEE Magn. Lett.* **2019**, *10*, 6104505. [CrossRef]
14. Sukhorukov, Y.P.; Telegin, A. Magnetic semiconductors as materials for spintronics. *Magnetochemistry* **2022**, *8*, 173. [CrossRef]
15. Ning, S.; Kumar, A.; Klyukin, K.; Cho, E.; Kim, J.H.; Su, T.; Kim, H.-S.; LeBeau, J.M.; Bilge Yildiz, B.; Ross, C.A. An antisite defect mechanism for room temperature ferroelectricity in orthoferrites. *Nat. Commun.* **2021**, *12*, 4298. [CrossRef]
16. Jeong, Y.K.; Lee, J.-H.; Ahn, S.-J.; Song, S.-W.; Jang, H.M.; Choi, H.; Scott, J.F. Structurally Tailored Hexagonal Ferroelectricity and Multiferroism in Epitaxial $YbFeO_3$ Thin-Film Heterostructures. *J. Am. Chem. Soc.* **2012**, *134*, 1450–1453. [CrossRef]
17. Seo, J.W.; Fullerton, E.E.; Nolting, F.; Scholl, A.; Fompeyrine, J.; Locquet, J.-P. Antiferromagnetic $LaFeO_3$ thin films and their effect on exchange bias. *J. Phys. Condens. Matter.* **2008**, *20*, 264014. [CrossRef]
18. Harris, V.G.; Chen, Z.; Chen, Y.; Yoon, S.; Sakai, T.; Gieler, A.; Yang, A.; He, Y.; Ziemer, K.S.; Sun, N.X.; et al. Ba-hexaferrite films for next generation microwave devices (invited). *J. Appl. Phys.* **2006**, *99*, 08M911. [CrossRef]
19. Pullar, R.C. Hexagonal ferrites: A review of the synthesis, properties and applications of hexaferrite ceramics. *Prog. Mater. Sci.* **2012**, *57*, 1191–1334. [CrossRef]
20. Harris, V.G. Modern Microwave Ferrites. *IEEE Trans. Magn.* **2012**, *48*, 1075–1104. [CrossRef]
21. Geiler, A.L.; Yoon, S.D.; Chen, Y.; Chinnasamy, C.N.; Chen, Z.; Geiler, M.; Harris, V.G.; Vittoria, C. $BaFe_{12}O_{19}$ thin films grown at the atomic scale from $BaFe_2O_4$ and α-Fe_2O_3 targets. *Appl. Phys. Lett.* **2007**, *91*, 162510. [CrossRef]
22. Nie, Y.; Harward, I.; Balin, K.; Beaubien, A.; Celinski, Z. Preparation and characterization of barium hexagonal ferrite thin films on a Pt template. *J. Appl. Phys.* **2010**, *107*, 073903. [CrossRef]
23. Chen, Z.; Yang, A.; Cai, Z.; Yoon, S.; Ziemer, K.; Vittoria, C.; Harris, V. Structure and Magnetism of Ba-Hexaferrite Films Grown on Single Crystal 6-H SiC With Graduated Interfacial MgO Buffer Layers. *IEEE Trans. Magn.* **2006**, *42*, 2855–2857. [CrossRef]
24. Song, Y.-Y.; Kalarickal, S.; Patton, C.E. Optimized pulsed laser deposited barium ferrite thin films with narrow ferromagnetic resonance linewidths. *J. Appl. Phys.* **2003**, *94*, 5103. [CrossRef]
25. Yoon, S.D.; Shi, P.; Zuo, X.; Oliver, S.A.; Vittoria, C. Ferrimagnetic resonance linewidths of thick barium hexaferrite films on MgO (111). *IEEE Trans. Magn.* **2001**, *37*, 2383. [CrossRef]
26. Wang, S.G.; Yoon, S.D.; Vittoria, C. Microwave and magnetic properties of double-sided hexaferrite films on (111) magnesium oxide substrates. *J. Appl. Phys.* **2002**, *92*, 6728. [CrossRef]
27. Yoon, S.D.; Vittoria, C. Microwave and magnetic properties of barium hexaferrite films having the c-axis in the film plane by liquid phase epitaxy technique. *J. Appl. Phys.* **2003**, *93*, 8597. [CrossRef]
28. Yoon, S.D.; Vittoria, C. Thick M-type barium hexaferrite films grown on garnet substrates. *J. Appl. Phys.* **2004**, *96*, 2131. [CrossRef]
29. Zhang, X.; Meng, S.; Song, D.; Zhang, Y.; Yue, Z.; Harris, V.G. Epitaxially grown BaM hexaferrite films having uniaxial axis in the film plane for self-biased devices. *Sci. Rep.* **2017**, *7*, 44193. [CrossRef]
30. Chen, Y.; Sakai, T.; Chen, T.; Yoon, S.D.; Geiler, A.L.; Vittoria, C.; Harris, V.G. Oriented barium hexaferrite thick films with narrow ferromagnetic resonance linewidth. *Appl. Phys. Lett.* **2006**, *88*, 062516. [CrossRef]
31. Jakubisová, L.E.; Višňovský, Š.; Široký, P.; Hrabovský, D.; Pištora, J.; Vávra, I.; Dobročka, E.; Krišťan, P.; Štěpánková, H.; Harward, I.; et al. Magneto-optical studies of $BaFe_{12}O_{19}$ films grown by metallo-organic decomposition. *Opt. Mater. Express* **2015**, *5*, 1323–1330. [CrossRef]
32. Khoo, M.K.; Kameli, P. Structure and Magnetization of Strontium Hexaferrite (SrFe12O19) Films Prepared by Pulsed Laser Deposition. *Front. Mater.* **2021**, *8*, 717251. [CrossRef]
33. Nishkala, K.R.; Rao, R.R.; Mutalik, S.; Daivajna, M. Effect of La substitution on the structural and chemical properties of Barium hexaferrite via Mossbauer spectroscopy. *Hyperfine Interact.* **2023**, *244*, 7. [CrossRef]
34. Manglam, M.K.; Kumari, S.; Mallick, J.; Kar, M. Crystal structure and magnetic properties study on barium hexaferrite of different average crystallite size. *Appl. Phys. A* **2021**, *127*, 138. [CrossRef]
35. Zhu, Q.; Tang, R.; Peng, F.; Xu, S.; Liang, G.; Zhao, R.; Fang, Y.; You, L.; Su, X. Mechanical Regulation of the Magnetic Properties of Uniaxial Anisotropic Hexaferrite Thin Films. *Phys. Rev. Applied.* **2021**, *16*, 054006. [CrossRef]
36. Balashova, E.; Zolotarev, A.; Levin, A.A.; Davydov, V.; Pavlov, S.; Smirnov, A.; Starukhin, A.; Krichevtsov, B.; Zhang, H.; Li, F.; et al. Crystal Structure, Raman, FTIR, UV-Vis Absorption, Photoluminescence Spectroscopy, TG–DSC and Dielectric Properties of New Semiorganic Crystals of 2-Methylbenzimidazoleium Perchlorate. *Materials* **2023**, *16*, 1994. [CrossRef] [PubMed]
37. Bruker AXS. *Diffrac.Suite, Eva*, Version 5.1.0.5. User Manual. Bruker AXS: Karlsruhe, Germany, 2019.
38. International Centre for Diffraction Data (ICDD). *Powder Diffraction File-2 Release 2014*; ICDD: Newton Square, PA, USA, 2014.
39. Maunders, C.; Etheridge, J.; Wright, N.; Whitfield, H.J. Structure and microstructure of hexagonal $Ba_3Ti_2RuO_9$ by electron diffraction and microscopy. *Acta Crystallogr. B.* **2005**, *61*, 154–159. [CrossRef]
40. Terlan, B.; Levin, A.A.; Börrnert, F.; Simon, F.; Oschatz, M.; Schmidt, M.; Cardoso-Gil, R.; Lorenz, T.; Baburin, I.A.; Joswig, J.-O.; et al. Effect of Surface Properties on the Microstructure, Thermal, and Colloidal Stability of VB_2 Nanoparticles. *Chem. Mater.* **2015**, *27*, 5106–5115. [CrossRef]

41. Terlan, B.; Levin, A.A.; Börrnert, F.; Zeisner, J.; Kataev, V.; Schmidt, M.; Eychmüller, A. A Size-Dependent Analysis of the Structural, Surface, Colloidal, and Thermal Properties of $Ti_{1-x}B_2$ (x = 0.03–0.08) Nanoparticles. *Eur. J. Inorg. Chem.* **2016**, *6*, 3460–3468. [CrossRef]
42. Levin, A.A. Program SizeCr for Calculation of the Microstructure Parameters from X-Ray Diffraction Data. 2022; *Preprint*. [CrossRef]
43. Langford, J.I.; Cernik, R.J.; Louer, D. The Breadth and Shape of Instrumental Line Profiles in High-Resolution Powder Diffraction. *J. Appl. Phys.* **1991**, *24*, 913–919. [CrossRef]
44. Stokes, A.R.; Wilson, A.J.C. The diffraction of X-rays by distorted crystal aggregates. *Proc. Phys. Soc. London* **1944**, *56*, 174–181. [CrossRef]
45. Scherrer, P. Bestimmung der Grösse und der inneren Struktur von Kolloidteilchen mittels Röntgenstrahlen. *Nachr. Königl. Ges. Wiss. Göttingen.* **1918**, *26*, 98–100. (In German)
46. Janni, J.F. Energy loss, range, path length, time-of-flight, straggling, multiple scattering, and nuclear interaction probability: Intwo Parts. Part 1. For Elements $1 \leq Z \leq 92$. *At. Data Nucl. Data Tables* **1982**, *27*, 341–529. [CrossRef]
47. Bird, J.R.; Williams, J.S. (Eds.) *Ion Beam for Materials Analysis*; Academic Press: Cambridge, MA, USA, 1989. [CrossRef]
48. Ziegler, J.F. The Stopping and Range of Ions in Matter. Available online: www.srim.org (accessed on 1 February 2023).
49. Lisfi, A.; Lodder, J.C.; de Haan, P.; Haast, M.A.M.; Roesthuis, F.J.G. Magnetic and magneto-optical effects in barium ferrite films grown by pulsed laser deposition. *J. Magn. Soc. Jpn.* **1998**, *22*, 159–162. [CrossRef] [PubMed]
50. Atkinson, R.; Papakonstantinou, P.; Salter, I.W.; Gerber, R. Optical and magneto-optical properties of Co-Ti-substituted barium hexaferrite single crystals and thin films produced by laser ablation deposition. *J. Magn. Magn. Mater.* **1994**, *138*, 222–231. [CrossRef]
51. Kahn, F.J.; Pershan, P.S.; Remeika, J.P. Ultraviolet Magneto-Optical Properties of Single-Crystal Orthoferrites, Garnets, and Other Ferric Oxide Compounds. *Phys. Rev.* **1969**, *186*, 891–918. [CrossRef]
52. Višňovský, Š.; Široký, P.; Krishnan, R. Complex Polar Kerr Effect Spectra of Magnetoplumbite. *Czech. J. Phys. B* **1986**, *36*, 1434–1442. [CrossRef]
53. Visnovský, S.; Nývlt, M.; Krishnan, R.; Ramamurthy Acharya, B.; Prasad, S.; Venkatramani, N. Magnetooptical Kerr Spectra in Sputtered Strontium Ferrite Films. *Phys. IV Fr.* **1997**, *7*, C1-721–C1-722. [CrossRef]
54. Krinchik, G.S.; Gushchin, V.S. Magnetooptical Effect of Change of Electronic Structure of a Ferromagnetic Metal Following Rotation of the Magnetization Vector. *JETP Lett.* **1969**, *10*, 24.
55. Krinchik, G.S.; Gan'shina, E.A.; Gushchin, V.S. Orientational magneto-optic effect in nickel and ferrosilicon nanocrystals. *Sov. Phys. JETP* **1971**, *33*, 115–120.
56. Gan'shina, E.A. Magneto-Optical Spectroscopy of Ferro and Ferrimagnets. Ph.D. Thesis, Moscow State University, Moscow, Russia, 21 April 1994.
57. Geiler, A.L.; Yang, A.; Zuo, X.; Yoon, S.D.; Chen, Y.; Harris, V.G. Structural, Magnetic, and Microwave Properties of BaFe. *IEEE Trans. Magn.* **2008**, *44*, 2966–2969. [CrossRef]

materials

MDPI

Article

Magnetic Charge Model for Leakage Signals from Surface Defects in Ferromagnetic Material

Xinyu Li *, Guangming Sheng, Zimin Meng, Fan Qin and Zhifeng Liu

School of Mechanical Engineering, Hefei University of Technology, Hefei 230009, China; shenggm2022@163.com (G.S.); mengzimin1027@163.com (Z.M.); peak.liu@263.net (Z.L.)
* Correspondence: lixinyu@hfut.edu.cn

Abstract: A novel three-dimensional theoretical model of magnetic flux leakage (MFL) is proposed in this paper based on the magnetic dipole model. The magnetic dipole model assumes that a ferromagnetic specimen with defects is exposed to a uniform external magnetic field that causes a uniform magnetization around the defect surface. Under this assumption, the MFL can be regarded as arising from magnetic charges on the defect surface. Previous theoretical models were mostly used to analyze simple crack defects such as cylindrical and rectangular cracks. In this paper, we developed a magnetic dipole model for more complex defect shapes such as circular truncated holes, conical holes, elliptical holes, and double-curve-shaped crack holes to complement the existing defect shapes. Experimental results and comparisons with previous models demonstrate that the proposed model provides a better approximation of complex defect shapes.

Keywords: magnetic dipole; round hole defects; metal magnetic memory; numerical integration

Citation: Li, X.; Sheng, G.; Meng, Z.; Qin, F.; Liu, Z. Magnetic Charge Model for Leakage Signals from Surface Defects in Ferromagnetic Material. *Materials* **2023**, *16*, 3750. https://doi.org/10.3390/ma16103750

Academic Editor: Dana Popescu

Received: 3 April 2023
Revised: 28 April 2023
Accepted: 3 May 2023
Published: 15 May 2023

1. Introduction

Non-destructive testing technology is used to detect stress defects in many industries, especially in important industrial fields such as energy, automotive, shipping, and aerospace. Among them, metal magnetic memory testing is a widely used, non-destructive testing technology for ferromagnetic components. Leakage detection technology exposes the specimen to a constant size and known direction magnetic field, while metal magnetic memory testing technology utilizes the earth's magnetic field. When there is a crack defect on the surface, there is a leakage of magnetic flux near the defecting surface. Therefore, the leakage magnetic field generated by the defect contains important shape information regarding the surface damage defect. Therefore, the geometric shape of surface fracture defects in ferromagnetic materials can be accurately evaluated by metal magnetic memory testing [1]. Therefore, metal magnetic memory testing can be used in places where conventional detection methods such as rails, pipelines, and pressure vessels are difficult to detect. Shi associated magnetic dipole theory with stress and established a force–magnetic coupling magnetic dipole theory, which allows the magnetic dipole model to analyze the effects of various influencing factors on leakage magnetic field signals. Based on the rectangular groove defect, he also provided the first analytical expression for the magnetic dipole of trapezoidal groove defects, further expanding the form of the defect leakage magnetic field signal [2,3]. Mandache, Taniguchi, and Suresh analyzed the leakage magnetic field signals of cylindrical defects, associated them with defect size, and experimentally verified the effectiveness of the model [4–6]. Trevino et al., established an improved analytical magnetic dipole model to represent the 3D magnetic leakage field caused by surface-breaking defects on ferromagnetic specimens and verified the accuracy of the improved magnetic dipole model using finite element simulations [7]. Shi described a magnetic charge model for metal magnetic memory signals and simulated specimens with stress concentration zones and long elliptical defects based on this model [8]. Okolo et al., simulated the distribution characteristics of surface and far-field leakage fields using axial magnetization technology and detected and characterized the leakage distribution caused by surface and far-field hairline cracks on rectangular specimens [9]. Leng and Han separately established magnetic dipole integral models to describe the leakage fields generated by the plastic zone at the tip of a V-shaped notch in ferromagnetic materials due to dislocation accumulation and magnetic dipole models to evaluate the stress concentration caused by local plastic deformation [10,11]. Wu et al., analyzed the variation in surface magnetic charge density for defects in different directions and effectively described the directional influence of defects on the distribution of leakage magnetic field [12,13]. Li et al., proposed a "near-field" magnetic leakage method to quantify defect width, indicating that the distance between the two peaks of the "near-field" leakage field is closely related to the defect width and lift-off value, but not to the defect depth [14]. Xu et al., used the equivalent magnetic charge method to establish a self-leakage magnetic model for buried defects, which successfully evaluated the characteristics of buried defects [15]. He et al., established a mathematical model of magnetic field signals, providing theoretical and experimental evidence to identify the stress state of pipeline circumferential welds [16]. Suresh et al., proposed an analytical model for predicting the leakage magnetic field signals of surface defects in ferromagnetic pipes, which can be used for the rapid prediction of leakage magnetic field signals and for inputting data regarding defect reconstruction into leakage inverse problems [17]. Yang et al., derived and established a specific mathematical model of low-frequency magnetic leakage field based on the magnetic dipole theory, considering four types of crack defects with an equal length and width but different bottom shapes [18]. Liu et al., developed a numerical model for detecting axial cracks in pipelines based on electromagnetic theory and calculated the electromagnetic detection signals of cracks of different sizes and orientations [19]. Long et al., discussed the corresponding relationship between the defect leakage field signal and the defect opening contour model based on the fundamental principles of electromagnetic fields and developed an approximate method to detect the defect edge [20].

Although many experts and scholars have conducted extensive research and analyses of rectangular groove-type cracks on metal surfaces and have obtained the expression of the surface crack leakage magnetic field signal, there are not many analyses of various circular hole defects on the surface. Based on the magnetic dipole model of rectangular and cylindrical crack defects, this paper obtains the integral expression of the leakage magnetic defect signals of conical holes, elliptical holes, hyperbolic crack holes, and truncated cone holes, and verifies the rationality of the expression through experiments, providing a new method for quantitatively analyzing the variation in leakage magnetic signals on the surface of defects. At the same time, it also provides a new analysis method for difficult-to-detect locations and regions, such as stress corrosion defects with elliptical shapes in boilers and saddle-shaped defects in reactor pressure vessels [21,22].

2. Rectangular Magnetic Dipole Theoretical Model

Based on the theory of magnetic dipole for cylindrical hole defects in the literature [3], it is believed that the presence of defects on the surface of ferromagnetic materials will generate positive and negative opposite magnetic charges at the defect wall, which, in turn, will generate a leakage magnetic field on the defect surface. Taking the rectangular groove crack defect as an example, combined with the magnetic dipole model analysis, as shown in Figure 1, where l denotes the length of the crack, d denotes the width of the crack, and h the depth of the crack, there exists a magnetic dipole band with equal surface magnetic charge density on both sides of the rectangular groove, and a pair of equal positive and negative differential element magnetic charges dp^+ and dp^- are taken on the face of the rectangular groove crack defect, where r_1 and r_2 are the distances from the positive and negative magnetic charges to point p, respectively. The magnitudes of the magnetic fields generated at point p are dH_p^+ and dH_p^-, their expressions are given by Equations (1) and (2), respectively, so that the total magnetic field dH_p generated by the equivalent magnetic dipole at point p is the sum of dH_p^+ and dH_p^-. Meanwhile, the micro-magnetic charge can be expressed by the product of surface magnetic charge density and area, i.e., dp, and dp can be expressed by Equation (3), where μ_0 is the vacuum permeability, σ_s is the magnetic charge density of the surface, and *ds* is the area of the micro-element in which the magnetic charge is located. Based on the analytical model of magnetic dipole for rectangular groove defects, the foundation for the establishment of magnetic dipole model for cylindrical hole-type defects is laid.

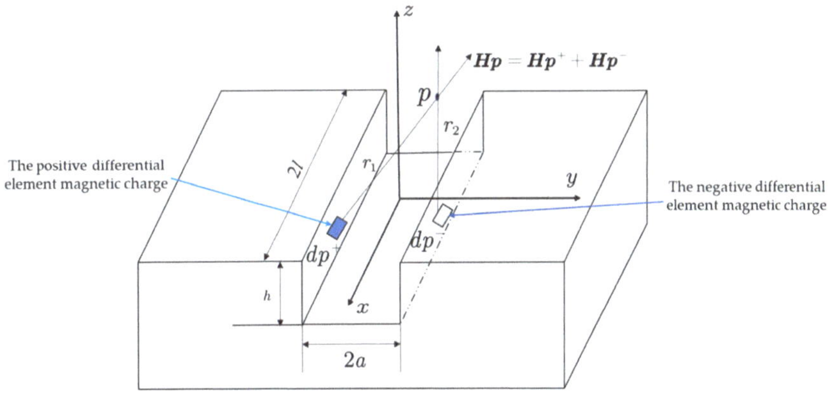

Figure 1. Rectangular groove defective magnetic dipole model.

$$dH_p^+ = \frac{dp^+}{4\pi\mu_0 r_1^3}\vec{r_1} \tag{1}$$

$$dH_p^+ = \frac{dp^-}{4\pi\mu_0 r_2^3}\vec{r_2} \tag{2}$$

$$dp = \sigma_s ds \tag{3}$$

3. Modeling of Cracking Defects

3.1. Cylindrical Hole-Shaped Defect

Assuming that a positive and negative equal surface magnetic charge also exists on the surface of the cylindrical hole defect, the interaction between the ambient magnetic field and the surface cracked defect leads to a spontaneous leakage magnetic field on the defect surface of the ferromagnetic material, which forms a magnetic charge distribution on its defect surface similar to the form of a uniform charge distribution. Consider a cylindrical hole-shaped defect, whose model schematic is shown in Figure 2, assuming that the cylindrical hole defect surface is filled by the surrounding air domain and embedded in the ferromagnetic medium. The cylindrical hole-shaped defect is divided into two parts, the left part with a positive magnetic charge and the right part with a negative magnetic charge; both positive and negative magnetic charges are uniformly distributed around the inner surface of the cylindrical hole-shaped defect. The main surface parameters of the cylindrical hole defect are the radius of the defect is *R*; the depth of the defect is *h*; the distance of any differential element magnetic charge on the defect surface to the point *p* is *r*. The top center point of the cylindrical defect is selected as the coordinate origin; the *z*-axis is perpendicular to the upper surface of the cylinder and passes the coordinate origin; the top region of the cylindrical defect is centered at (0, 0, 0) and the bottom region is centered at (0, 0, $-h$). Since the magnetic memory probe is measured in the region of $z > 0$ during the actual signal measurement, the analysis focuses on the leakage magnetic field signal generated in the region of $z > 0$ for the type of cylindrical hole defect. *dp* for the differential element magnetic charge on the defect surface, and *ds* for the differential element area selected for the surface magnetic charge, which is expressed in column coordinates as *ds*, with du being the length of the differential element area along the axis direction. *dp* is still expressed by Equation (3) and ds is expressed by Equation (4). Since the cylinder is symmetric at both the xoz plane and the yoz plane, the defect spontaneous leakage magnetic field signal along the x-axis and along the y-axis is practically the same. For the convenience of analysis, it is assumed that the positive and negative magnetic charges are symmetrically and uniformly distributed about the xoz plane. The defect surface with a positive magnetic charge is taken as an example to analyze the leakage magnetic field signal at the measurement point p. Assuming that the measurement trajectory is measured along the y-axis, the coordinates of the measurement point p are $(0, y, z)$ and the coordinates of the positive magnetic charge differential element are $(-Rcos\theta, -Rsin\theta, u)$; then, the distance r_1 from the positive magnetic charge differential element to the measurement point can be obtained, where r_1^z denotes the component of vector $\vec{r_1}$ along the z-axis direction $r_1^z = z - u$, r_1^y denotes the component of vector $\vec{r_1}$ along the y-axis direction; that is, $r = y + Rsin\theta$. The distance r_2 from the negative magnetic charge differential element to the measurement point can be similarly obtained; the z-axis vector component $r = z - u$, and the y-axis vector component is $r = y - Rsin\theta$, so the differential element of positive and negative magnetic charge surface to the differential element of leakage magnetic field signal at the measurement point p can be expressed by Equation (5). Equations (5) and (6) can be integrated, and the corresponding integral expressions can be summed up to obtain Equation (7), which provides the integral expressions for the total tangential and normal directions of the cylindrical hole-shaped defect at measurement point p.

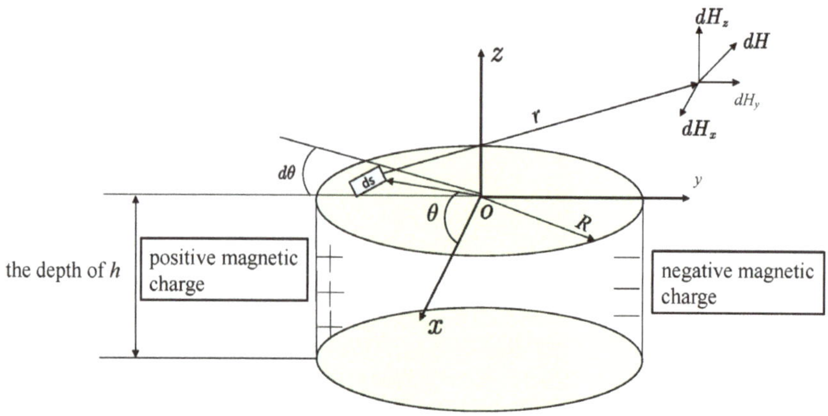

Figure 2. Cylindrical hole-shaped defective magnetic dipole model.

$$ds = Rd\theta du \tag{4}$$

$$\begin{cases} dH_z^+ = \frac{dp^+}{4\pi\mu_0 r_1^2} r_1^z = \frac{\sigma_s R}{4\pi\mu_0 r_1^2} r_1^z d\theta du \\ dH_z^- = -\frac{dp^+}{4\pi\mu_0 r_2^2} r_2^z = -\frac{\sigma_s R}{4\pi\mu_0 r_2^2} r_2^z d\theta du \\ dH_y^+ = \frac{dp^+}{4\pi\mu_0 r_1^2} r_1^y = \frac{\sigma_s R}{4\pi\mu_0 r_1^2} r_1^y d\theta du \\ dH_y^- = -\frac{dp^+}{4\pi\mu_0 r_2^2} r_2^y = -\frac{\sigma_s R}{4\pi\mu_0 r_2^2} r_2^y d\theta du \end{cases} \tag{5}$$

$$\begin{cases} H_z^+ = \frac{\sigma_s R}{4\pi\mu_0}\int_0^\pi d\theta \int_{-h}^0 \frac{(z-u)}{\left[(R\cos\theta)^2+(y+R\sin\theta)^2+(z-u)^2\right]^{\frac{3}{2}}}du \\ H_z^- = -\frac{\sigma_s R}{4\pi\mu_0}\int_0^\pi d\theta \int_{-h}^0 \frac{(z-u)}{\left[(R\cos\theta)^2+(y-R\sin\theta)^2+(z-u)^2\right]^{\frac{3}{2}}}du \\ H_y^+ = \frac{\sigma_s R}{4\pi\mu_0}\int_0^\pi d\theta \int_{-h}^0 \frac{(y+R\sin\theta)}{\left[(R\cos\theta)^2+(y+R\sin\theta)^2+(z-u)^2\right]^{\frac{3}{2}}}du \\ H_y^- = -\frac{\sigma_s R}{4\pi\mu_0}\int_0^\pi d\theta \int_{-h}^0 \frac{(y-R\sin\theta)}{\left[(R\cos\theta)^2+(y-R\sin\theta)^2+(z-u)^2\right]^{\frac{3}{2}}}du \end{cases} \quad (6)$$

$$\begin{cases} H_z = H_z^+ + H_z^- \\ H_y = H_y^+ + H_y^- \end{cases} \quad (7)$$

3.2. Circular Truncated Hole Defect

Because the radius R of the cylindrical defective hole does not change with depth, while the radius R of the cylindrical surface of the circular truncated hole defect is changed linearly with depth, $R' = R + b + ub/h$, R' is the radius of the defective hole at depth u. The depth h of the circular truncated hole is shown in Figure 3; therefore, the difference in the size of the depth h of the circular truncated hole will also have a different effect on the defect leakage magnetic field signal. On the basis of the cylindrical hole defect, the R in its integral expression is replaced by R' to obtain the integral expression of the tangential and directional leakage magnetic field signal of the circular truncated hole defect at measurement point p. This is expressed by Equation (8).

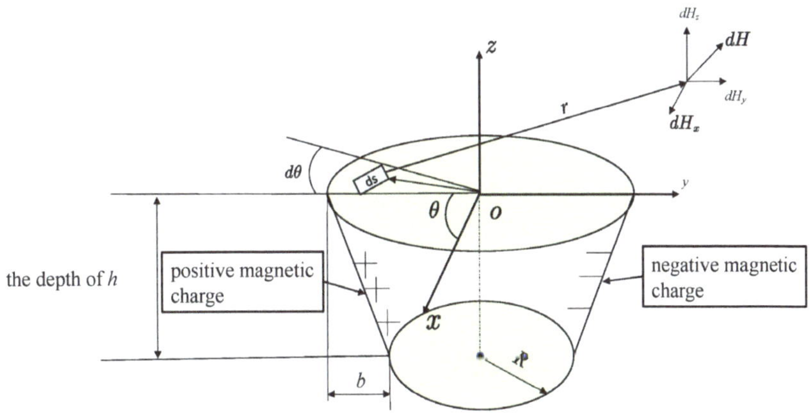

Figure 3. Circular truncated hole defective magnetic dipole model.

$$\begin{cases} H_z^+ = \frac{\sigma_s}{4\pi\mu_0}\int_0^\pi d\theta \int_{-h}^0 (R+b+ub/h)\cdot \\ \frac{(z-u)}{\left[((R+b+ub/h)\cos\theta)^2+(y+(R+b+ub/h)\cos\theta)^2+(z-u)^2\right]^{\frac{3}{2}}}du \\ H_z^- = -\frac{\sigma_s}{4\pi\mu_0}\int_0^\pi d\theta \int_{-h}^0 (R+b+ub/h)\cdot \\ \frac{(z-u)}{\left[((R+b+ub/h)\cos\theta)^2+(y-(R+b+ub/h)\sin\theta)^2+(z-u)^2\right]^{\frac{3}{2}}}du \\ H_y^+ = \frac{\sigma_s}{4\pi\mu_0}\int_0^\pi d\theta \int_{-h}^0 (R+b+ub/h)\cdot \\ \frac{[y+(R+b+ub/h)\sin\theta]}{\left[((R+b+ub/h)\cos\theta)^2+(y+(R+b+ub/h)\sin\theta)^2+(z-u)^2\right]^{\frac{3}{2}}}du \\ H_y^- = -\frac{\sigma_s}{4\pi\mu_0}\int_0^\pi d\theta \int_{-h}^0 (R+b+ub/h)\cdot \\ \frac{[y-(R+b+ub/h)\sin\theta]}{\left[((R+b+ub/h)\cos\theta)^2+(y-(R+b+ub/h)\sin\theta)^2+(z-u)^2\right]^{\frac{3}{2}}}du \end{cases} \quad (8)$$

3.3. Conical Hole-Shaped Defect

When R = 0, the circular truncated hole defect can be transformed into a conical hole-shaped defect, as shown in Figure 4, so that the integrated expression of the circular truncated hole defect leakage magnetic field signal R = 0 can be obtained as conical hole-shaped defect leakage magnetic field signal integral expression (9).

$$\begin{cases} H_z^+ = \frac{\sigma_s}{4\pi\mu_0}\int_0^\pi d\theta \int_{-h}^0 (b+ub/h)\cdot \\ \frac{(z-u)}{\left[((b+ub/h)\cos\theta)^2+(y+(b+ub/h)\cos\theta)^2+(z-u)^2\right]^{\frac{3}{2}}}du \\ H_z^- = -\frac{\sigma_s}{4\pi\mu_0}\int_0^\pi d\theta \int_{-h}^0 (b+ub/h)\cdot \\ \frac{(z-u)}{\left[((b+ub/h)\cos\theta)^2+(y-(b+ub/h)\sin\theta)^2+(z-u)^2\right]^{\frac{3}{2}}}du \\ H_y^+ = \frac{\sigma_s}{4\pi\mu_0}\int_0^\pi d\theta \int_{-h}^0 (b+ub/h)\cdot \\ \frac{(y+(b+ub/h)\sin\theta)}{\left[((b+ub/h)\cos\theta)^2+(y+(b+ub/h)\sin\theta)^2+(z-u)^2\right]^{\frac{3}{2}}}du \\ H_y^- = -\frac{\sigma_s}{4\pi\mu_0}\int_0^\pi d\theta \int_{-h}^0 (b+ub/h)\cdot \\ \frac{(y-(b+ub/h)\sin\theta)}{\left[((b+ub/h)\cos\theta)^2+(y-(b+ub/h)\sin\theta)^2+(z-u)^2\right]^{\frac{3}{2}}}du \end{cases} \quad (9)$$

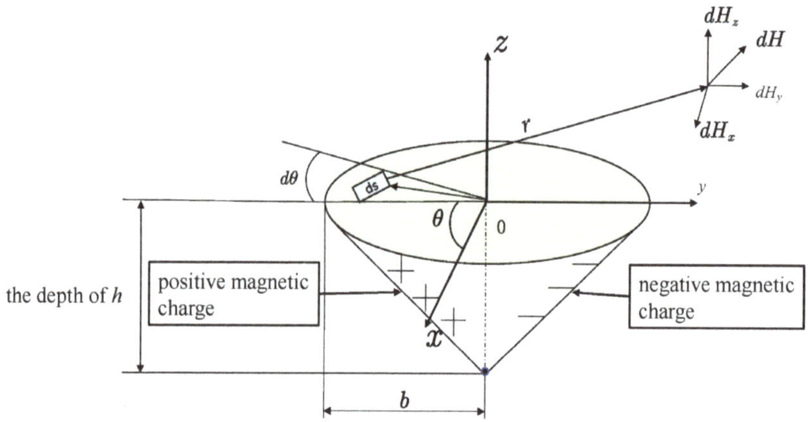

Figure 4. Conical hole-shaped defective magnetic dipole model.

3.4. Elliptical Hole-Shaped Defect

The established magnetic dipole model of the elliptical hole-shaped defect is shown in Figure 5. Converting the right-angle coordinate system into a spherical coordinate system, the angle between the magnetic charge differential element and the x-axis is θ, and the angle between the magnetic charge differential element and the z-axis is φ. Since the general ellipsoidal formula is $x^2/a^2+y^2/a^2+u^2/c^2=1$, the following formula can be obtained by converting it into a spherical coordinate system.

$$\begin{cases} x_0 = a\cos\theta\sin\varphi \\ y_0 = b\sin\theta\sin\varphi \\ u_0 = c\cos\varphi \end{cases} \quad (10)$$

The expression of the magnetic charge differential element $dp = \sigma_s R^2 \sin\theta d\theta dz$ when $R = R_t$ is:

$$R_t = \sqrt{x_0^2 + y_0^2 + u_0^2} = \sqrt{(a\cos\theta\sin\varphi)^2 + (b\sin\theta\sin\varphi)^2 + (c\cos\varphi)^2} \tag{11}$$

In turn, the expression for the magnetic charge differential element in the spherical coordinate system is obtained as:

$$dp = \sigma_s ds = \sigma_s R_t^2 \sin\theta d\theta d\varphi = \sigma_s \left[(a\cos\theta\sin\varphi)^2 + (b\sin\theta\sin\varphi)^2 + (c\sin\varphi)^2\right] \sin\theta d\theta d\varphi \tag{12}$$

The position of the magnetic charge differential element on the elliptical surface is:

$$\begin{cases} x = R_t\cos\theta\sin\varphi \\ y = R_t\sin\theta\cos\varphi \\ u = R_t\cos\varphi \end{cases} \tag{13}$$

In turn, the integral expression of the leakage field signal in the tangential and normal directions for elliptical hole-type defects is obtained as:

$$\begin{cases} H_z^+ = \frac{\sigma_s}{4\pi\mu_0}\int_0^\pi d\theta \int_{\frac{\pi}{2}}^{\pi} \frac{R_t^2 \sin\varphi\cdot(z-R_t\cos\varphi)}{\left[(R_t\cos\theta\sin\varphi)^2+(y+R_t\sin\theta\sin\varphi)^2+(z-R_t\cos\varphi)^2\right]^{\frac{3}{2}}} \sin\theta d\varphi d\theta \\ H_z^- = -\frac{\sigma_s}{4\pi\mu_0}\int_0^\pi d\theta \int_{\frac{\pi}{2}}^{\pi} \frac{R_t^2 \sin\varphi\cdot(z-R_t\cos\varphi)}{\left[(R_t\cos\theta\sin\varphi)^2+(y-R_t\sin\theta\sin\varphi)^2+(z-R_t\cos\varphi)^2\right]^{\frac{3}{2}}} \sin\theta d\varphi d\theta \\ H_y^+ = \frac{\sigma_s}{4\pi\mu_0}\int_0^\pi d\theta \int_{\frac{\pi}{2}}^{\pi} \frac{R_t^2 \sin\varphi\cdot(y+R_t\sin\theta\sin\varphi)}{\left[(R_t\cos\theta\sin\varphi)^2+(y+R_t\sin\theta\sin\varphi)^2+(z-R_t\cos\varphi)^2\right]^{\frac{3}{2}}} \sin\theta d\varphi d\theta \\ H_y^- = -\frac{\sigma_s}{4\pi\mu_0}\int_0^\pi d\theta \int_{\frac{\pi}{2}}^{\pi} \frac{R_t^2 \sin\varphi\cdot(y-R_t\sin\theta\sin\varphi)}{\left[(R_t\cos\theta\sin\varphi)^2+(y-R_t\sin\theta\sin\varphi)^2+(z-R_t\cos\varphi)^2\right]^{\frac{3}{2}}} \sin\theta d\varphi d\theta \end{cases} \tag{14}$$

When $a = b = c = R$, Equation (14) is the same as the equation of the magnetic dipole model of the spherical hole in the literature [23], which verifies the reasonableness of the change in the model equation from the side.

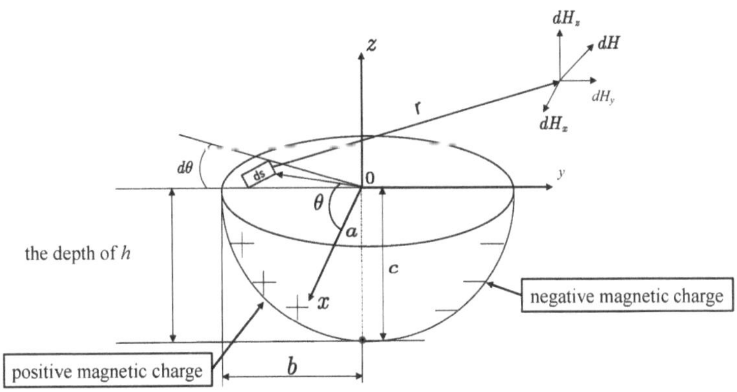

Figure 5. Elliptical hole-shaped defective magnetic dipole model.

3.5. Double-Curve-Shaped Crack Hole Defect

The established magnetic dipole model of the double-curve-shaped crack hole defect is shown in Figure 6. The right-angle coordinate system is transformed into a column coordinate system. The formula for the hyperbola on the xoy plane is $y^2/a^2 - x^2/b^2 = 1$, and the parametric coordinates of the hyperbola are $x = b\tan\theta$, $y = a\sec\theta$; the magnetic charge area differential element $ds = dldu$, where dl can be expressed by Equation (11). Then, the integral expression of the leakage magnetic field signal in the form of the hyperbola cracked hole is obtained, as shown in Equation (16).

$$dl = \sqrt{(dx)^2 + (dy)^2} = \sqrt{b^2 \sec^4\theta + a^2 \sec^2\theta \tan^2\theta}d\theta \tag{15}$$

$$\begin{aligned}
H_z^+ &= \frac{\sigma_s}{4\pi}\int_{-h}^{0} du \int_{-a\sec\theta}^{a\sec\theta} \frac{(z-u)\sqrt{b^2\sec^4\theta + a^2\sec^2\theta\tan^2\theta}}{\left[(b\tan\theta)^2 + (y+a\sec\theta)^2 + (z-u)^2\right]^{\frac{3}{2}}} d\theta \\
H_z^- &= \frac{\sigma_s}{4\pi}\int_{-h}^{0} du \int_{-a\sec\theta}^{a\sec\theta} \frac{(z-u)\sqrt{b^2\sec^4\theta + a^2\sec^2\theta\tan^2\theta}}{\left[(b\tan\theta)^2 + (y-a\sec\theta)^2 + (z-u)^2\right]^{\frac{3}{2}}} d\theta \\
H_y^+ &= \frac{\sigma_s}{4\pi}\int_{-h}^{0} du \int_{-a\sec\theta}^{a\sec\theta} \frac{(y+a\sec\theta)\sqrt{b^2\sec^4\theta + a^2\sec^2\theta\tan^2\theta}}{\left[(b\tan\theta)^2 + (y+a\sec\theta)^2 + (z-u)^2\right]^{\frac{3}{2}}} d\theta \\
H_y^- &= \frac{\sigma_s}{4\pi}\int_{-h}^{0} du \int_{-a\sec\theta}^{a\sec\theta} \frac{(y-a\sec\theta)\sqrt{b^2\sec^4\theta + a^2\sec^2\theta\tan^2\theta}}{\left[(b\tan\theta)^2 + (y-a\sec\theta)^2 + (z-u)^2\right]^{\frac{3}{2}}} d\theta
\end{aligned} \tag{16}$$

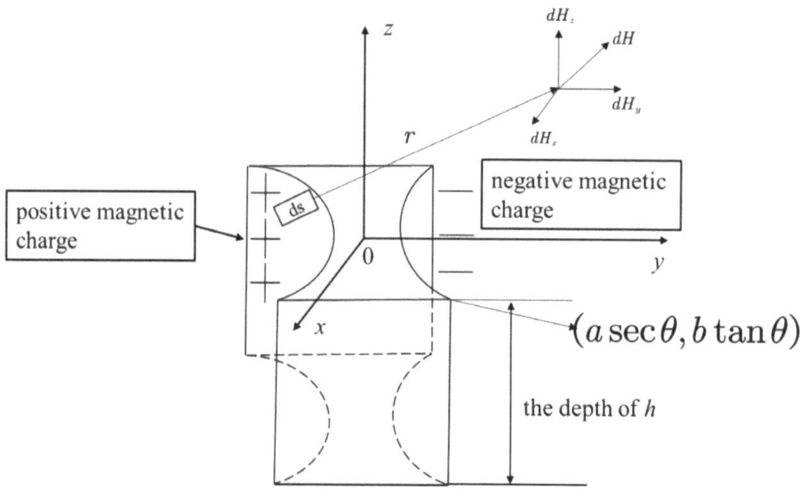

Figure 6. Double-curve-shaped crack hole defective magnetic dipole model.

4. Comparison and Validation of Theoretical Models

Since the model-derived dual integral $F = \int_a^b dx \int_c^d f(x,y)dy$ is non-integrable, combined with the binary complexified Simpson's product formula, the integral region of the rectangle is divided equally into $n \times m$ small rectangles. Four vertices are taken in the middle of each small rectangle in equal parts, and the Simpson's product formula for each small rectangle can be obtained by combining the four vertices.

$$\begin{aligned}
F = \frac{(b-a)(d-c)}{36mn}\{f(x_0,y_0) + f(x_2,y_0) - f(x_0,y_2) + f(x_2,y_2) + \\
4[f(x_0,y_1) + f(x_1,y_0) + f(x_2,y_1) + f(x_1,y_2)] + 16f(x_1,y_1)\}
\end{aligned}$$

After summing them, we can obtain:

$$\int_a^b dx \int_c^d f(x,y)dy \approx G$$

where the G equation is equal to:

$$G = \frac{(b-a)(d-c)}{36mn}\sum_{j=0}^{m-1}\sum_{i=0}^{n-1}\{f(x_{2i},y_{2j}) + f(x_{2i+2},y_{2j}) + f(x_{2i+1},y_{2j+2}) + f(x_{2i},y_{2j+2})$$
$$+4[f(x_{2i+1},y_{2j}) + f(x_{2i+2},y_{2j+1}) + f(x_{2i+1},y_{2j+2}) + f(x_{2i},y_{2j+1}) + 16f(x_{2i+1},y_{2j+1})]\}$$
$$= \frac{(b-a)(d-c)}{36mn}\sum_{j=0}^{2m}\sum_{i=0}^{2n}\lambda_{ij}f(x_i,y_j)$$

λ_{ij} are the coefficients of the complexified Simpson product formula, and λ_{ij} are the elements of the following λ-matrices.

$$\lambda = \begin{bmatrix} 1 & 4 & 2 & 4 & 2 & \cdots & 4 & 2 & 4 & 1 \\ 4 & 16 & 8 & 16 & 8 & \cdots & 16 & 8 & 16 & 4 \\ 2 & 8 & 4 & 8 & 4 & \cdots & 8 & 4 & 8 & 2 \\ & & \cdots & & & \cdots & & & \cdots & \\ & & \cdots & & & \cdots & & & \cdots & \\ & & \cdots & & & \cdots & & & \cdots & \\ 4 & 16 & 8 & 16 & 8 & \cdots & 16 & 8 & 16 & 4 \\ 2 & 8 & 4 & 8 & 4 & \cdots & 8 & 4 & 8 & 2 \\ 4 & 16 & 8 & 16 & 8 & \cdots & 16 & 8 & 16 & 4 \\ 1 & 4 & 2 & 4 & 2 & \cdots & 4 & 2 & 4 & 2 \end{bmatrix}$$

and the error of the complexified Simpson product formula is:

$$E_r = -\frac{(b-a)(d-c)}{180}\left[\left(\frac{b-a}{2}\right)^4\frac{\partial^4 f(\alpha,\beta)}{\partial x^4} + \left(\frac{d-c}{2}\right)^4\frac{\partial^4 f(\overline{\alpha},\overline{\beta})}{\partial y^4}\right]$$

where $\alpha,\overline{\alpha} \in (a,b)$, $\beta,\overline{\beta} \in (c,d)$.

The original reference [3] expresses the rectangular slot defect as:

$$H_x = \frac{\sigma_s}{2\pi\mu_0}\left[\arctan\frac{h+y}{a-x} + \arctan\frac{h+y}{a+x} - \arctan\frac{y}{a-x} - \arctan\frac{y}{a+x}\right] \quad (17)$$

$$H_y = \frac{\sigma_s}{4\pi\mu_0}\ln\left[\frac{(a+x)^2+(h+y)^2}{(a+x)^2+y^2}\cdot\frac{(a-x)^2+y^2}{(a-x)^2+(y+h)^2}\right] \quad (18)$$

Take the magnetic charge density as $\sigma_s = 4\pi\mu_0\delta_s$; that used to obtain rectangular groove defects; magnetic charge density is $\delta_s = \frac{2.65}{2\pi}\left(\frac{h/a+1}{h/(\mu a)+1}\right)$, the magnetic charge density of cylindrical hole-shaped defects is $\delta_s = \frac{2.65}{2\pi}\left(\frac{h/R+1}{h/(\mu R)+1}\right)H_a$; the magnetic charge density of the conical hole-shaped defect is $\delta_s = \frac{2.65}{2\pi}\left(\frac{h/(R+2b)+1}{h/\mu(R+2b)+1}\right)H_a$; the magnetic charge density of the elliptical hole-shaped defect is $\delta_s = \frac{2.65}{2\pi}\left(\frac{c/[(a+b)/2]+1}{c/\mu[(a+b)/2]+1}\right)H_a$, where μ_0 is the vacuum magnetic permeability, $\mu_0 = 4\pi \times 10^{-7}$, $H_a = 50$ A/m.

At different defect depths, the tangential and normal leakage magnetic field signals are calculated by the circular truncated hole defective magnetic dipole model, and the results are shown in Figures 7 and 8.

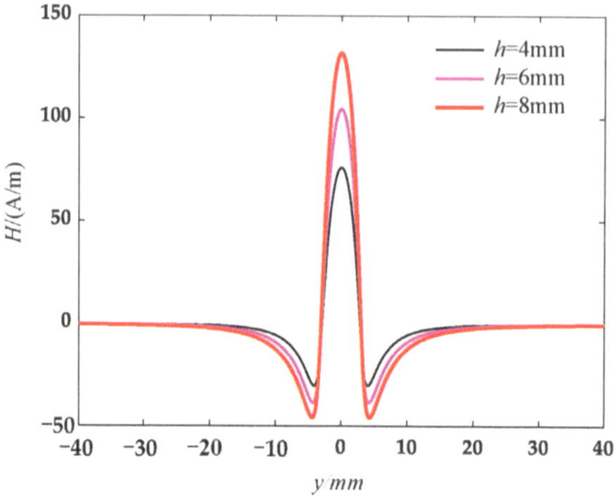

Figure 7. Distribution of tangential leakage field signals of circular truncated hole defects at different depths.

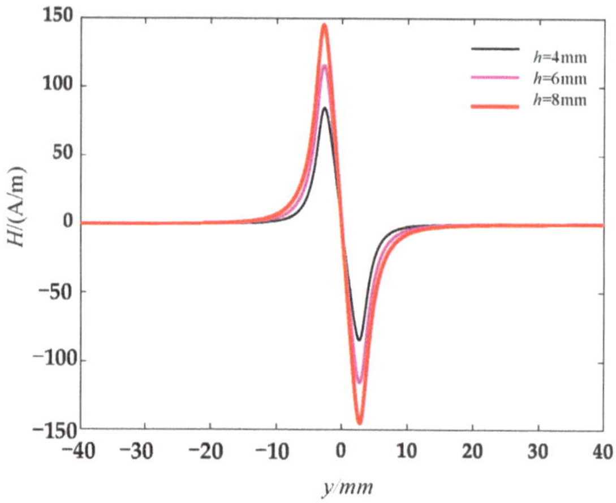

Figure 8. Distribution of normal leakage magnetic field signals of circular truncated hole defects at different depths.

The analysis of circular truncated hole defects using the complexified Simpson product formula is similar to the results of the damage signal analysis of rectangular slot defects derived in the literature [3] and can thus expand the existing defects.

5. Materials and Methods

5.1. Metal Magnetic Memory Simulation

A three-dimensional finite element model of the magnetic memory detection shown in Figure 9 is established, and the black arrow represents the scanning trajectory and direction. The simulation coordinate system selects the position in the middle of the surface of the three-dimensional model. The geometric model material is selected as 45-gauge steel, and the circular truncated hole defects with different depths are set on the surface, where the length and width of the material are 100 mm, 100 mm, and 15 mm, respectively. Select the scanning line close to the surface along the y-axis in the positive direction above the symmetrical center of each defect, and calculate the electromagnetic field using the magnetic field without current module, and the intrinsic relationship in the air domain satisfies $B = \mu_0 H$. Since the magnetic memory detection is simulated in the geomagnetic field, the geomagnetic field used the length, width and height dimensions of 500 mm, 300 mm and 300 mm, respectively, and the magnetic field size is set to 50 A/m, with the magnetic field direction along the positive direction of the y-axis to make the model's internal magnetic field strength uniform. The intrinsic relationship of the specimen material is calculated using the magnetization relationship of $B = \mu_0(H + M)$. Figure 10 shows a schematic diagram of the magnetic field simulation results of the circular truncated cone defect. The red arrow in the figure indicates the direction of the magnetic field at different positions in space. It can be seen that under the condition of uniform external magnetic field, ferromagnetic materials will undergo polarization. Its magnetization vector will start from one end and return to the other end, thus forming a complete loop. When there is a defect in the material, it causes a magnetic signal to be generated in the defect area, which increases the local magnetization strength and leads to local magnetic field distortion. Here, the leakage magnetic field signal on the surface directly above each circular truncated cone is scanned, and the results are obtained as shown in Figure 11. It is not difficult to see that the tangential component of the leakage magnetic field signal peaks at the cracked hole defect damage site, and the normal component of the leakage magnetic field signal crosses the zero point.

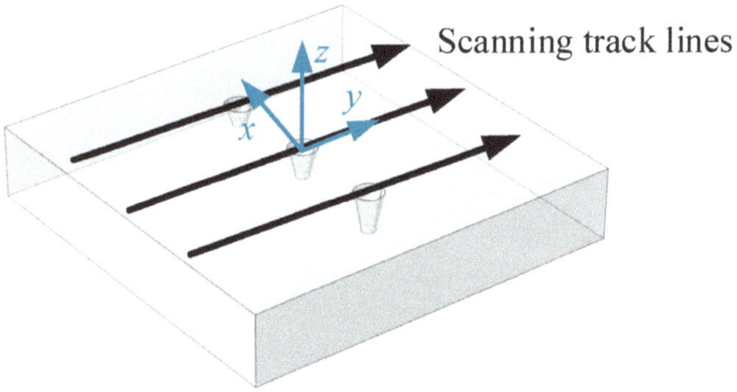

Figure 9. Three-dimensional model of circular truncated hole defects.

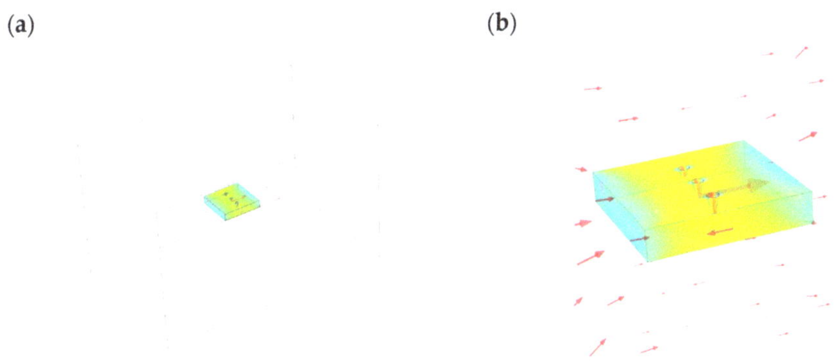

Figure 10. Schematic diagram of simulation results of metal magnetic memory: (**a**) full view of simulation; (**b**) specimen and surrounding magnetic field distribution map.

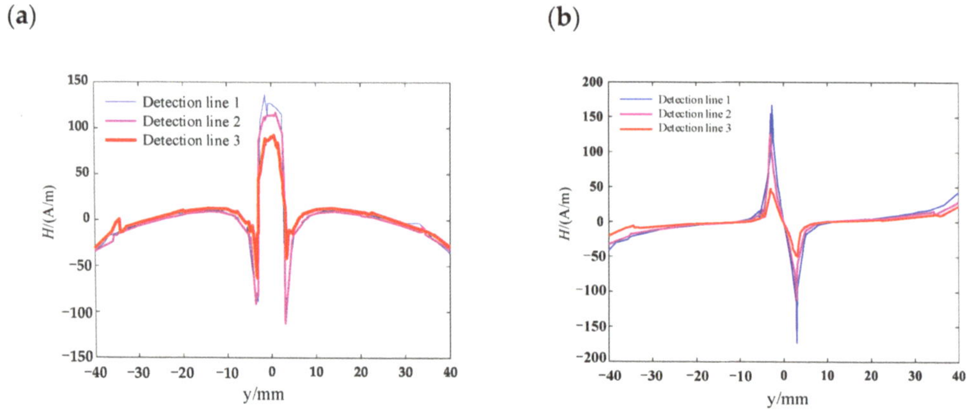

Figure 11. Schematic diagram of the simulation of magnetic field leakage signal from a circular truncated hole: (**a**) tangential leakage magnetic field signal; (**b**) normal leakage magnetic field signal.

5.2. Experimental Verification and Analysis

The experimental material used in this study is common engineering structural steel 45, with the chemical composition listed in Table 1. The 45 steel was processed into rectangular plates with a thickness of 15 mm and dimensions of 100 mm by 100 mm. This type of steel has good strength and machinability and is often used in engineering for wear-resistant and high-toughness parts.

Table 1. Chemical composition of experimental material 45# steel.

Element	C	Si	Mn	S	P
mass fraction	0.42~0.50%	0.17~0.37%	0.50~0.80%	≤0.035%	≤0.035%

Computer numerical control machine tools are used to process various surface defect shapes on the surface of the material specimen. The pictures of the processed patterns are shown in Figure 12 and mark one detection line along the length direction of the specimen in the defect detection area. The length of the detection line is 80 mm. Using the TSC-2M-8 metal magnetic memory detector, the detection probes are measured along the detection lines on the specimen surface, and the detection line 1, detection line 2 and detection line 3 corresponded to the crack depths from the maximum to the minimum. Each specimen is prefabricated with three different defect depths of 8 mm, 6 mm, and 4 mm, respectively. To reduce the influence of signals between defects, the spacing between prefabricated defects is set to 25 mm. The prefabricated defect specimens are shown in Figure 13. The measurement results of the surface leakage magnetic field signal for defects are shown in Figures 14–17. By analyzing the tangential component H_y and normal component H_z of the measured signal and combining the magnetic dipole model and spontaneous leakage magnetic theory, it can be concluded that the leakage magnetic field signal has a peak in the tangential component H_y and a characteristic of passing through zero in the normal component H_z. Moreover, the peak value on each detection line is different for each defect. Combining the magnetic dipole model and metal magnetic memory simulation results, it can be found that as the depth of defects increases, the amplitude of leakage magnetic field signals also increases. Therefore, it is possible to reflect the depth of defects with different shapes through leakage magnetic field signals, which is consistent with the expected results of the model.

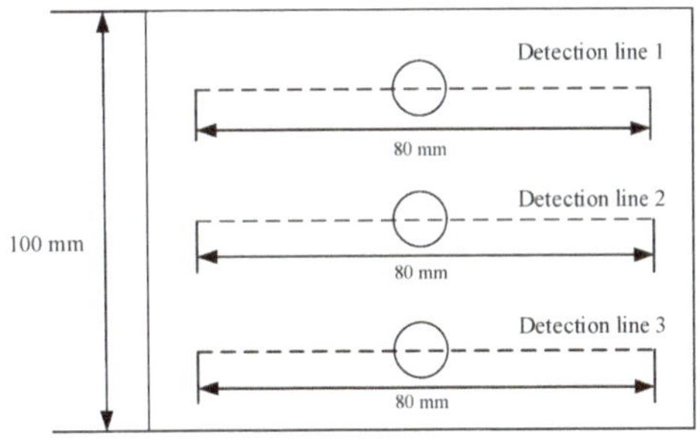

Figure 12. Schematic diagram of leakage magnetic field signal detection for the specimen.

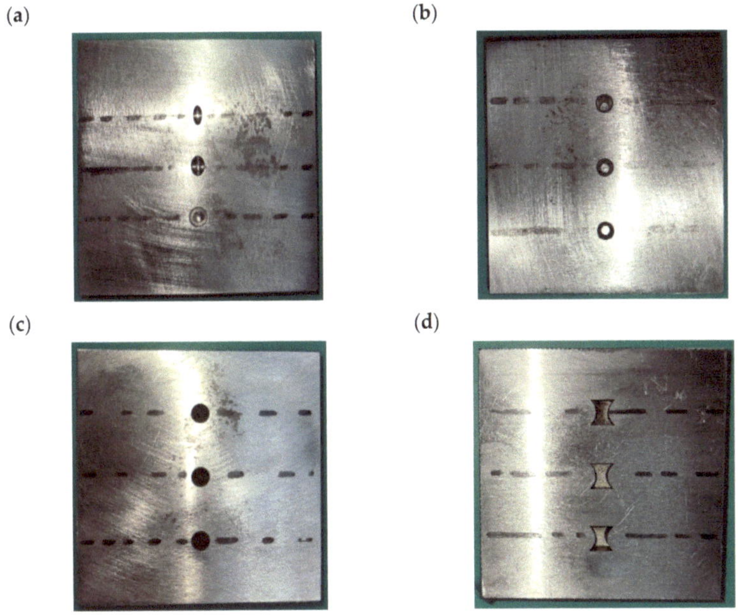

Figure 13. Processed defective specimens: (**a**) elliptical hole-shaped defect; (**b**) circular truncated hole-shaped hole defect; (**c**) conical hole-shaped defect; (**d**) double curve cracked hole defect.

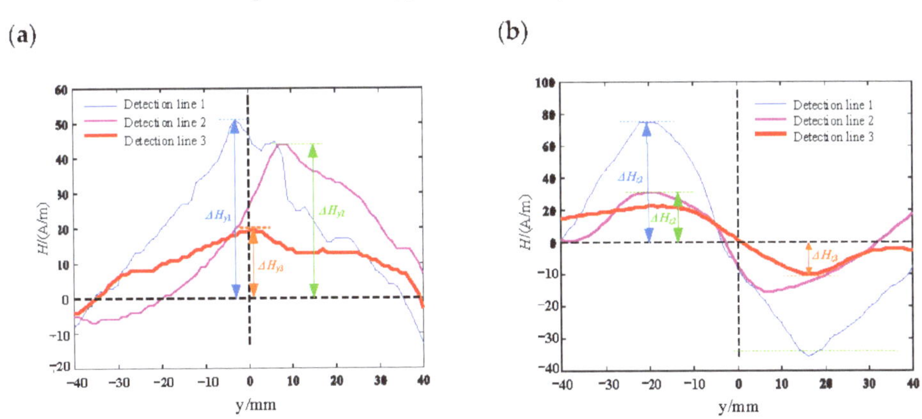

Figure 14. Leakage signal of elliptical hole-shaped defects: (**a**) tangential leakage magnetic field signal; (**b**) normal leakage magnetic field signal.

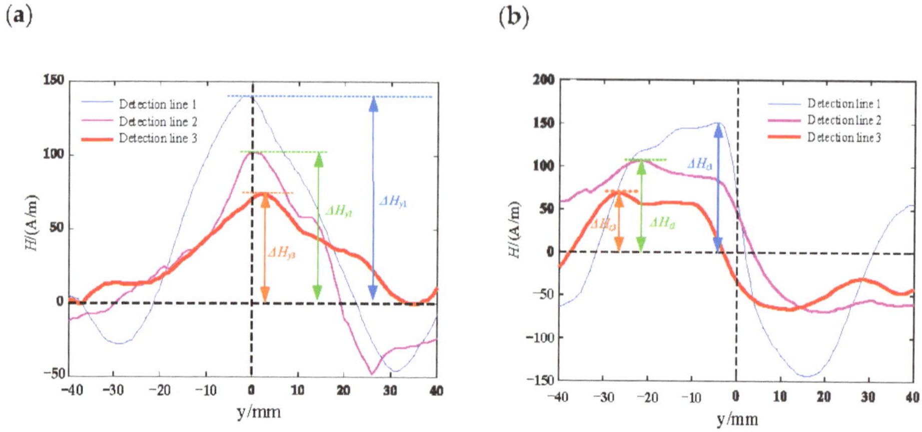

Figure 15. Leakage magnetic field signal of circular truncated hole defects: (**a**) tangential leakage magnetic field signal; (**b**) normal leakage magnetic field signal.

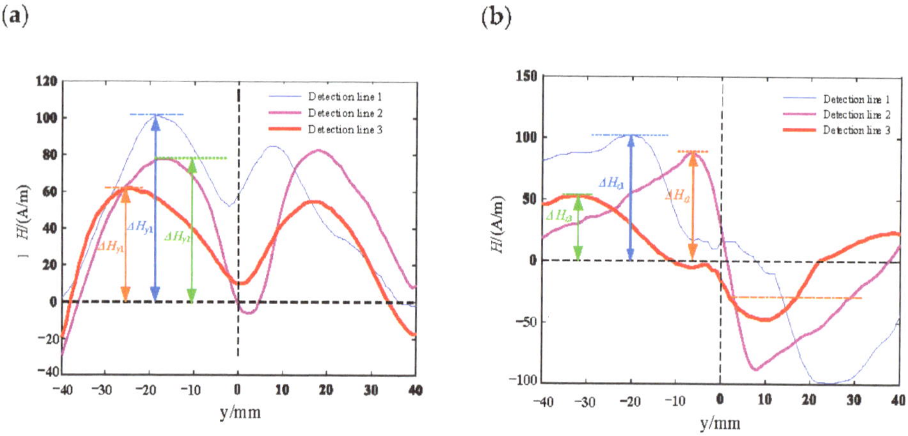

Figure 16. Leakage field signal of conical hole-shaped defects (**a**) tangential leakage magnetic field signal; (**b**) normal leakage magnetic field signal.

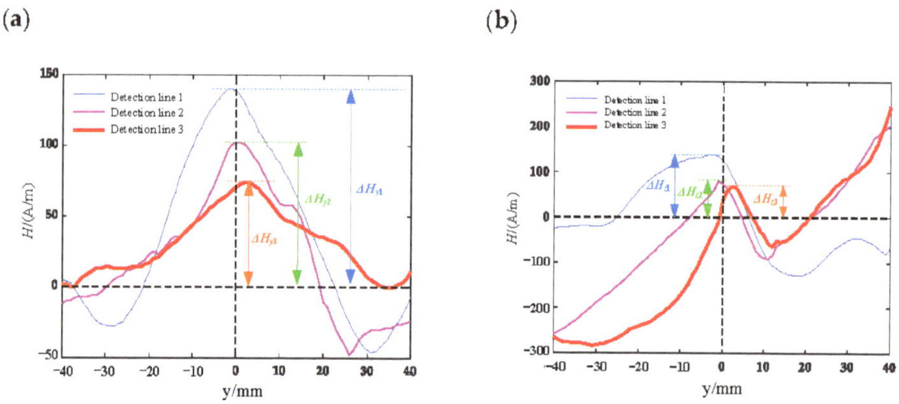

Figure 17. Leakage magnetic signal of double curved cracked hole defects: (**a**) tangential leakage magnetic field signal; (**b**) normal leakage magnetic field signal.

Combining the measured magnetic leakage signals with the theoretical models of the given defect shape and the existing simple rectangular crack model, the established model can better express the magnetic leakage signals of complex-shaped defects. Taking the circular truncated hole as an example, the circular truncated hole defect is approximated by a rectangular crack form and compared with the actual signal. Meanwhile, the magnetic dipole model of the established circular truncated hole defect is also compared with the actual measured signal. A rectangular slot defect is used to replace the circular truncated hole. The half-width a of the rectangular slot is approximately represented by half of the sum of the upper and lower radii of the circular truncated hole. The upper radius of the circular truncated hole is 3 mm and the lower radius is 1.5 mm. The height of the lift-off value is selected to be 1 mm. The theoretical value of the magnetic flux leakage signal is calculated by substituting these values into the formula. The results are shown in Figure 18, where Figure 18a,b shows the tangential and normal signal comparison diagrams of the circular truncated hole defect theoretical model and the actual magnetic leakage field signal, respectively, and Figure 18c,d shows the tangential and normal signal comparison diagrams of the theoretical model of the rectangular crack defect approximation and the actual magnetic leakage field signal, respectively. It can be seen that the theoretical model of the circular truncated hole defect can better express the relationship between the depth and the peak value of the magnetic leakage field signal, indicating that the theoretical model of the simple rectangular crack is not accurate enough to express the complex magnetic leakage field signal and cannot accurately reflect the depth of the defect.

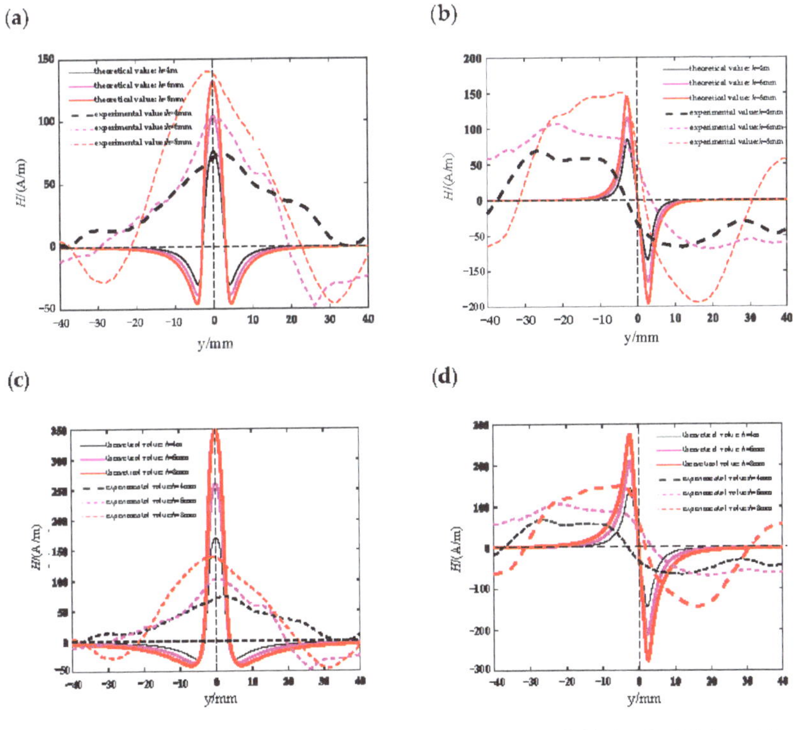

Figure 18. Comparison of theoretical and actual leakage magnetic field signals of defects. (**a**) comparison between the magnetic dipole model of a circular truncated hole and the actual tangential leakage magnetic field signal; (**b**) comparison between the magnetic dipole model of a circular truncated hole and the actual normal leakage magnetic field signal; (**c**) comparison between the rectangular crack model and the actual tangential leakage magnetic field signal; (**d**) comparison between the rectangular crack model and the actual normal leakage magnetic field signal.

6. Results

Along the detection line direction, although the tangential and normal leakage magnetic signal values of each specimen are different, the change trend is roughly the same, and the tangential leakage magnetic signal of each defective specimen is approximately mirror-symmetric, and the normal leakage magnetic signal is approximately symmetrical about the origin. As shown in Figure 14, the peak value ΔH_y of the tangential magnetic leakage signal of the elliptical hole increases with the depth of the defect; the peak value ΔH_z of the normal magnetic leakage signal of the elliptical hole defect also increases with the depth of the defect, and the signal is symmetric at about the zero point. In fact, the increase in the width of the elliptical hole defect will cause the peak value of the magnetic leakage signal to decrease, but the effect of the depth of the defect on the signal is much greater than that of the width. As shown in Figure 15, the peak value ΔH_y of the tangential magnetic leakage signal component of the circular truncated hole increases with the depth, and the curves of detection line 2 and detection line 3 are relatively close. The peak value ΔH_z of the normal magnetic leakage signal component of the circular truncated hole defect also increases with depth. As shown in Figure 16, the peak value ΔH_y of the tangential magnetic leakage signal component of the conical hole defect increases with the depth, and a low valley is found at the center position of the conical hole. The peak value ΔH_z of the normal magnetic leakage signal component of the conical hole defect also increases with depth. As shown in Figure 17, the peak value ΔH_y of the tangential magnetic leakage signal component of the double-curve-shaped crack hole defect increases with depth, and the peak value ΔH_z of the normal magnetic leakage signal component near the zero point also increases with depth. The model established based on the magnetic dipole theory for the elliptical, circular truncated cone and double-curve-shaped crack holes agrees with the experimental results, but the tangential magnetic leakage signal of the conical hole defect does not completely agree, while the experimental results of the normal magnetic leakage signal still agree. Therefore, there are differences in the leakage signals of different types of defects, and further improvement of the theoretical model is needed in the future.

7. Conclusions

Based on the magnetic dipole theory model, the leakage magnetic field signals of different types of circular hole defects are physically modelled and the corresponding integral expressions are derived, which make a useful addition to the existing types of defects. The integral expressions of the leakage field signal for different types of circular hole defects are given, and the dual integrals are numerically computed by compounding Simpson's formula. The leakage field signal images of the defects with different shapes are analyzed. The validity of the expression is experimentally verified.

8. Discussion

This paper mainly uses magnetic dipole theory to establish and analyze the magnetic flux leakage signals of complex defects without considering the deflection of defect direction, the defects located inside the material, and the influence of too close a defect spacing on the magnetic flux leakage signal. Since actual defects may deflect at different angles, it is impossible to ensure that the measurement path moves along the center of the defect symptoms during the measurement process. Therefore, it is necessary to consider the impact of defect deflection in different directions on magnetic flux leakage, and this can be solved by combining the coordinate transformation in reference [12]. For defects that exist inside the material and are a certain distance from the surface, this can be improved by combining the equivalent magnetic charge method in reference [15]. For multiple closely spaced defects, this can be analyzed by combining the superposition between multiple magnetic dipole models in reference [24]. If all these factors are taken into account, it will improve the application range of this paper's theoretical model.

Author Contributions: Conceptualization, X.L. and G.S.; methodology, G.S.; writing—original draft preparation, G.S.; writing—review and editing, X.L., F.Q. and Z.M.; project administration, X.L. and Z.L.; funding acquisition, Z.L. All authors have read and agreed to the published version of the manuscript.

Funding: This research was funded by the National Key R&D Program in China, grant number 2019YFC1908002.

Institutional Review Board Statement: Not applicable.

Informed Consent Statement: Not applicable.

Data Availability Statement: Data are available from the corresponding author upon reasonable request.

Acknowledgments: This research was funded by the National Key Research and Development Program in China [Grant number: 2019YFC1908002].

Conflicts of Interest: The authors declare no conflict of interest.

References

1. Wang, C.L.; Chen, Z.L.; Ma, X.L. *Visualization Technology of Defect in Magnetic Flux Leakage Testing;* National Defense Industry Press: Beijing, China, 2014.
2. Shi, P.P.; Hao, S. Analytical solution of magneto-mechanical magnetic dipole model for metal magnetic memory method. *Acta Pet. Sin.* **2021**, *70*, 34101. [CrossRef]
3. Pengpeng, S. Analytical solutions of magnetic dipole model for defect leakage magnetic fields. *NDT* **2015**, *91*, 399–404.
4. Mandache, C.; Clapham, L. A model for magnetic flux leakage signal predictions. *J. Phys. D Appl. Phys.* **2003**, *36*, 2427–2431. [CrossRef]
5. Suresh, V.; Abudhahir, A.; Daniel, J. Development of magnetic flux leakage measuring system for detection of defect in small diameter steam generator tube. *Measurement* **2017**, *95*, 273–279. [CrossRef]
6. Taniguchi, T. An analytical computation of magnetic field generated from a cylinder ferromagnet. *J. Magn. Magn. Mater.* **2018**, *452*, 464–472. [CrossRef]
7. Trevino, D.A.G.; Dutta, S.M.; Ghorbel, F.H.; Karkoub, M. An Improved Dipole Model of 3-D Magnetic Flux Leakage. *IEEE Trans. Magn.* **2015**, *52*, 1–7. [CrossRef]
8. Pengpeng, S.; Xiaojing, Z. Magnetic charge model for 3D MMM signals. *Nondestruct. Test. Eval.* **2015**, *31*, 45–60. [CrossRef]
9. Okolo, C. *Modelling and Experimental Investigation of Magnetic Flux Leakage Distribution for Hairline Crack Detection and Characterization;* Cardiff University: Wales, UK, 2018.
10. Leng, J.C.; Xing, H.Y.; Zhou, G.Q.; Gao, Y.T. Dipole modelling of metal magnetic memory for V-notched plates. *Insight Non-Destr. Test. Cond. Monit.* **2013**, *55*, 498–503. [CrossRef]
11. Han, G.; Huang, H. A dual-dipole model for stress concentration evaluation based on magnetic scalar potential analysis. *NDT E Int.* **2020**, *118*, 102394. [CrossRef]
12. De-Hui, W.; Zhi-Tian, L.; Xiao-Hong, W.; Ling-Xin, S. Mechanism analysis of influence of surface-breaking orientation on magnetic leakage field distribution. *Acta Phys. Sin.* **2017**, *66*, 48102. [CrossRef]
13. Dehui, W.; Lingxin, S.; Xiaohong, W.; Zhitian, L. A Novel Non-destructive Testing Method by Measuring the Change Rate of Magnetic Flux Leakage. *J. Nondestruct. Eval.* **2017**, *36*, 24. [CrossRef]
14. Li, E.; Chen, Y.; Chen, X.; Wu, J. Defect Width Assessment Based on the Near-Field Magnetic Flux Leakage Method. *Sensors* **2021**, *21*, 5424. [CrossRef] [PubMed]
15. Xu, K.; Qiu, X.; Tian, X. Theoretical investigation of metal magnetic memory testing technique for detection of magnetic flux leakage signals from buried defect. *Nondestruct. Test. Eval.* **2016**, *33*, 45–55. [CrossRef]
16. He, G.; He, T.; Liao, K.; Deng, S.; Chen, D. Experimental and numerical analysis of non-contact magnetic detecting signal of girth welds on steel pipelines. *ISA Trans.* **2022**, *125*, 681–698. [CrossRef]
17. Suresh, V.; Abudhahir, A. An Analytical Model for Prediction of Magnetic Flux Leakage from Surface Defects in Ferromagnetic Tubes. *Meas. Sci. Rev.* **2016**, *16*, 8–13. [CrossRef]
18. Yang, L.; Huang, P.; Gao, S.; Du, Z.; Bai, S. Research on the magnetic flux leakage field distribution characteristics of defect in low-frequency electromagnetic detection technique. *IEICE Electron. Express* **2021**, *18*, 20200362. [CrossRef]
19. Liu, B.; Ma, Z.; Liu, Z.; Luo, N.; Xu, X. Research on internal detection technology for axial crack of long-distance oil and gas pipeline based on micromagnetic method. *Struct. Health Monit.* **2019**, *19*, 1123–1136. [CrossRef]
20. Long, Y.; Huang, S.; Peng, L.; Wang, S.; Zhao, W. A Characteristic Approximation Approach to Defect Opening Profile Recognition in Magnetic Flux Leakage Detection. *IEEE Trans. Instrum. Meas.* **2021**, *70*, 1–12. [CrossRef]
21. Kalu, I.E.W.; Inglis, H.M.; Kok, S. The sensitivity of failure analysis of boiler tubes to the shape of elliptical external erosion flaws. *Eng. Fail. Anal.* **2021**, *119*, 104952. [CrossRef]

22. Ha, Y.; Tobita, T.; Takamizawa, H.; Katsuyama, J. Fracture Toughness Evaluation of the Heat-Affected Zone Under the Weld Overlay Cladding in Reactor Pressure Vessel Steel. *J. Press. Vessel. Technol.* **2022**, *145*, 21501. [CrossRef]
23. Huang, X.; Wu, J.; Sun, Y.; Kang, Y. 3D magnetic dipole models of magnetic flux leakage forconcave'andbump'defects. *Int. J. Appl. Electromagn. Mech.* **2019**, *59*, 1305–1312. [CrossRef]
24. Deng, Z.; Sun, Y.; Yang, Y.; Kang, Y. Effects of surface roughness on magnetic flux leakage testing of micro-cracks. *Meas. Sci. Technol.* **2017**, *28*, 45003. [CrossRef]

materials

Article

Relationship between the Synthesis Method and the Magnetoelectric Properties of Bismuth Sodium-Potassium Titanate/Nickel Cobalt Ferrite Lead-Free Composites

Javier Camargo *, Leandro Ramajo and Miriam Castro *

Institute of Research in Materials Science and Technology (INTEMA), Av. Colón 10850, Mar del Plata 7600, Argentina
* Correspondence: jcamargo@fi.mdp.edu.ar (J.C.); mcastro@fi.mdp.edu.ar (M.C.)

Abstract: In this work, the influence of the synthesis methods of piezoelectric and magnetostrictive phases on the final properties of the $Bi_{0.5}(Na_{0.8}K_{0.2})_{0.5}TiO_3$-$Ni_{0.5}Co_{0.5}Fe_2O_4$ composites was studied. Different routes were used to individually synthesize each phase, and the composites were prepared using different fractions for each phase. Composites were sintered, and the structural, microstructural, dielectric, and magnetoelectric properties were evaluated. According to the selected synthesis method employed for each phase, different particle sizes and reactivities of the individual phases were obtained. These differences determined the suitable sintering temperature for each set of composites and were responsible for the final properties. In fact, magnetoelectric properties were modulated by the combination of composition and synthesis routes.

Keywords: magnetoelectric; piezoelectric; magnetic; sol-gel; solid state; Pechini

Citation: Camargo, J.; Ramajo, L.; Castro, M. Relationship between the Synthesis Method and the Magnetoelectric Properties of Bismuth Sodium-Potassium Titanate/Nickel Cobalt Ferrite Lead-Free Composites. *Materials* **2023**, *16*, 2759. https://doi.org/10.3390/ma16072759

Academic Editor: Dana Popescu

Received: 8 March 2023
Revised: 24 March 2023
Accepted: 28 March 2023
Published: 30 March 2023

1. Introduction

Magnetoelectric materials are able to modify magnetization and polarization by the application of an external electric or magnetic field. These materials are extensively studied due to their scientific and technological relevance in several applications [1–3]. In most single-phase multiferroics, the magnetoelectric (ME) coupling between electrical and magnetic ordering is generally weak and operates at very low temperatures, making it difficult to use in practical applications [4–6]. For example, some compounds based on the $Bi_2LaNb_{1.5}Mn_{0.5}O_9$ Aurivillius phase [7], the $Ba_4(Sm_xLa_{1-x})_2Fe_2Nb_8O_{30}$ tungsten bronze phase [8] or the $(Ca,Sr)_3Mn_2O_7$ Ruddlesden-Popper perovskites [9,10] show low magnetoelectric coupling properties even at low temperatures. On the other hand, although single-phase $BiFeO_3$ ceramics are magnetoelectric at room temperature, it is difficult to obtain single-phase ceramics with acceptable properties. Therefore, the research focus has shifted towards the multiferroic compounds, which show a strong ME coupling at room temperature. These composites can be obtained through the direct mix of particles of both phases, the use of ceramic fibers, or the multilayer deposition [11–14].

Among the composites with magnetoelectric response, it is possible to find some ceramics that use lead-based materials in the piezoelectric phase, such as $PbZr_{0.52}Ti_{0.48}O_3$ (PZT) or $Pb(Mg_{1/3}Nb_{2/3})_{0.67}Ti_{0.33}O_3$ (PMN-PT) and, for example, cobalt and nickel ferrites as the magnetostrictive phase [15]. However, in many countries, the use of lead-based materials is prohibited or limited due to their known toxicity. Consequently, in recent years different lead-free alternatives have been developed for both piezoelectric and magnetoelectric materials [16,17]. Lead-free ceramics, such as $BaTiO_3$ (BT), $Bi_{0.5}Na_{0.5}TiO_3$ (BNT), $Sr_{0.5}Ba_{0.5}Nb_2O_6$ (SBN), $K_{0.5}Na_{0.5}NbO_3$ (KNN), and $82BaTiO_3$-$10BaZrO_3$-$8CaTiO_3$ (BZT-BCT), have been proposed for the piezoelectric phase [18–25].

$Bi_{0.5}Na_{0.5}TiO_3$ (BNT) is a widely studied perovskite with high Curie temperature (T_c~320 °C) and remanent polarization (P_r~38 μC/cm^2). Nevertheless, its high coercive field and low piezoelectric coefficient compared to other piezoceramics hinder its application [26,27]. A possible solution to considerably enhance the piezoelectric response is the introduction of BNT solid solutions. Specifically, the $Bi_{0.5}Na_{0.5}TiO_3$ (BNT)-$Bi_{0.5}K_{0.5}TiO_3$ (BKT) solid solution presents a morphotropic phase boundary (MPB) for the 80BNT-20BKT composition where the maximum value of the piezoelectric coefficient (d_{33}), and a diminution in the value of the coercive field are reached [28–31].

In the case of magnetic materials with AB_2O_4 spinel structure, cobalt ferrite is a well-known hard magnetic material with relatively high coercivity (H_c), and magnetic saturation (M_s), whereas nickel ferrite is a soft magnetic material with low coercivity and magnetic saturation. Therefore, the combination of these hard and soft ferrites allows their use to be extended to a wide variety of applications [32–34]. Indeed, Ni-Co ferrites are commonly employed in electronic devices in the area of telecommunications. For example, these ferrites are used in high-quality filters, radio frequency circuits, transformer cores, and read/write heads for high-speed digital tapes. Moreover, they can be used in the miniaturization of high-frequency application devices due to the reached electromagnetic properties [35].

Another way to improve the functional properties of these ceramics is by optimizing the size of the starting powders and the size ratio between the magnetic and piezoelectric phases. It has been reported that the size of the powders plays a very important role in obtaining a certain crystalline phase and the electrical properties of ceramic pieces when these powders are in the submicron size range [36,37]. For this reason, powder synthesis methods have been extensively studied. In general, ceramic powders are prepared using the conventional solid-state reaction [38,39]. However, to ensure the complete formation of the desired phases, by this type of reaction, a long time and a high temperature (above 1100 °C) are generally required. Consequently, the synthesized powders typically have a large particle size. In contrast, chemical routes (i.e., co-precipitation technique, sol-gel process, or hydrothermal) are effective in reducing the synthesis temperature and the size of crystals in the nanometer range [38,40,41]. Additionally, it has been determined that the direct interaction between the piezoelectric and magnetic phases, without reaching the percolation limit, is essential for obtaining improvements in the magnetoelectric coefficient.

In this work, magnetoelectric compounds corresponding to the $Bi_{0.5}(Na_{0.8}K_{0.2})_{0.5}TiO_3$ (BNKT)-$Ni_{0.5}Co_{0.5}Fe_2O_4$ (NCF) system were prepared. The $Bi_{0.5}(Na_{0.8}K_{0.2})_{0.5}TiO_3$ (BNKT) piezoelectric phase was synthesized by two different routes: the solid-state reaction method with a mechanochemical activation step and the sol-gel method. Nevertheless, the $Ni_{0.5}Co_{0.5}Fe_2O_4$ (NCF) magnetostrictive phase was prepared by Pechini's method. The compounds were prepared using different fractions of the previously synthesized phases. The dielectric, piezoelectric, and magnetic properties, as well as the magnetoelectric (ME) coefficient, were investigated. Furthermore, the ME coupling was improved by controlling the grain size to facilitate the contact between the phases and minimize the reactivity of the phases.

2. Materials and Methods

Two alternative routes were followed for the synthesis of the piezoelectric phase (the solid-state reaction with a mechanochemical activation step of the reagents (SS) and the sol-gel (SG) methods), whereas the magnetostrictive phase was synthesized through Pechini's method (PE).

2.1. Synthesis of BNKT through the Solid-State Method (SS)

The reagents used for the synthesis were bismuth oxide, Bi_2O_3 (Aldrich 99.8%; Saint Louis, MO, USA), sodium carbonate, Na_2CO_3 (Aldrich 99.5%; USA), potassium carbonate, K_2CO_3 (Aldrich 99.5%; USA), and titanium oxide, TiO_2 (Aldrich 99.9%; USA). The reagents were weighted in the stoichiometric ratio, ground in a planetary mill (Fritsch, Pulverisette 7, 1050 rpm, Idar-Oberstein, Germany) for 6 h, and then calcined at 700 °C for 2 h.

2.2. Synthesis of BNKT through the Sol-Gel Method (SG)

The reagents used for the synthesis were bismuth nitrate pentahydrate, $Bi(NO_3)_3{\cdot}5H_2O$ (Aldrich 98%; USA), sodium acetate tetrahydrate, $C_2H_3NaO_2{\cdot}4H_2O$ (Anedra 98%; Los Troncos del Tala, BA, Argentina), potassium acetate tetrahydrate, $C_2H_3KO_2{\cdot}4H_2O$ (Biopack 98%; Argentina), and titanium butoxide, $Ti(C_4H_9O)_4$ (Aldrich 97%; USA). Specifically, 5 mM bismuth nitrate pentahydrate, 4 mM sodium acetate tetrahydrate, and 1 mM potassium acetate tetrahydrate were separately dissolved in 25 mL acetic acid. On the other hand, 10 mM titanium butoxide, and 20 mM acetylacetone were dissolved in 12 mL 2-methoxy ethanol. All the solutions were stirred for 30 min, mixed, and finally, 2.5 mL ammonium hydroxide was added to reach pH 3. This solution was also kept at 150 °C for 24 h to obtain a fine powder. Subsequently, a heat treatment at 700 °C for 30 min was carried out to obtain the desired BNKT phase.

2.3. Synthesis of NCF Phase

The reagents were nickel nitrate, $Ni(NO_3)_2{\cdot}6H_2O$ hexahydrate (Baker 99.8%; Phillipsburg, NJ, USA), cobalt nitrate hexahydrate, $Co(NO_3)_3.6H_2O$ (Biopack 98%, Buenos Aires, Argentina), iron nitrate nonahydrate, $Fe(NO_3)_3{\cdot}9H_2O$ (Aldrich 98%; USA), and citric acid (Cicarelli 99%; Santa Fe, Argentina). The reactants were separately dissolved in a 1:1:4:6 stoichiometric ratio in water in a 1.25 molar ratio. All the solutions were stirred for 30 min, then they were mixed, and a final pH 7 was obtained by the addition of ammonium hydroxide. The final solution was reserved at 200 °C until self-combustion to obtain a fine black powder. Afterwards, in order to find the desired phase, the black powder was thermally treated at 850 °C for 2 h.

2.4. Composites Preparation

BNKT and NCF powders were mixed in different weight proportions according to $xBi_{0.5}(Na_{0.8}K_{0.2})_{0.5}TiO_3$-(100-x)$Ni_{0.5}Co_{0.5}Fe_2O_4$, x = 0, 60, 70, 80, 90, 95, and 100 in a planetary mill for 15 min. The powders were uniaxially pressed into disks 1 cm in diameter and 0.2 mm thick. Taking into account previous studies carried out on xBNKT-(100-x)NCF composites [38], the sintering temperature of the compounds was established from an analysis of the densification degree of the samples with composition 70BNKT-30NCF. For this composition, the sintering temperature was varied between 950 °C and 1125 °C. Based on this analysis, the sintering cycle for the composites was determined. BNKT (SS)-NCF (PE) samples were sintered at 1075 °C, whereas BNKT (SG)-NCF (PE) samples were sintered at 975 °C.

2.5. Characterization

The obtained ceramics were evaluated using several techniques (X-ray diffraction (XRD, PANalytical, X'pert Pro, CuKα), Raman spectroscopy (Renishaw in Via microscope through the 514 nm Ar-ion laser line, Wotton-under-Edge, Glos, UK), and Field Emission Scanning Electron Microscopy (Zeiss Crossbeam 350, Jena, Germany) equipped with Energy-Dispersive Spectroscopy (EDS Oxford Ulti Max 100, High Wycombe, Bucks, UK). The apparent density of the sintered composites was obtained by means of Archimedes' method using distilled water as the immersion medium at room temperature. The theoretical density (δ) for BNKT-NCF composites was calculated from the mixture rule, using the following equation:

$$\delta = (W_{BNKT} + W_{NCF})/(W_{BNKT}/\delta_{BNKT} + W_{NCF}/\delta_{NCF}) \quad (1)$$

where W_{BNKT} and W_{NCF}, are the weight percentages of the BNKT and NCF in the mixture, respectively; δ_{BNKT} and δ_{NCF} are the theoretical densities of BNKT (5.97 g/cm^3) and NCF phases (5.334 g/cm^3). The relative density was calculated by the quotient between the apparent density and the theoretical density.

Dielectric measurements were carried out on the sintered discs painted with silver/palladium electrodes on the plain surfaces. Real permittivity and dielectric loss tangent were determined employing an impedance analyzer (HP4284A LCR meter, Agilent Technologies, Englewood, CO, USA) in a temperature range between 20 to 500 °C while the piezoelectric coefficient d_{33} was measured using a quasi-static piezoelectric d_{33} meter (YE2730—Sinoceramics, Shanghai, China).

Magnetization was measured, at room temperature and as a function of the magnetic field between +13 and −13 kOe, using a vibrating sample magnetometer (Lakeshore 7300, Westerville, OH, USA).

The magnetoelectric voltage coefficient (α_{33}) of the composites was obtained from the magnetic field-induced voltage measured across the sample using a lock-in amplifier (model NF LI5640) and employing a bias magnetic field up to 15 kOe with an ac magnetic field of 2 Oe at 1 kHz.

3. Results

Based on a previous work carried out on BNT(SS)-NCF(SS) composites [38], the density of the 70BNT(SS)-30NCF (70-30) sample was taken as a parameter to determine the sintering temperature of each set of samples. Table 1 shows the density values of the sintered samples at different temperatures. For the SS-PE samples the maximum density value was observed when samples were sintered at 1075 °C, whereas for the SG-PE samples the maximum density value was obtained in the samples sintered at 975 °C. Consequently, all sets of samples were sintered at the temperature where the maximum density was registered. Thus, the complete SS-PE set was sintered at 1075 °C for 2 h, whereas the SG-PE set was sintered at 975 °C for 2 h.

Table 1. Experimental and relative density values of 70BNKT-30NCF (SS-PE and SG-PE) samples sintered at different temperatures.

70BNKT-30NCF SS-PE Samples			**70BNKT-30NCF SG-PE Samples**		
Temperature (°C)	**Density (g/cm^3)**	**Relative Density (%)**	**Temperature (°C)**	**Density (g/cm^3)**	**Relative Density (%)**
975	5.15 ± 0.10	89.3	950	5.16 ± 0.02	89.5
1000	5.13 ± 0.07	89.0	975	5.23 ± 0.04	90.7
1025	5.13 ± 0.08	89.0	1000	5.04 ± 0.08	87.4
1050	5.17 ± 0.08	89.6	1025	4.98 ± 0.08	86.3
1075	5.18 ± 0.07	89.8	1050	4.60 ± 0.20	80.4
1100	4.40 ± 0.20	76.8	1075	4.39 ± 0.01	76.2
1125	4.20 ± 0.10	73.1	1100	4.10 ± 0.10	70.4

In Figures 1 and 2 the XRD patterns of the sintered samples are shown. All patterns of the composite samples exhibit the peaks assigned to the BNKT [42] and NCF (JCPDS card 04-002-0422) phases without the presence of secondary phases. Analyzing in detail the main peak of NCF (located at ~35.5°), a displacement to lower angles as the amount of BNKT increases in SS-PE samples is registered. This change can be attributed to the expansion of lattice parameters due to the diffusion of larger radius ions into smaller ion-site in the NCF phase. However, in the SG-PE samples (Figure 2), as they were sintered at a lower temperature, the ferrite remains stable, and the peak displacement is less perceptible.

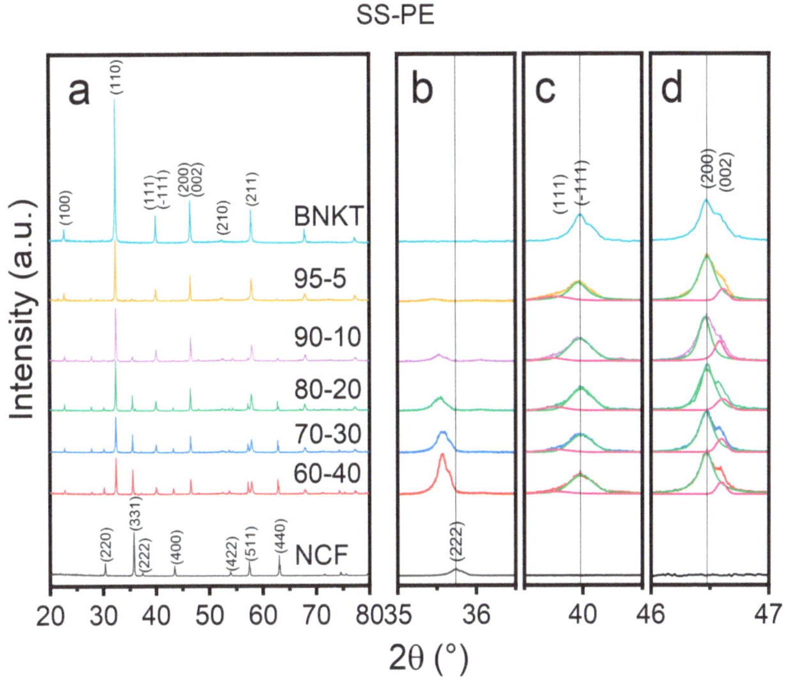

Figure 1. XRD patterns of the xBNKT-(100-x)NCF (x = 100, 95, 90, 80, 70, 60, 0) SS-PE composites (**a**). Enlargements of 35–36.5° (**b**), 39–41° (**c**), and 46–47° (**d**) angles.

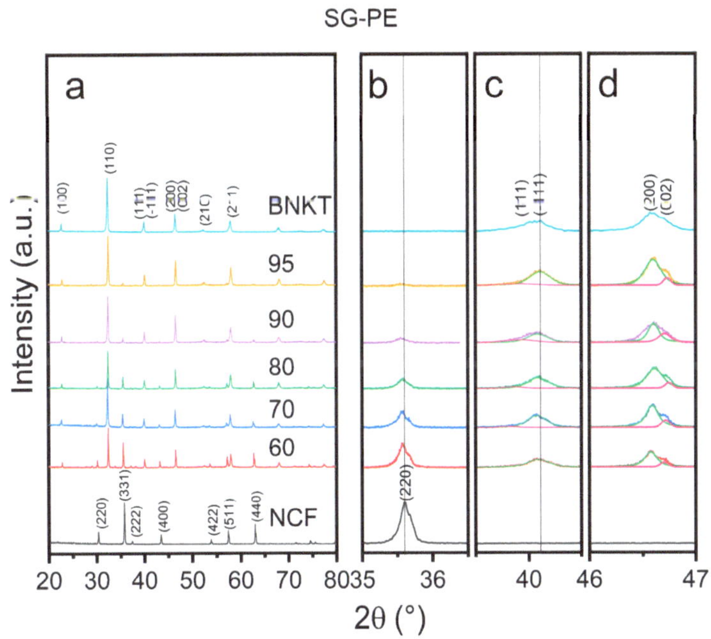

Figure 2. XRD patterns of the xBNKT-(100-x)NCF (x = 100, 95, 90, 80, 70, 60, 0) SG-PE composites (**a**) Enlargements of 35–36.5° (**b**), 39–41° (**c**), and 46–47° (**d**) angles.

For the BNKT phase, structural phase evolution can be examined by the peak splitting phenomenon of the (111) and (200) peaks. In these composite samples, the rhombohedral stabilization can be confirmed by the presence of the (−111) and (111) peaks near 41° and the combination of the (002)/(200) peaks into only one peak at 46.5°, indicating the tetragonal phase absence, for both methods. It must be pointed out that the small peak located at 46.7° belongs to $K_{\alpha 2}$ [43].

Figure 3 shows the SEM images of the sintered composites. Firstly, taking into account the grain size of pure BNKT ceramics, synthesized by the solid-sate (0.7 ± 0.2 μm) or the sol-gel (0.22 ± 0.10 μm) methods [44,45], and the grain size of the BNKT phase in the 95-5 composites (BNKT SS-PE 7 ± 4 μm and BNKT SG-PE 2.4 ± 1.4 μm), considerable grain growth is registered in these composites. Prasertpalichat et al. studied the influence of Fe addition on the final properties of $0.93Bi_{0.5}Na_{0.5}TiO_3$-$0.07BaTiO_3$ and they attributed the grain size increasing to the oxygen vacancies generated from doping [46]. Oxygen vacancies facilitate mass transportation during the sintering process and promote grain growth. Consequently, the initial grain growth of the BNKT phase by the NCF addition in the composites can be attributed to the iron diffusion into the BNKT phase. Secondly, the increasing ferrite concentration does not produce a continuous grain growth of the piezoelectric phase. Apparently, as the amount of the ferrite phase increases, the Fe diffusion into the BNKT is reduced due to the stabilization of the ferrite phase into the composites. In addition, the ferrite phase could impede the grain growth of the BNKT phase due to steric issues. Finally, the lower grain growth observed in the SG-PE than in the SS-PE set of samples can be assigned to the inferior sintering temperature required for the SG-PE set of samples and to the smaller initial particle size of the SG powder.

In Figure 4 the atomic mapping of the different elements in the 70BNKT-30NCF composites can be observed. Additionally, the atomic content corresponding to the ferrite and piezoelectric phases is reported in Table 2. Taking into account the detection limit of the technique, the final compositions obtained by both synthesis processes were in good agreement with the expected ones. Moreover, in both cases, the small iron content detected in the perovskite phase corroborates the influence of this element on the BNKT grain growth.

Table 2. Atomic content determined by EDS of the 70BNKT-30NCF SS-PE and SG-PE composites.

						70BNKT-30NCF SS-PE							
Point 1							**Point 2**						
Bi	Na	K	Ti	O	Fe	Ni	Co	Fe	O	Bi	Na	K	Ti
0.23	0.16	0.02	0.42	1.54	0.04	0.21	0.21	0.74	1.72	0.01	0.00	0.00	0.06
						70BNKT-30NCF SG-PE							
Point 1							**Point 2**						
Bi	Na	K	Ti	O	Fe	Ni	Co	Fe	O	Bi	Na	K	Ti
0.23	0.15	0.02	0.43	1.55	0.03	0.22	0.22	0.73	1.77	0.01	0.02	0.00	0.07

Table 3 shows the apparent and theoretical densities, and the densification degree values for both sets of samples. Taking into account that the sintering temperature of each set of samples was optimized for the 70-30 composition, relevant variations with composition are not observed. Although the synthesis method of the piezoelectric phase is changing (solid-state reaction or sol-gel) and therefore so is its reactivity, this effect is compensated with the different sintering temperatures (975 °C for the SG-PE and 1075 °C for the SS-PE).

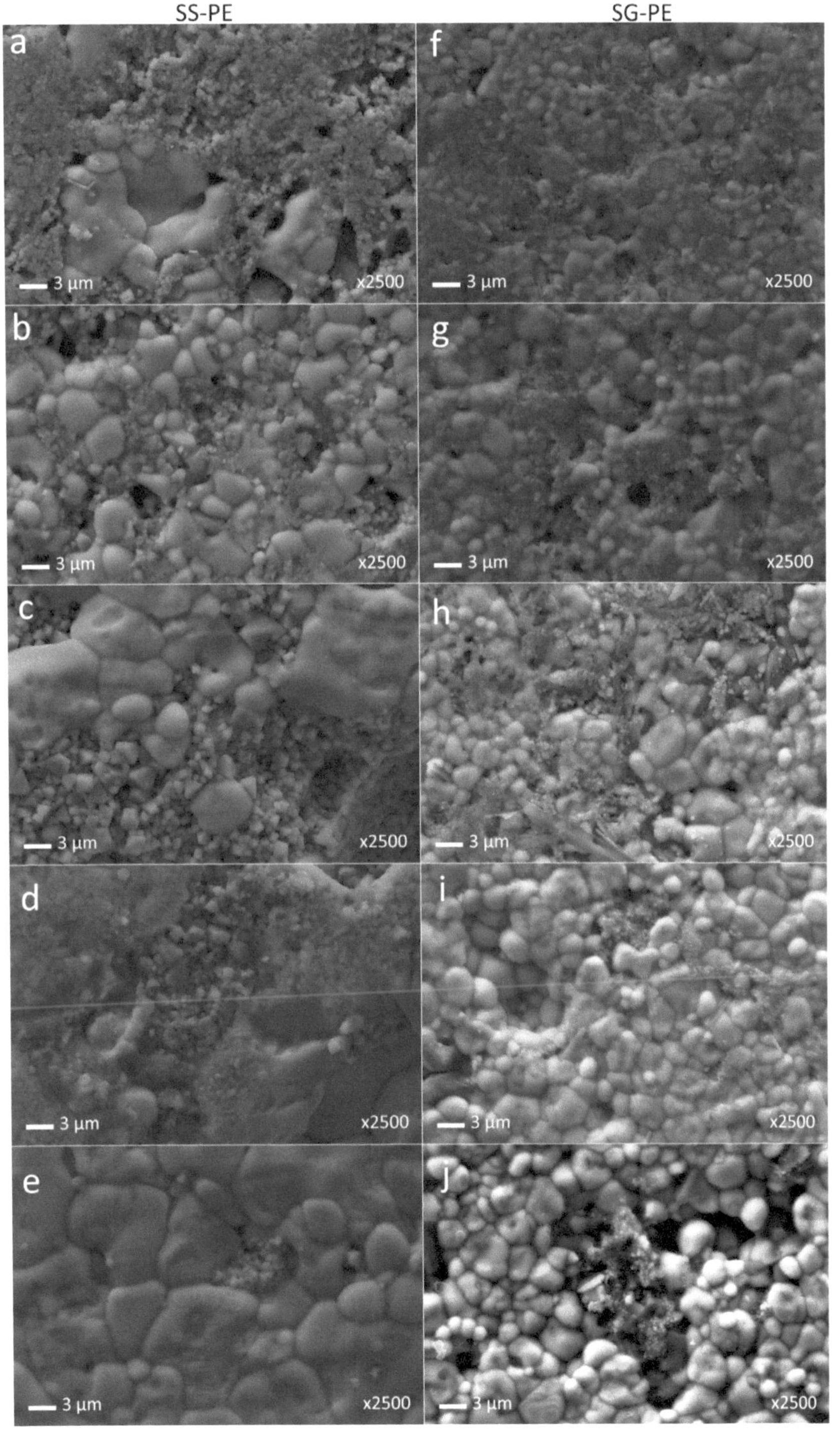

Figure 3. SEM images of the xBNKT-(100-x)NCF (x = 60 (**a**,**f**), 70 (**b**,**g**), 80 (**c**,**h**), 90 (**d**,**i**), 95 (**e**,**j**)) sintered composites.

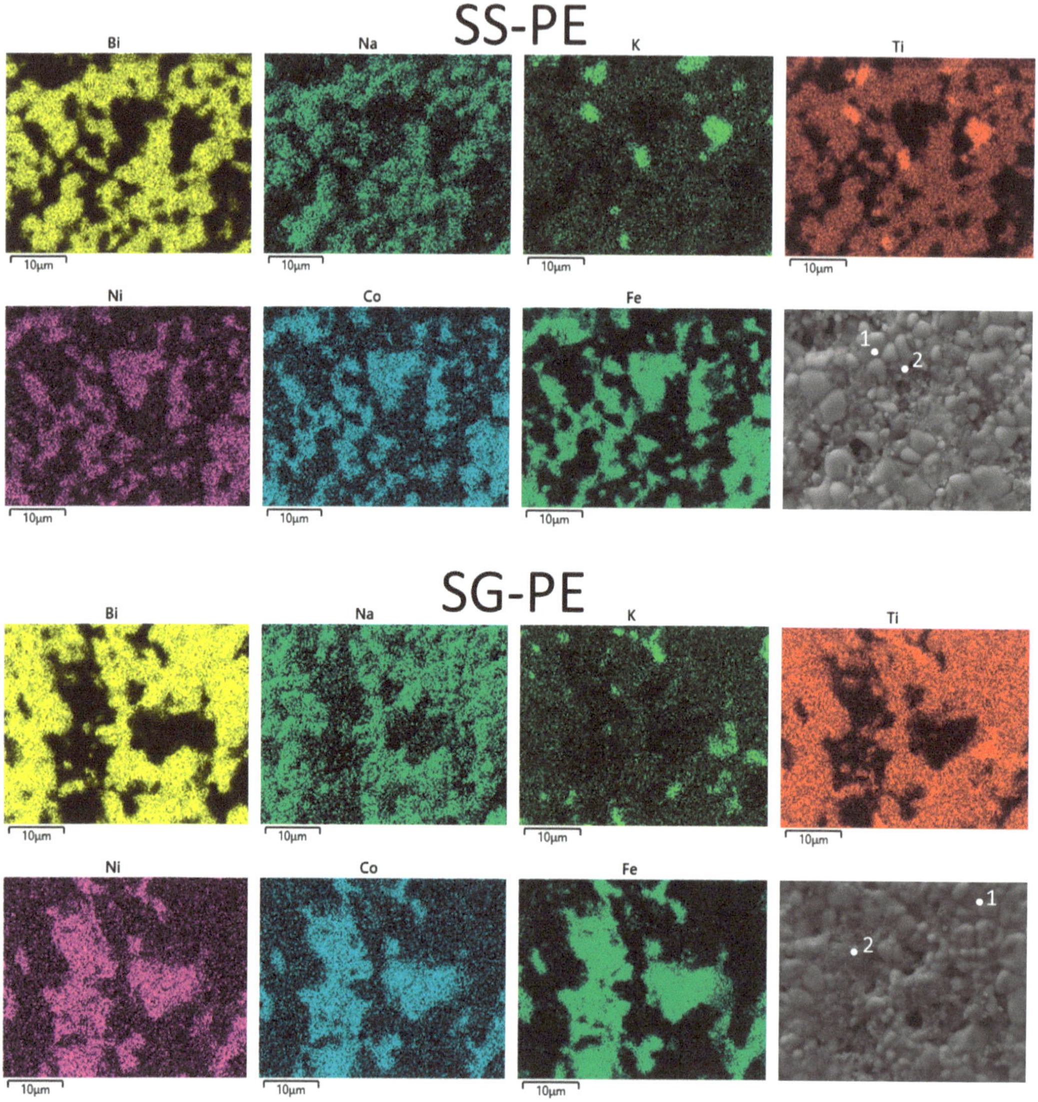

Figure 4. EDS mapping of the different elements corresponding to the 70BNKT-30NCF composites. Points 1 and 2 indicate the places where the atomic content was analyzed.

Table 3. Apparent, and theoretical densities, and densification degree values of BNKT-NCF (SS-PE and SG-PE) samples.

xBNKT-(100-x)NCF (x)	Theoretical Density (g/cm^3)	SS-PE (Sintered a 1075 °C)		SG-PE (Sintered at 975 °C)	
		Density (g/cm^3)	Densification Degree (%)	Density (g/cm^3)	Densification Degree (%)
60	5.70	5.14 ± 0.04	90.2	5.07 ± 0.07	89.0
70	5.76	5.18 ± 0.07	89.8	5.23 ± 0.04	90.7
80	5.83	5.27 ± 0.06	90.4	5.26 ± 0.04	90.8
90	5.90	5.27 ± 0.10	89.3	5.34 ± 0.09	90.5
95	5.93	5.51 ± 0.02	92.9	5.33 ± 0.08	89.8

Figure 5 shows the Raman spectra of both sets of BNKT-NCF sintered composites, where the spectra of BNKT and NCF phases sintered at the corresponding sintering temperatures are also presented. For BNKT samples, four characteristic bands are observed. The band at 130 cm^{-1} is associated with the A-site vibration in the perovskite structure, the band around 240 and 400 cm^{-1} is associated with the Ti-O vibration, the band at 430–700 cm^{-1} is connected to the TiO_6 octahedra vibration, and finally, the band above 700 cm^{-1} is related to A_1 and E longitudinal optical overlapping bands [47]. For pure NCF samples, five bands in the Raman spectra are identified. The bands located at 700, 650, and 620 cm^{-1}, associated with the symmetric stretch of oxygen atoms along the (Fe/M)-O bond, can be assigned to Fe-O, Ni-O, and Co-O cations, respectively. The band at 485 cm^{-1} is associated with the anti-symmetric stretch of oxygen atoms along the (Fe/M)-O bond. The bands at 327 and 570 cm^{-1} are symmetric and antisymmetric bending modes of oxygen concerning Fe(M), respectively, and finally, the band at 213 cm^{-1} is assigned to the translational movement of tetrahedron MO_4 [48].

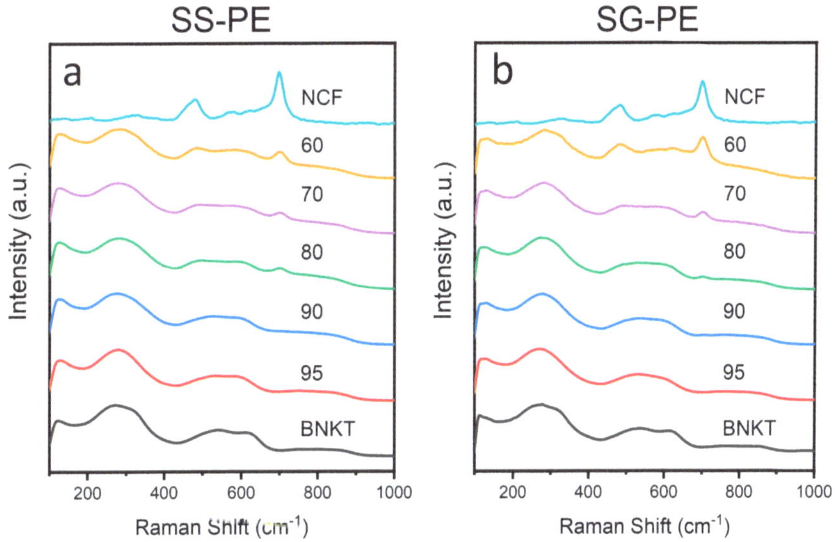

Figure 5. Raman spectra of xBNKT-(100-x)NCF (**a**) SS-PE and (**b**) SG-PE composites.

To study in detail the influence of ferrite on the Raman spectrum of the piezoelectric phase, the spectra of the compounds were analytically calculated from the algebraic sum of the experimentally obtained spectra of the individual phases (BNKT and NCF) corresponding to the different composites. After comparing the experimental Raman spectra of the composites with the corresponding algebraic sum (see, Figure A1 in Appendix A), it is observed that the BNKT bands below 400 cm^{-1} are not affected by the ferrite addition. In the zone between 400 and 800 cm^{-1}, ferrite addition modifies the TiO_6 octahedral vibration of the BNKT phase. In addition, from the analysis of the experimental and calculated Raman spectra, it is observed that the ferrite bands are only individualized in NCF compositions of 20% or higher, in agreement with other published papers [49,50].

In Figure 6 real permittivity and loss tangent vs. temperature curves at different frequencies of pure phases and composites, with 5% of NCF addition, obtained by the different methods are shown. From the figure, a strong influence of the ferrite conductivity process at higher temperatures for frequencies lower than 100 kHz is observed. This effect was also observed in the previously published results corresponding to BNKT(SS)-NCF(SS) samples [51]. These curves show two dielectric anomalies, the depolarization temperature (T_d) existing at around 130 °C and a maximum permittivity temperature (T_m) close to 280 °C. Moreover, the typical relaxor behavior with frequency is clearly observed in sample

BNKT-SS (Figure 6). Broad peaks at T_d and T_m indicate the characteristics of a diffuse phase transition [52,53]. Interestingly, BNKT samples synthesized by the sol-gel method rendered high dielectric losses at low frequencies which could be associated with the low densification level, and the small grain size obtained after sintering at 975 °C.

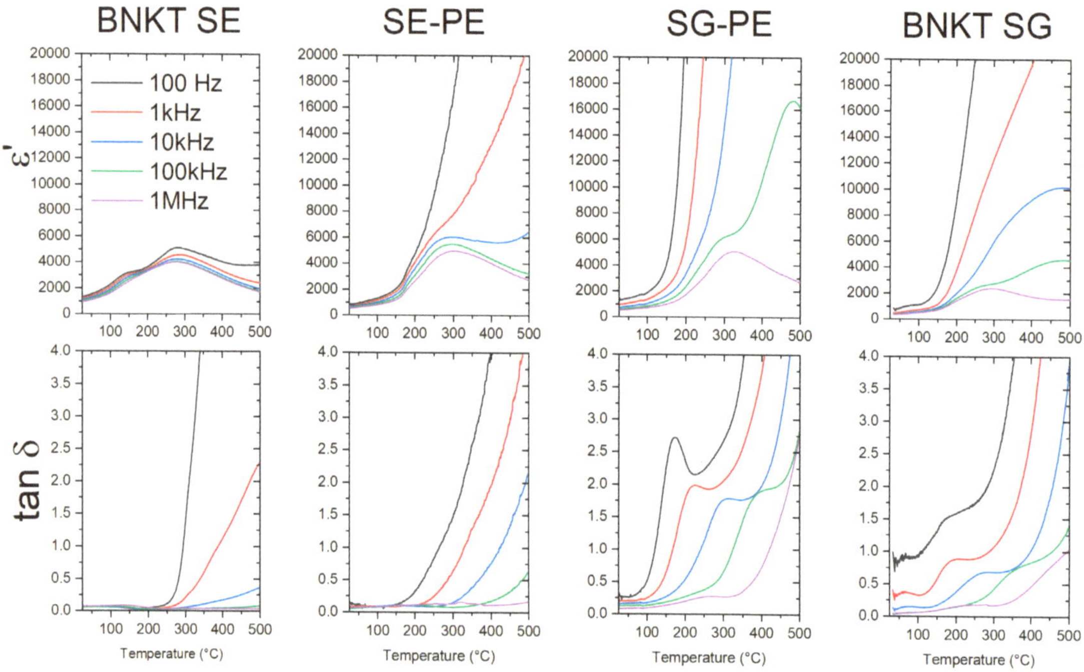

Figure 6. Real permittivity (ε′) and loss tangent (tanδ) vs. temperature curves at different frequencies of BNKT (SS and SG), and 95BNKT-5NCF (SS-PE and SG-PE) sintered samples.

To avoid the frequency dependence and to be able to compare the influence of the ferrite addition on dielectric properties, Figure 7 only shows the dielectric properties of both sets of samples at 1 MHz. Considering that the dielectric permittivity of the ferrite phase is lower than that of the piezoelectric phase, then the dielectric permittivity decreases when the ferrite addition increases. However, in SG-PE samples, the notable increase in the real permittivity with the addition of NCF to the pure BNKT phase can be associated with the observed grain size increment. Indeed, the low dielectric properties of the BNKT SG sample can be related to the low sintering temperature and, consequently, the small grain growth reached. In addition, the temperature corresponding to the maximum dielectric permittivity moves to higher values when the NCF content is increased, particularly for the SS-PE set of samples. On the contrary, P Gupta et al. observed a reduction in the T_m with Fe addition to $BaTiO_3$ and they attributed the observed behavior to the substitution of the lower valence Fe^{2+}/Fe^{3+} at the Ti^{4+} site creating oxygen vacancies [46]. In these sets of samples, two possible effects influence the dielectric properties: the possible ions diffusion between both phases and the variation in the grain growth of the BNKT phase as the NCF content is increased. Additionally, for the highest temperatures, the dielectric loss of all composites increases due to the ferrite conductivity.

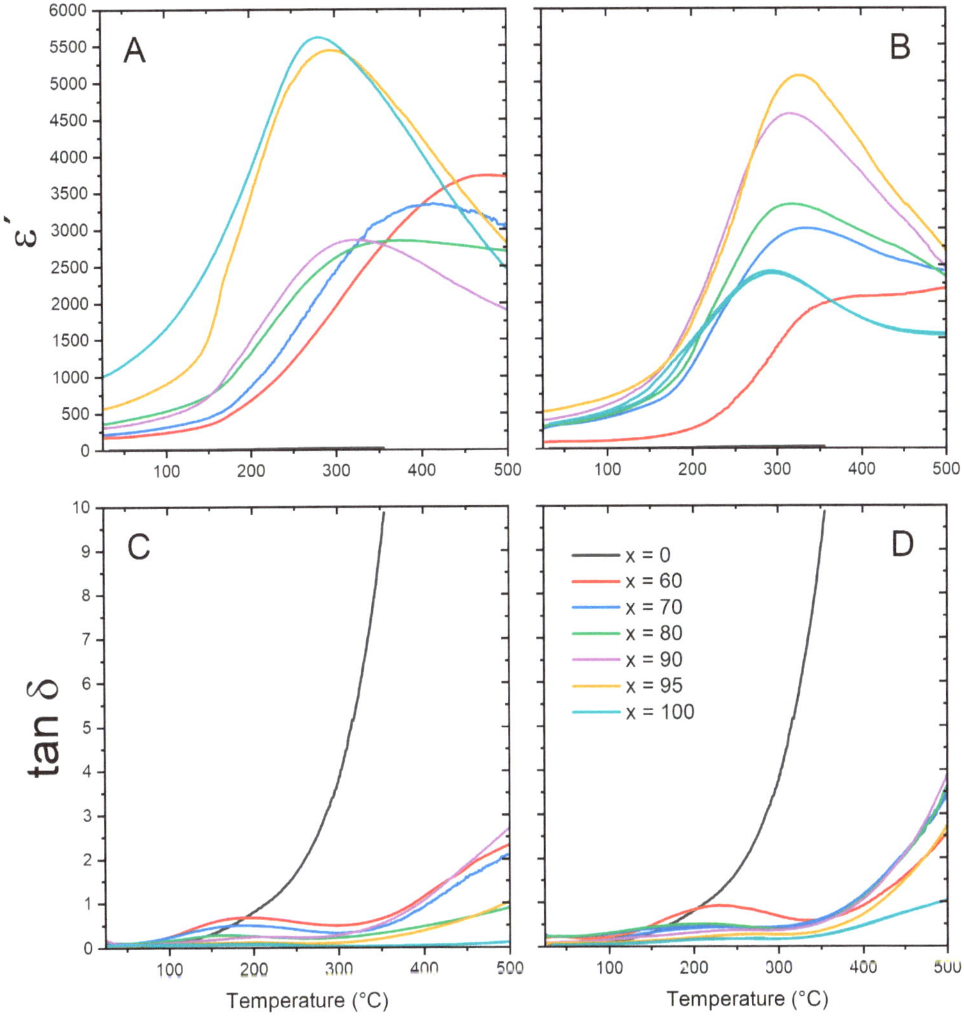

Figure 7. Real permittivity (ε') and loss tangent ($\tan\delta$) vs. temperature curves for BNKT, NCF, and BNKT-NCF sintered samples obtained by the different methods at 1 MHz. SS-PE (**A**, **C**) and SG-PE (**B**, **D**).

In Figure 8, the magnetic hysteresis loops and the saturation magnetization and coercive field evolution with composition are presented. In both systems, as the BNKT phase amount increases, the saturation magnetization decreases according to Vegard's Law. Nevertheless, the BNKT phase increment produces a different effect in the coercive magnetic field depending on the selected synthesis and processing conditions (see Table 4). These variations are associated with compositional changes and the final grain size of the magnetic phase.

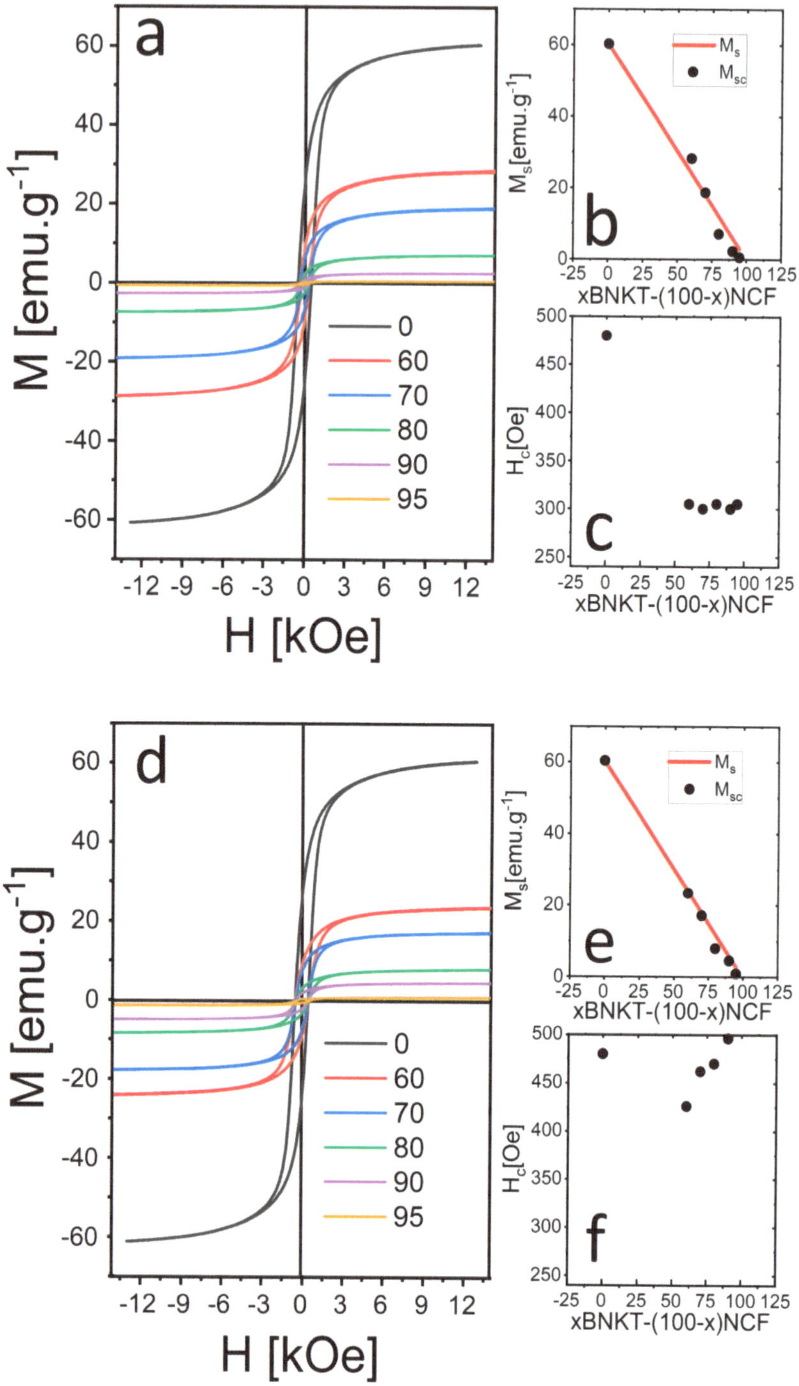

Figure 8. Magnetic hysteresis loops of the pure NCF sample and xBNKT-(100-x)NCF composites (**a**,**d**), Saturation magnetization (**b**,**e**) and coercive magnetic field (**c**,**f**) progress with composition.

Table 4. Experimental (M_s) and calculated (M_{sc}) saturation magnetization and coercive field (H_c) of xBNKT-(100-x)NCF (x = 95, 90, 80, 70, 60, 0) SS-PE and SG-PE composites.

Sample	SS-PE Samples			SG-PE Samples		
	M_s (emu/g)	M_{sc} (emu/g)	H_c (Oe)	M_s (emu/g)	M_{sc} (emu/g)	H_c (Oe)
0	60.3	-	480	60.3	-	480
60	28.3	24.2	260	23.5	24.1	426
70	18.8	18.1	300	17.3	18.1	462
80	7.1	12.1	305	8.1	12.1	470
90	2.3	6.0	300	4.7	6.0	496
95	0.6	3.0	305	1.1	3.0	512

Figure 9 shows the magnetoelectric voltage coefficient as a function of the magnetic field for the different SS-PE composites. Unfortunately, SG-PE composites were not completely polarized as a consequence of the small grain size of the magnetic phase due to the low sintering temperature required to avoid the possible reaction or the loss of the BNKT volatile elements. Consequently, the magnetoelectric voltage coefficient could not be obtained for this set of samples. From the figure, the highest magnetoelectric voltage coefficient is registered for the 70BNKT-30NCF composite. For higher NCF concentrations, the magnetoelectric voltage coefficient decreases. In composites, the ME voltage coefficient is determined by the conjunction between piezoelectric, magnetostrictive, and microstructural properties [51]. In addition, it can be seen that the highest value of the magnetoelectric coefficient is obtained for the composition with a 30% ferrite phase. Interestingly, comparing the magnetoelectric voltage coefficient values obtained in the SS-PE composites with those previously reported for compounds of the same composition but with both phases obtained by the solid-state reaction (SS) method and sintered at 1100 °C, it is observed that the maximum magnetoelectric voltage coefficient is also found at 30% of magnetic phase and the observed values are similar to those found here (Table 5). Considering that in SS-SS and SS-PE, the selected sintering temperature is not enough for the NCF grain growth, the characteristics of the BNKT phase and the overall microstructure determine the magnetoelectric properties. Additionally, as was expected, the higher the amount of the NCF phase, the smaller the value of the piezoelectric coefficient. From these results, it is possible to affirm that the selected synthesis method used for the piezoelectric or ferrite phases determines the final properties of the composites. In fact, magnetoelectric properties are strongly dependent on the possible diffusion or reaction between both phases. Specifically, comparing composites of the same composition but with both phases obtained from the solid-state reaction method Table 5, the use of ferrite powder obtained by Pechini's method allowed for reducing the calcination and sintering temperatures for NCF obtained from the solid-state reaction method T_{cal} = 1050 °C and for SS-SS composites T_s = 1100 °C during 5 h [51], whereas for NCF obtained from Pechini's method T_{cal} = 700 °C and for SS-PE composites T_s = 1075 °C during 2 h, maintaining the magnetoelectric properties. Moreover, an improvement in the magnetoelectric properties for x = 60 composites is registered for SS-PE samples. This variation could be associated with a better distribution of the phases. Finally, it is necessary to pointed out that the magnetoelectric coupling coefficients here presented are in good agreement with those found for other lead-free piezoelectric-ferrite particulate composites (2.02 $mVcm^{-1}Oe^{-1}$ for 20 $K_{0.5}Na_{0.5}NbO_3$/80 $MnFe_2O_4$ [54], and 4.875 $mVcm^{-1}Oe^{-1}$ for 60$Na_{0.5}Bi_{0.5}TiO_3$-40$CoCr_{0.4}Fe_{1.6}O_4$ composite [55]), where for better magnetoelectric coupling coefficient values laminates composites [56] or films [57] are required.

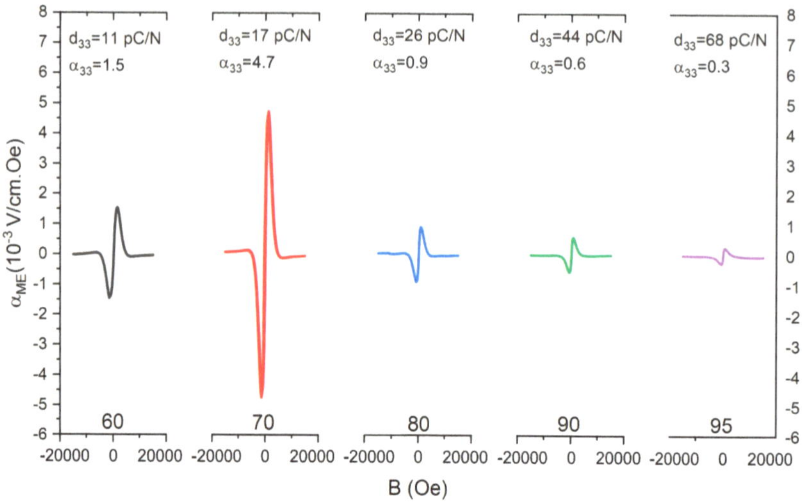

Figure 9. Magnetoelectric voltage coefficient (α_{33}) vs. dc magnetic field (B) of SS-PE sintered composites employing an ac magnetic field of 2 Oe at 1 kHz. The piezoelectric coefficient (d_{33}) of the composites is also reported in the figure.

Table 5. Magnetoelectric voltage coefficient (α_{33}) and piezoelectric coefficient (d_{33}) of SS-SS [51] and SS-PE composites.

Sample	α_{33} (mV/cmOe)		d_{33} (pC/N)	
	SS-SS	SS-PE	SS-SS	SS-PE
60	0.69	1.5	10	11
70	4.81	4.7	24	17
80	2.3	0.9	28	26

4. Conclusions

$Bi_{0.5}(Na_{0.8}K_{0.2})_{0.5}TiO_3$ (BNKT) powders were successfully synthesized by two alternative methods (solid-state reaction (SS) and sol-gel (SG)), whereas $Ni_{0.5}Co_{0.5}Fe_2O_4$ (NCF) powders were obtained by the Pechini's method (PE). Composites containing different amounts of the ferrite phase, and free of secondary phases, were obtained for both sets of samples (SS-PE and SG-PE). Trying to avoid the possible reaction between both phases and the decomposition of the piezoelectric phase, the sintering temperature of each set of samples was selected according to the final density of the sintered 70-30 composites. Consequently, in order to obtain similar final densities, SG-PE set samples were sintered at 975 °C, while samples of the SS-PE set were sintered at 1075 °C. By XRD and Raman spectroscopy, only the piezoelectric and ferrite phases were confirmed. Nevertheless, in the samples sintered at the highest temperature (SS-PE set) changes in the position of the XRD peaks can be assigned to variations in the lattice parameters of the ferrite phase due to the diffusion of the ions between both phases. In both sets of samples, an increase in the average grain size was registered for the lowest ferrite amount due to the possible iron diffusion in the piezoelectric phase. However, for the other compositions, this rise in the BNKT grain size was not increased. Although magnetization values of both sets of samples follow Vegard's Law, the magnetoelectric coefficient was strongly dependent on the selected synthesized method. SS-PE composites rendered similar magnetoelectric properties than composites, previously reported, made off both powders obtained by the solid-state reaction method, whereas for SG-PE composites the polarization of the samples was reduced due to the highest samples conductivity. In summary, to improve magnetoelectric properties, the possible diffusion or reaction between both phases must be avoided. Consequently, the selected synthesis method for the obtention of the piezoelectric or ferrite phases determines the final properties of the composites. In this case, when comparing with composites of the same composition but with both phases obtained from the solid-state reaction method, the use of ferrite powder obtained by Pechini's method was a valuable alternative to reduce the calcination and sintering temperatures, without reducing the magnetoelectric properties.

Author Contributions: Conceptualization, J.C. and M.C.; Methodology, J.C.; Validation, J.C.; Formal analysis, J.C. and M.C.; Investigation, J.C. and M.C; Writing—original draft, J.C. and M.C; Writing—review & editing, J.C., L.R. and M.C.; Visualization, J.C.; Supervision, L.R. and M.C.; Project administration, M.C.; Funding acquisition, M.C. All authors have read and agreed to the published version of the manuscript.

Funding: This research was funded by ANPCyT (grant number PICT 2014-1314), CONICET (grant number 112-201101-00432), and UNMdP (grant number 15/G527).

Institutional Review Board Statement: Not applicable.

Informed Consent Statement: Not applicable.

Data Availability Statement: Not applicable.

Conflicts of Interest: The authors declare no conflict of interest. The funders had no role in the design of the study; in the collection, analyses, or interpretation of data; in the writing of the manuscript; or in the decision to publish the results.

Appendix A

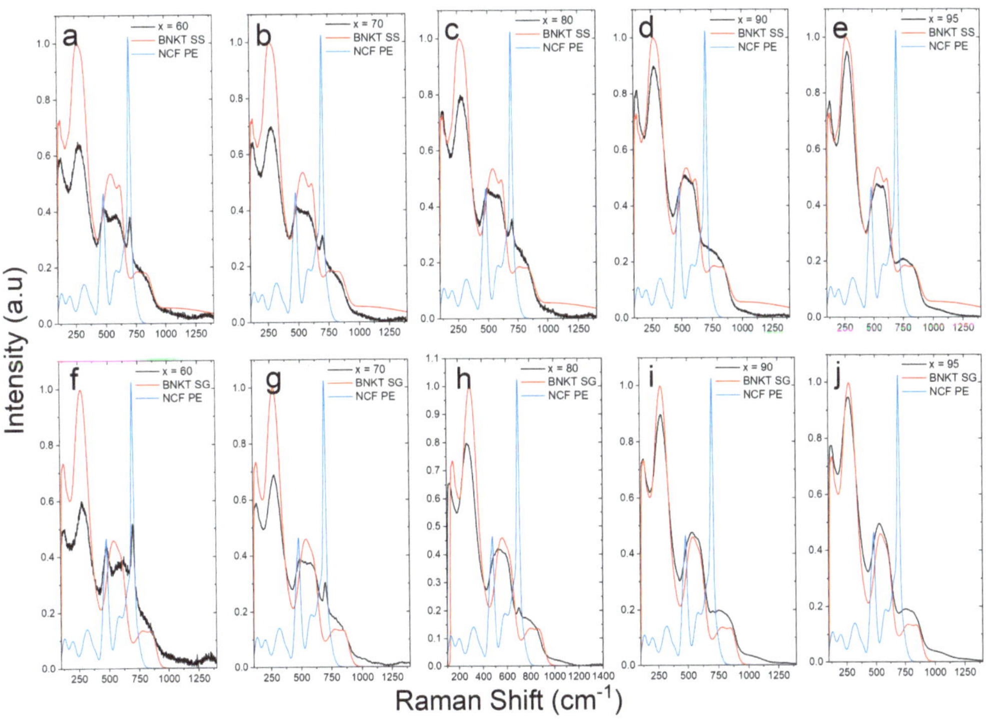

Figure A1. Experimental and calculated Raman spectra of xBNKT-(100-x)NCF (x = 100, 95, 90, 80, 70, 60, 0) SS-SS (**a–e**) and SG-PE (**f–j**) sintered samples.

References

1. Srinivas, A.; Krishnaiah, R.V.; Karthik, T.; Suresh, P.; Asthana, S.; Kamat, S.V. Observation of Direct and Indirect Magnetoelectricity in Lead Free Ferroelectric ($Na_{0.5}Bi_{0.5}TiO_3$)–Magnetostrictive ($CoFe_2O_4$) Particulate Composite. *Appl. Phys. Lett.* **2012**, *101*, 082902. [CrossRef]
2. Ramesh, R.; Spaldin, N.A. Multiferroics: Progress and Prospects in Thin Films. In *Nanoscience and Technology: A Collection of Reviews from Nature Journals*; World Scientific Publishing Co.: Singapore, 2009; pp. 20–28, ISBN 978-981-4287-00-5.
3. Oanh, L.T.M.; Do, D.B.; Van Minh, N. A Study of Structural Phase Transition, Vibration, Optical, and Magnetic Properties of Fe-Doped $PbTiO_3$ Nanostructured Powders. *Appl. Phys. A* **2016**, *122*, 660. [CrossRef]
4. Zhang, S.T.; Kounga, A.B.; Aulbach, E.; Granzow, T.; Jo, W.; Kleebe, H.J.; Rödel, J. Lead-Free Piezoceramics with Giant Strain in the System $Bi_{0.5}Na_{0.5}TiO_3$-$BaTiO_3$-$K_{0.5}Na_{0.5}NbO_3$. I. Structure and Room Temperature Properties. *J. Appl. Phys.* **2008**, *103*, 034107. [CrossRef]
5. Wang, B.Y.; Wang, H.T.; Singh, S.B.; Shao, Y.C.; Wang, Y.F.; Chuang, C.H.; Yeh, P.H.; Chiou, J.W.; Pao, C.W.; Tsai, H.M.; et al. Effect of Geometry on the Magnetic Properties of $CoFe_2O_4$–$PbTiO_3$ Multiferroic Composites. *RSC Adv.* **2013**, *3*, 7884. [CrossRef]
6. Ahn, C.W.; Kim, H.S.; Woo, W.S.; Won, S.S.; Seog, H.J.; Chae, S.A.; Park, B.C.; Jang, K.B.; Ok, Y.P.; Chong, H.H.; et al. Low-Temperature Sintering of $Bi_{0.5}(Na,K)_{0.5}TiO_3$ for Multilayer Ceramic Actuators. *J. Am. Ceram. Soc.* **2015**, *98*, 1877–1883. [CrossRef]
7. Wendari, T.P.; Arief, S.; Mufti, N.; Blake, G.R.; Baas, J.; Suendo, V.; Prasetyo, A.; Insani, A.; Zulhadjri, Z. Lead-Free Aurivillius Phase $Bi_2LaNb_{1.5}Mn_{0.5}O_9$: Structure, Ferroelectric, Magnetic, and Magnetodielectric Effects. *Inorg. Chem.* **2022**, *61*, 8644–8652. [CrossRef] [PubMed]
8. Li, C.; Huang, Y.H.; Hong, J.S.; Liu, X.Q.; Li, J.; Wu, Y.J. Magnetoelectric Effect in Sm-Substituted Tungsten Bronze Structure $Ba_4(SmLa_{1-x})_2Fe_2Nb_8O_{30}$ Ceramics. *J. Alloys Compd.* **2019**, *786*, 126–133. [CrossRef]
9. Saini, N.; Jindal, R.; Tripathi, A. Study of Lattice Dynamics of Bilayered Tetragonal Ruddlesden-Popper Compounds $(Ca,Sr)_3Mn_2O_7$. *Mater. Today Commun.* **2023**, *34*, 105190. [CrossRef]
10. Liu, M.; Zhang, Y.; Lin, L.-F.; Lin, L.; Yang, S.; Li, X.; Wang, Y.; Li, S.; Yan, Z.; Wang, X.; et al. Direct Observation of Ferroelectricity in $Ca_3Mn_2O_7$ and Its Prominent Light Absorption. *Appl. Phys. Lett.* **2018**, *113*, 022902. [CrossRef]
11. Baji, A.; Mai, Y.-W.; Yimnirun, R.; Unruan, S. Electrospun Barium Titanate/Cobalt Ferrite Composite Fibers with Improved Magnetoelectric Performance. *RSC Adv.* **2014**, *4*, 55217–55223. [CrossRef]
12. Yang, F.; Zhang, F.; Dong, C.; Liu, F.; Tang, M. Multiferroic Properties of $Bi_{3.15}Nd_{0.85}Ti_3O_{12}$–$NiFe_2O_4$–$Bi_{3.15}Nd_{0.85}Ti_3O_{12}$ Trilayer Composite Thin Films Prepared by a Sol–Gel Process. *J. Sol-Gel Sci. Technol.* **2014**, *73*, 469–475. [CrossRef]
13. Martin, L.W.; Crane, S.P.; Chu, Y.H.; Holcomb, M.B.; Gajek, M.; Huijben, M.; Yang, C.H.; Balke, N.; Ramesh, R. Multiferroics and Magnetoelectrics: Thin Films and Nanostructures. *J. Phys. Condens* **2008**, *20*, 434220. [CrossRef]
14. Chen, Y.; Fitchorov, T.; Geiler, A.L.; Gao, J.; Vittoria, C.; Harris, V.G. Dynamic Response of Converse Magnetoelectric Effect in a PMN-PT-Based Multiferroic Heterostructure. *Appl. Phys. A* **2010**, *100*, 1149–1155. [CrossRef]
15. Nan, C.W.; Bichurin, M.I.; Dong, S.; Viehland, D.; Srinivasan, G. Multiferroic Magnetoelectric Composites: Historical Perspective, Status, and Future Directions. *J. Appl. Phys.* **2008**, *103*, 031101. [CrossRef]
16. Etier, M.; Shvartsman, V.V.; Salamon, S.; Gao, Y.; Wende, H.; Lupascu, D.C. The Direct and the Converse Magnetoelectric Effect in Multiferroic Cobalt Ferrite-Barium Titanate Ceramic Composites. *J. Am. Ceram. Soc.* **2016**, *99*, 3623–3631. [CrossRef]
17. Priya, S.; Islam, R.; Dong, S.; Viehland, D. Recent Advancements in Magnetoelectric Particulate and Laminate Composites. *J. Electroceramics* **2007**, *19*, 149–166. [CrossRef]
18. Krishnaiah, R.V.; Srinivas, A.; Kamat, S.V.; Karthik, T.; Asthana, S. Effect of $CoFe_2O_4$ Mole Percentage on Multiferroic and Magnetoelectric Properties of $Na_{0.5}Bi_{0.5}TiO_3/CoFe_2O_4$ Particulate Composites. *Ceram. Int.* **2014**, *40*, 7799–7804. [CrossRef]
19. Chen, X.M.; Tang, Y.H.; Chen, I.-W.; Xu, Z.C.; Wu, S.Y. Dielectric and Magnetoelectric Characterization of $CoFe_2O_4/Sr_{0.5}Ba_{0.5}Nb_2O_6$ Composites. *J. Appl. Phys.* **2004**, *96*, 6520–6522. [CrossRef]
20. Röbisch, V.; Yarar, E.; Urs, N.O.; Teliban, I.; Knöchel, R.; McCord, J.; Quandt, E.; Meyners, D. Exchange Biased Magnetoelectric Composites for Magnetic Field Sensor Application by Frequency Conversion. *J. Appl. Phys.* **2015**, *117*, 17B513. [CrossRef]
21. Sreenivasulu, G.; Fetisov, L.Y.; Fetisov, Y.K.; Srinivasan, G. Piezoelectric Single Crystal Langatate and Ferromagnetic Composites: Studies on Low-Frequency and Resonance Magnetoelectric Effects. *Appl. Phys. Lett.* **2012**, *100*, 052901. [CrossRef]
22. Yang, H.; Zhang, J.; Lin, Y.; Wang, T. High Curie Temperature and Enhanced Magnetoelectric Properties of the Laminated $Li_{0.058}(Na_{0.535}K_{0.48})_{0.942}NbO_3/Co_{0.6}Zn_{0.4}Fe_{1.7}Mn_{0.3}O_4$ Composites. *Sci. Rep.* **2017**, *7*, 44855. [CrossRef]
23. Kumar, A.S.; Lekha, C.S.C.; Vivek, S.; Saravanan, V.; Nandakumar, K.; Nair, S.S. Multiferroic and magnetoelectric properties of $Ba_{0.85}Ca_{0.15}Zr_{0.1}Ti_{0.9}O_3$–$CoFe_2O_4$ core–shell nanocomposite. *J. Magn. Magn. Mater.* **2016**, *418*, 294–299. [CrossRef]
24. Islam, R.A.; Priya, S. Magnetoelectric Properties of the Lead-Free Cofired $BaTiO_3$-$(Ni_{0.8}Zn_{0.2})Fe_2O_4$ Bilayer Composite. *Appl. Phys. Lett.* **2006**, *89*, 152911. [CrossRef]
25. Mane, S.M.; Pawar, S.A.; Patil, D.S.; Kulkarni, S.B.; Tayade, N.T.; Shin, J.C. Magnetoelectric, Magnetodielectric Effect and Dielectric, Magnetic Properties of Microwave-Sintered Lead-Free $x(Co_{0.9}Ni_{0.1}Fe_2O_4)$-$(1-x)[0.5(Ba_{0.7}Ca_{0.3}TiO_3)$-$0.5(BaZr_{0.2}Ti_{0.8}O_3)]$ Particulate Multiferroic Composite. *Ceram. Int.* **2020**, *46*, 3311–3323. [CrossRef]
26. Prado-Espinosa, A.; Camargo, J.; Ramajo, L.; Castro, M. Improvement on Dielectric and Microstructural Properties of Lead Free $Bi_{0.5}Na_{0.5}TiO_3$ Ceramics through Processing Conditions. *J. Mater. Sci. Mater. Electron.* **2017**, *28*, 16836–16841. [CrossRef]
27. Takenaka, T.; Maruyama, K.; Sakata, K. $(Bi_{1/2}Na_{1/2})TiO_3$-$BaTiO_3$ System for Lead-Free Piezoelectric Ceramics. *Jpn. J. Appl. Phys.* **1991**, *30*, 2236–2239. [CrossRef]

28. Sasaki, A.; Chiba, T.; Mamiya, Y.; Otsuki, E. Dielectric and Piezoelectric Properties of $(Bi_{0.5}Na_{0.5})TiO_3$–$(Bi_{0.5}K_{0.5})TiO_3$ Systems. *Jpn. J. Appl. Phys.* **1999**, *38*, 5564–5567. [CrossRef]
29. Camargo, J.; Ramajo, L.; Rubio-Marcos, F.; Castro, M. Ferroelectric Properties of $Bi_{0.5}(Na_{0.8}K_{0.2})_{0.5}TiO_3$ Ceramics. *AMR* **2014**, *975*, 3–8. [CrossRef]
30. Moosavi, A.; Bahrevar, M.A.; Aghaei, A.R.; Ramos, P.; Algueró, M.; Amorín, H. High-Field Electromechanical Response of $Bi_{0.5}Na_{0.5}TiO_3$-$Bi_{0.5}K_{0.5}TiO_3$ across Its Morphotropic Phase Boundary. *J. Phys. D* **2014**, *47*, 055304. [CrossRef]
31. Camargo, J.; Ramajo, L.; Rubio-Marcos, F.; Castro, M. Estudio de Las Condiciones de Procesamiento de $Bi_{0.5}(Na_{0.8}K_{0.2})_{0.5}TiO_3$. *Bol. Soc. Esp. Ceram. Vidr.* **2014**, *53*, 27–31. [CrossRef]
32. Gao, X.; Liu, L.; Birajdar, B.; Ziese, M.; Lee, W.; Alexe, M.; Hesse, D. High-Density Periodically Ordered Magnetic Cobalt Ferrite Nanodot Arrays by Template-Assisted Pulsed Laser Deposition. *Adv. Funct. Mater.* **2009**, *19*, 3450–3455. [CrossRef]
33. Ahalya, K.; Suriyanarayanan, N.; Ranjithkumar, V. Effect of Cobalt Substitution on Structural and Magnetic Properties and Chromium Adsorption of Manganese Ferrite Nano Particles. *J. Magn. Magn. Mater.* **2014**, *372*, 208–213. [CrossRef]
34. Bensebaa, F.; Zavaliche, F.; L'Ecuyer, P.; Cochrane, R.W.; Veres, T. Microwave Synthesis and Characterization of Co-Ferrite Nanoparticles. *J. Colloid Interface Sci.* **2004**, *277*, 104–110. [CrossRef]
35. Mohamed, R.M.; Rashad, M.M.; Haraz, F.A.; Sigmund, W. Structure and Magnetic Properties of Nanocrystalline Cobalt Ferrite Powders Synthesized Using Organic Acid Precursor Method. *J. Magn. Magn. Mater.* **2010**, *322*, 2058–2064. [CrossRef]
36. Park, M.B.; Hwang, S.J.; Cho, N.H. Effect of the Grain Size and Chemical Features on the Phase Transition and Physical Characteristics of Nano-Grained $BaTiO_3$ Ceramics. *Mater. Sci. Eng. B* **2003**, *99*, 155–158. [CrossRef]
37. Pazik, R.; Hreniak, D.; Strek, W.; Kessler, V.G.; Seisenbaeva, G.A. Photoluminescence Investigations of Eu^{3+} Doped $BaTiO_3$ Nanopowders Fabricated Using Heterometallic Tetranuclear Alkoxide Complexes. *J. Alloys Compd.* **2008**, *451*, 557–562. [CrossRef]
38. Li, H.; Feng, C.; Yao, W. Some Effects of Different Additives on Dielectric and Piezoelectric Properties of $(Bi_{1/2}Na_{1/2})TiO_3$–$BaTiO_3$ Morphotropic-Phase-Boundary Composition. *Mater. Lett.* **2004**, *58*, 1194–1198. [CrossRef]
39. Su, B.; Holmes, J.E.; Cheng, B.L.; Button, T.W. Processing Effects on the Microstructure and Dielectric Properties of Barium Strontium Titanate (BST) Ceramics. *J. Electroceramics* **2002**, *9*, 111–116. [CrossRef]
40. Zhang, H.; Jiang, S. Effect of Repeated Composite Sol Infiltrations on the Dielectric and Piezoelectric Properties of a $Bi_{0.5}(Na_{0.82}K_{0.18})_{0.5}TiO_3$ Lead Free Thick Film. *J. Eur. Ceram. Soc.* **2009**, *29*, 717–723. [CrossRef]
41. Fang, T.-T.; Lin, H.-B.; Hwang, J.-B. Thermal Analysis of Precursors of Barium Titanate Prepared by Coprecipitation. *J. Am. Ceram. Soc.* **1990**, *73*, 3363–3367. [CrossRef]
42. Jones, G.O.; Thomas, P.A. Investigation of the Structure and Phase Transitions in the Novel A-Site Substituted Distorted Perovskite Compound $Na_{0.5}Bi_{0.5}TiO_3$. *Acta Cryst. B* **2002**, *58*, 168–178. [CrossRef] [PubMed]
43. van Quyet, N.; Huu Bac, L.; Odkhuu, D.; Duc Dung, D. Effect of Li_2CO_3 Addition on the Structural, Optical, Ferroelectric, and Electric-Field-Induced Strain of Lead-Free BNKT-Based Ceramics. *J. Phys. Chem. Solids* **2015**, *85*, 148–154. [CrossRef]
44. Camargo, J.E.; Parra, R.; Ramajo, L.A.; Castro, M.S. Synthesis and Characterization of $Bi_{0.5}(Na_{0.8}K_{0.2})_{0.5}TiO_3$-Based Ceramics Obtained through the Sol-Gel Method. *Ferroelectrics* **2019**, *545*, 62–69. [CrossRef]
45. Camargo, J.; Prado Espinosa, A.; Ramajo, L.; Castro, M. Influence of the Sintering Process on Ferroelectric Properties of $Bi_{0.5}(Na_{0.8}K_{0.2})_{0.5}TiO_3$ Lead-Free Piezoelectric Ceramics. *J. Mater. Sci. Mater. Electron.* **2018**, *29*, 5427–5432. [CrossRef]
46. Prasertpalichat, S.; Khengkhatkan, S.; Siritanon, T.; Jutimoosik, J.; Kidkhunthod, P.; Bongkarn, T.; Patterson, E.A. Comparison of Structural, Ferroelectric, and Piezoelectric Properties between A-Site and B-Site Acceptor Doped $0.93Bi_{0.5}Na_{0.5}TiO_3$-$0.07BaTiO_3$ Lead-Free Piezoceramics. *J. Eur. Ceram. Soc.* **2021**, *41*, 4116–4128. [CrossRef]
47. Kushvaha, D.K.; Rout, S.K.; Tiwari, B. Structural, Piezoelectric and Highdensity Energy Storage Properties of Lead-Free BNKT-BCZT Solid Solution. *J. Alloys Compd.* **2019**, *782*, 270–276. [CrossRef]
48. Nandan, B.; Bhatnagar, M.C.; Kashyap, S.C. Cation Distribution in Nanocrystalline Cobalt Substituted Nickel Ferrites: X-ray Diffraction and Raman Spectroscopic Investigations. *J. Phys. Chem. Solids* **2019**, *129*, 298–306. [CrossRef]
49. Mane, S.M.; Nimbalkar, A.R.; Kim, H.; Kulkarni, S.B.; Tayade, N.T.; Thombare, J.V.; Dhasade, S.S.; Shin, J.C. Magnetoelectric and Magnetodielectric Coupling in Partially Ni-Doped $CoFe_2O_4$ and $0.15(Ba_{0.7}Ca_{0.3}TiO_3)$–$0.85(BaZr_{0.2}Ti_{0.8}O_3)$ Composites Prepared via Clean Microwave Sintering. *J. Alloys Compd.* **2020**, *849*, 156599. [CrossRef]
50. Ramana, E.V.; Figueiras, F.; Graça, M.P.F.; Valente, M.A. Observation of Magnetoelectric Coupling and Local Piezoresponse in Modified $(Na_{0.5}Bi_{0.5})TiO_3$–$BaTiO_3$–$CoFe_2O_4$ Lead-Free Composites. *Dalton Trans.* **2014**, *43*, 9934–9943. [CrossRef]
51. Camargo, J.; Prado Espinosa, A.; Zabotto, F.; Ramajo, L.; Castro, M. Magnetoelectric Interactions in Bismuth Sodium-Potassium Titanate-Nickel Cobalt Ferrite Lead-Free Composite Ceramics. *J. Alloys Compd.* **2020**, *826*, 154129. [CrossRef]
52. Manotham, S.; Butnoi, P.; Jaita, P.; Tunkasiri, T. Structure and Electrical Properties of BNKT-Based Ceramics. *Solid State Phenom.* **2018**, *283*, 147–153. [CrossRef]
53. Hussain, A.; Ahn, C.W.; Lee, J.S.; Ullah, A.; Kim, I.W. Large Electric-Field-Induced Strain in Zr-Modified Lead-Free $Bi_{0.5}(Na_{0.78}K_{0.22})_{0.5}TiO_3$ Piezoelectric Ceramics. *Sens. Actuator A Phys.* **2010**, *158*, 84–89. [CrossRef]
54. Kiran; Thakur, K.; Kumar, S.; Kumar, S.; Thakur, N. Investigation of Structural, Dielectric, and Magnetoelectric Properties of $K_{0.5}Na_{0.5}NbO_3$–$MnFe_2O_4$ Lead Free Composite System. *J. Alloys Compd.* **2021**, *857*, 158251. [CrossRef]
55. Goel, R.; Syal, R.; Sharma, G.; Singh, R.K.; Dhiman, S.; Singh, A.K.; Kumar, S. Magnetoelectric Coupling Susceptibility in Novel Lead-Free 0–3 Type Multiferroic Particulate Composites of $(1\text{-}x)Na_{0.5}Bi_{0.5}TiO_3$ -$(x)CoCr_{0.4}Fe_{1.6}O_4$. *Mater. Chem. Phys.* **2022**, *282*, 126004. [CrossRef]

56. Pandey, A.H.; Varade, P.; Miriyala, K.; Sowmya, N.S.; Kumar, A.; Arockiarajan, A.; Kulkarni, A.R.; Venkataramani, N. Magnetoelectric Response in Lead-Free $Na_{0.4}K_{0.1}Bi_{0.5}TiO_3/NiFe_2O_4$ Laminated Composites. *Mater. Today Commun.* **2021**, *26*, 101898. [CrossRef]
57. Tang, Z.; Zhang, Z.; Chen, J.; Zhao, S. Magnetoelectric Effect of Lead-Free Perovskite $BiFeO_3/Bi_{0.5}(Na_{0.85}K_{0.15})_{0.5}TiO_3$ Composite Films. *J. Alloys Compd.* **2017**, *696*, 1–8. [CrossRef]

materials MDPI

Article

Nonlinear Optical Properties in an Epitaxial $YbFe_2O_4$ Film Probed by Second Harmonic and Terahertz Generation

Hongwu Yu [1], Yoichi Okimoto [1,*], Atsuya Morita [1], Shuhei Shimanuki [1], Kou Takubo [1], Tadahiko Ishikawa [1], Shin-ya Koshihara [1], Ryusei Minakami [2], Hirotake Itoh [2], Shinichiro Iwai [2], Naoshi Ikeda [3], Takumi Sakagami [4], Mayu Nozaki [4] and Tatsuo Fujii [4]

1 Department of Chemistry, Tokyo Institute of Technology, 2-12-1, Meguro, Tokyo 152-8551, Japan
2 Department of Physics, Tohoku University, Sendai 980-8578, Japan
3 Department of Physics, Okayama University, 3-1-1, Tsushimanaka, Okayama 700-8530, Japan
4 Department of Applied Chemistry, Okayama University, 3-1-1, Tsushimanaka, Okayama 700-8530, Japan
* Correspondence: okimoto.y.aa@m.titech.ac.jp

Abstract: An epitaxial film of $YbFe_2O_4$, a candidate for oxide electronic ferroelectrics, was fabricated on yttrium-stabilized zirconia (YSZ) substrate by magnetron sputtering technique. For the film, second harmonic generation (SHG), and a terahertz radiation signal were observed at room temperature, confirming a polar structure of the film. The azimuth angle dependence of SHG shows four leaves-like profiles and is almost identical to that in a bulk single crystal. Based on tensor analyses of the SHG profiles, we could reveal the polarization structure and the relationship between the film structure of $YbFe_2O_4$ and the crystal axes of the YSZ substrate. The observed terahertz pulse showed anisotropic polarization dependence consistent with the SHG measurement, and the intensity of the emitted terahertz pulse reached about 9.2% of that emitted from ZnTe, a typical nonlinear crystal, implying that $YbFe_2O_4$ can be applied as a terahertz wave generator in which the direction of the electric field can be easily switched.

Keywords: ferroelectrics; nonlinear optics; second harmonic generation; strongly correlated systems; thin film

Citation: Yu, H.; Okimoto, Y.; Morita, A.; Shimanuki, S.; Takubo, K.; Ishikawa, T.; Koshihara, S.-y.; Minakami, R.; Itoh, H.; Iwai, S.; et al. Nonlinear Optical Properties in an Epitaxial $YbFe_2O_4$ Film Probed by Second Harmonic and Terahertz Generation. *Materials* **2023**, *16*, 1989. https://doi.org/10.3390/ma16051989

Academic Editor: Dana Popescu

Received: 8 February 2023
Revised: 23 February 2023
Accepted: 24 February 2023
Published: 28 February 2023

1. Introduction

Ferroelectrics are a class of materials in which crystals have a spontaneous polarization that an external electric field can control. With this characteristic feature, ferroelectrics have been applied in various industries, such as information storage devices, capacitors, actuators, and nonlinear crystals. Particularly for memory devices, it is desirable to achieve ferroelectrics that operate at ultra-high speeds and conserve energy. For that purpose, several studies on ultrafast control in organic and inorganic ferroelectrics have been conducted [1–3]. Also, attention has been paid to developing novel ferroelectrics that operate at higher speeds and lower energies than conventional ferroelectrics.

To achieve these goals, the concept of "electronic ferroelectrics" has been proposed experimentally [4–6] and theoretically [7,8] since the beginning of the 21st century. The characteristics of electronic ferroelectrics are that the order of ions with different valances originates polarization and the direction of polarization can be reversed by electron transfer between those ions. While in conventional ferroelectrics, spontaneous displacement of the constituent atoms in real space from their equilibrium points is important to reverse polarization. Therefore, electronic ferroelectrics are expected to have a significantly faster response time of polarization reversal and a smaller coercive electric field than conventional ferroelectrics. Research on developing electronic ferroelectrics is attracting attention in materials science and ferroelectric applications owing to their unique properties such as low coercive field, excellent durability, and ultrafast response.

While looking for such electronic ferroelectrics, Ikeda et al. proposed an iron oxide complex, RFe_2O_4 (R = rare earth ions), as a novel electronic ferroelectric material [4]. Figure 1a shows the crystal structure of $YbFe_2O_4$ without oxygen ions reported in a previous study [9]. This crystal is composed of an alternate stacking of the R-layer and W-layer along the c_m direction, as depicted in the schematic image. The R-layer comprises Yb^{3+} and oxygen ions, and the W-layer comprises two layers composed of Fe (Fe^{2+} and Fe^{3+}) and oxygen ions stacked along the c_m-axis direction [4,9]. Fe^{2+}, Fe^{3+}, and oxygen ions form a triangular lattice in the W-layer, as shown in Figure 1b, and this crystal has inversion symmetry ($R\bar{3}m$) above 500 K [10]. Furthermore, Fe^{2+} and Fe^{3+} in the W-layer show charge ordering with a three-fold periodic superstructure below 500 K [4]. Under these circumstances, the total amount of Fe^{2+} and Fe^{3+} is not equal in the upper and lower layers of the W-layer; In the upper layer, the number of Fe^{3+} is twice as many as that of Fe^{2+} and vice versa in the lower layer. This indicates the total charge of the upper layer is greater than that of the lower layer. Such charge disproportionation within the W-layer results in electric polarization in the direction of the yellow arrow in Figure 1a. Thus, polarization by electronic ordering occurs in this system at room temperature.

Although there have been several opinions regarding the presence or absence of polar structure in RFe_2O_4 [11,12], recent measurements of second harmonic generation (SHG) in chemically equivalent R = Yb crystals have revealed the presence of electrical polarization [9]. In addition, analysis of the azimuthal dependence of SHG has shown that $YbFe_2O_4$ crystals are polarized and belong to the monoclinic C_m point group. In this symmetry, the electric polarization exists in the $a_m c_m$ plane, as shown in Figure 1a, supporting that the electric polarization forms owing to electron disproportionation in the upper and lower sheets of the W-layers, as proposed by Ikeda et al. [4].

To apply electronic ferroelectricity observed in a bulk $YbFe_2O_4$ to future ferroelectric devices, thin film crystals have been fabricated [13–17]. Several studies have succeeded in the epitaxial growth of $YbFe_2O_4$ thin films on various substrates, including yttrium-stabilized zirconia (YSZ) [13–15], MgO [16], and α-Al_2O_3 [17,18], whose structures are compatible with the hexagonal lattice of RFe_2O_4. Furthermore, based on transmission electron microscopy (TEM) observations, Fujii et al. [14,17] first discovered that the resulting $YbFe_2O_4$ films have a superlattice structure owing to the order of Fe ions having the same three-fold periodicity as single crystals [4,19]. Thus, thin film crystals are expected to exhibit a polar structure like a single crystal. However, the X-ray diffraction analyses of RFe_2O_4 have not determined the crystal symmetry consistent with the TEM measurement. Hence, it is yet to be determined whether these thin film crystals are polar or not.

In this report, we investigated two types of second-order nonlinear effects (SHG and terahertz generation) on a film of $YbFe_2O_4$ with a three-fold Fe ordering. The former experiment revealed the symmetry of the polarized film structure and the relationship between the polar direction and the crystal axes of the substrate. The latter demonstrated that $YbFe_2O_4$ film radiates a strong terahertz pulse owing to the polar structure, proposing a possible application of electronic ferroelectric oxide for nonlinear optical devices.

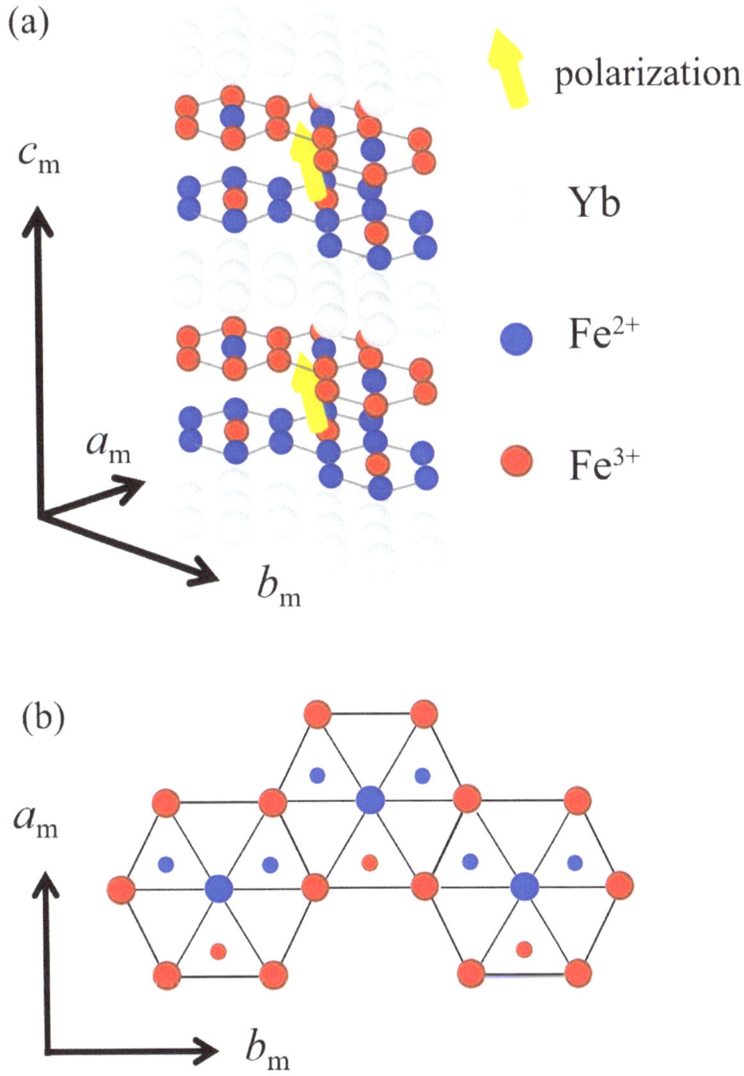

Figure 1. (**a**) Crystal structure of $YbFe_2O_4$ after Ref. [9]. Oxygens are not depicted for clarity. The yellow arrows denote the polarization direction. (**b**) Schematic view of the *W*-layer in a plane perpendicular to the c_m axis. Red and blue circles denote Fe^{3+} and Fe^{2+}, respectively and small ones Fe^{3+} and Fe^{2+} in the lower layer.

2. Experiment

We synthesized $YbFe_2O_4$ thin film crystal on YSZ (111) single crystal substrate. The films were deposited using the RF magnetron sputtering; the details are described elsewhere [14]. Table 1 shows the O_2 partial pressure during sputtering and the retention time before deposition in which the substrate temperature is kept at ~1400 K. Conventional X-ray diffraction (XRD) and TEM measurements were performed to determine the crystallinity of the film. The c_m-axis lattice constant obtained from XRD and the film thickness are also shown in Table 1.

Table 1. Experimental conditions for sputtering and the structural parameters of the fabricated $YbFe_2O_4$ thin film.

O_2 Partial Pressure (10^{-6} Pa)	Retention Time (min)	*c*-axis Lattice Constant (Å)	Film Thickness (nm)
1.36	8	25.125(2)	~50

Figure 2b shows the azimuth angle dependence of the $(10\bar{1}4)_h$ Bragg reflection for the film, which shows the in-plane crystal orientation. For comparison, Figure 2a shows the azimuth angle dependence of the Bragg peak for the YSZ crystal, the substrate used for the film deposition. The appearance of the peaks at every 120° rotation of the crystal shows that the crystal has a three-fold symmetric structure. Like the YSZ substrate, the film shows a three-fold symmetry, although residues of a six-fold peak are observed owing to the crystalline domains (60° domains). The ratio of the six-fold to the three-fold component is about 8%, which is smaller than that observed in other RFe_2O_4 films reported so far [14,17], indicating a good crystallization of $YbFe_2O_4$ film.

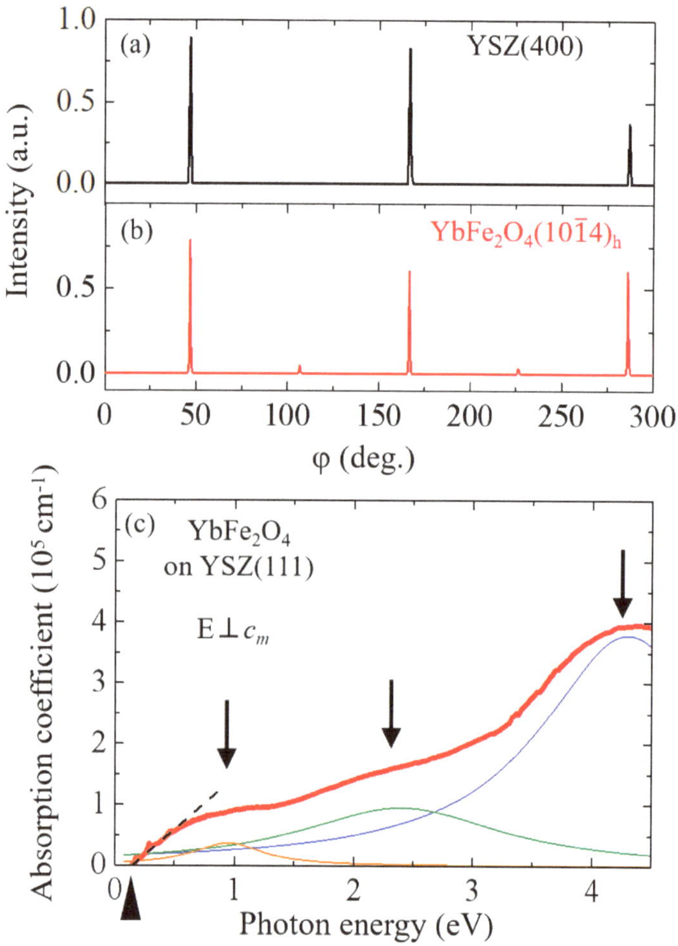

Figure 2. (**a**) Azimuth angle dependence of the (400) Bragg reflection for YSZ substrate. (**b**) Azimuth angle dependence of the $(10\bar{1}4)_h$ Bragg reflection for $YbFe_2O_4$ thin film. (**c**) In-plane absorption coefficient spectrum of $YbFe_2O_4$ thin film.

The normal incident transmission spectrum ($T(\omega)$) was measured using a Fourier transform interferometer in the mid- and near-infrared region (0.01–1.0 eV). Above the near-infrared energy region (0.9–3.3 eV), we used grating-type monochromators and detected the transmitted monochromatic light with InGaAs (0.7–1.4 eV) and Si photodiodes (1.2–3.1 eV). The absorption coefficient spectrum ($\alpha(\omega)$) was calculated by the relation that $\alpha(\omega) = -\ln [T(\omega)]/l$, l being the thickness of the film.

SHG and terahertz radiation were measured using a regenerate amplified mode-locked Ti: a sapphire laser source (pulse width: 35 fs; repetition rate: 1 kHz; center wavelength: 800 nm). Both measurement systems are depicted in Figure 3a,b. The thin film was irradiated with fundamental 800 nm pulses in normal incidence to avoid SHG or terahertz waves generated by the surface. The polarization of the 800 nm pulses to generate SHG and terahertz pulse is controlled using a half-wavelength plate. In the SHG measurement, an SHG pulse of approximately 400 nm was detected with a photomultiplier tube after removing the strong fundamental light using a high-pass filter and a grating-type monochromator. The terahertz pulse radiating in the direction of the a_m axis of the film was detected with a conventional electro-optical sampling method. We carefully confirmed that neither SHG nor terahertz signal was observed by irradiating the YSZ substrate with the amplified fundamental pulses.

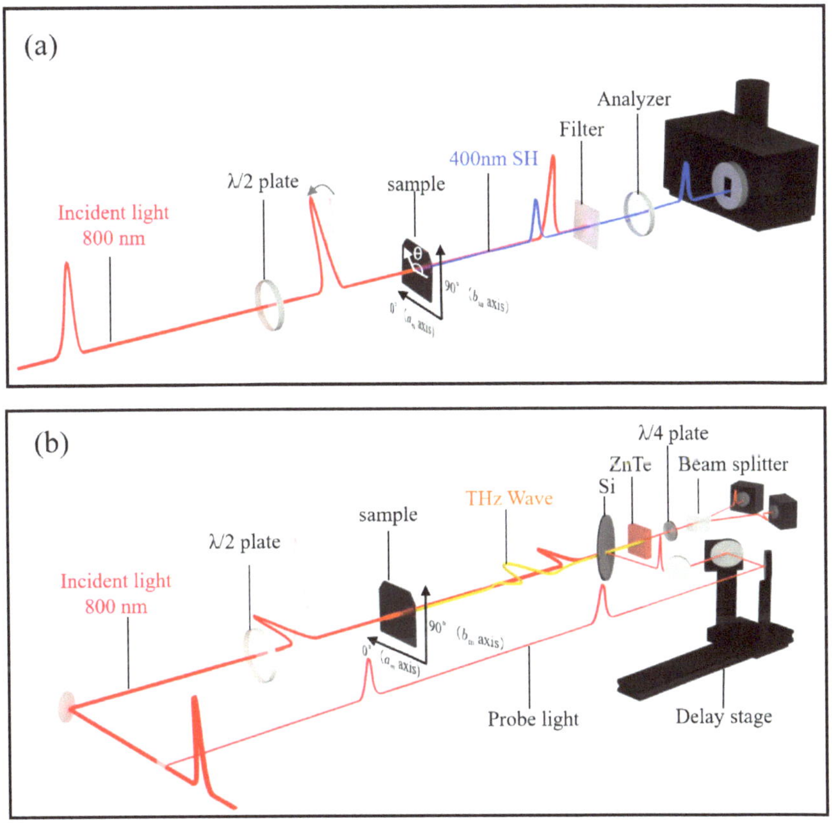

Figure 3. Schematic figures of the measurement systems of SHG (**a**) and terahertz radiation (**b**).

3. Results and Discussion

The absorption spectrum was measured to observe the electronic structure of the fabricated thin film. In Figure 2c, we show $\alpha(\omega)$ of the film by a black line. As shown by the arrows in the figure, $\alpha(\omega)$ contains three broad absorption bands at approximately 4.0 eV, 2.3 eV and 1.0 eV, which are assigned to the O 2$p \rightarrow$ Fe 3d charge-transfer transition, the Fe^{2+} on site transition, and the $Fe^{2+} \rightarrow Fe^{3+}$ d–d transition, respectively [15,18,20–22]. (In the figure, those absorption components extracted by Lorentz fitting are displayed by thin lines.)

An important feature in the electronic structure of the $YbFe_2O_4$ film has a larger spectral weight in the mid-infrared region than that of the single crystal of $LuFe_2O_4$ [22] or $YbFe_2O_4$ [23]. This discrepancy may be caused by the lattice mismatch between the film and the YSZ substrate [14]. This makes the electronic structure different from that in a single crystal of $YbFe_2O_4$, resulting in a reduction in the d–d transition energy and an increase in the spectral weight in the mid-infrared region. We extrapolated the rising part of the $\alpha(\omega)$ of the thin film crystal as a dashed line and estimated the optical gap to be about 0.1 eV from its intersection with the abscissa (see the closed triangle), which is smaller than that of $LuFe_2O_4$ single crystal [22].

Then, the nonlinear optical response of the thin film was investigated. By irradiating the $YbFe_2O_4$ film with an 800 nm pulse, a finite signal of 400 nm pulse was observed, as shown by red circles in Figure 4a. (Both the polarizations of the incident 800 nm pulse and the detected 400 nm pulse are parallel to YSZ [11$\bar{2}$] axis.) A fitting analysis of the obtained data, assuming that the signal is proportional to the square of the incident intensity, is shown by the solid black line in Figure 4a. This result indicates that the film exhibits an SHG signal and has a polar structure with broken inversion symmetry.

To further investigate SHG in $YbFe_2O_4$ film, the azimuth angle dependence of SHG was investigated. Red and blue circles in Figure 4b,c show the angular dependence of the SHG intensity polarized along the YSZ [11$\bar{2}$] axis and the [1$\bar{1}$0] axis, respectively. (See also Figure 3a.) The numbers on the circumference denote the angle between the polarization of the incident laser light (800 nm) and the [11$\bar{2}$] direction of the substrate. There is a significant SHG in both polarizations, and the angular profiles of four clover-shaped leaves are observed, almost identical to that of a single bulk crystal [9].

In general, the SHG intensity (I_{SH}) is related to the incident optical electric field (E) using the equation, $I_{SH} \propto \left|P^{(2)}\right|^2 \propto \left|\varepsilon_0 \chi^{(2)} E E\right|^2$, where $\chi^{(2)}$ is the second-order nonlinear susceptibility and ε_0 is the dielectric constant in a vacuum. We assumed that the thin film crystal has the monoclinic C_m symmetry (Exactly speaking, we can consider three different point groups to reproduce the observed SHG profile in Figure 4b,c; i.e., 1 in triclinic, C_m in monoclinic, and mm2 in orthorhombic. However, there is no reason of triclinic structure from XRD in the films investigated. In addition, the crystal analysis based on neutron diffraction measurement for a single crystal of $YbFe_2O_4$ [9], indicates the symmetry should be monoclinic as long as the charge ordering of iron ions are involved. Thus, we chose C_m among the three) and defined the substrate's [11$\bar{2}$] (0°) and [1$\bar{1}$0] (90°) directions as a_m and b_m in the assumed monoclinic lattice, respectively. Then, the relationship between SHG intensity and θ, an angle between the polarization direction of the incident pulse and YSZ [11$\bar{2}$], is described as follows [24]:

$$I_{SH}^{(a)} \propto \varepsilon_0^2 E^4 \left(d_{11} \cos^2\theta + d_{12} \sin^2\theta \right)^2 \quad (1)$$

$$I_{SH}^{(b)} \propto \varepsilon_0^2 E^4 (d_{26} \sin 2\theta)^2 \quad (2)$$

where $I_{SH}^{(a)}$ and $I_{SH}^{(b)}$ denotes the SHG component observed along a_m and b_m axis, respectively, assuming the monoclinic C_m symmetry. The black lines in Figure 4b,c are the results of fitting the SHG angle profiles based on Equations (1) and (2) and agree with the experimental results. These fitting results show that the $YbFe_2O_4$ thin film deposited on YSZ substrates is monoclinic C_m, whose b_m and a_m axes correspond to the substrate's [1$\bar{1}$0] and [11$\bar{2}$] directions, respectively. The ratio of the obtained contracted tensor components is d_{11}: d_{12}: d_{26} = 1: –1.1:1.5, which almost corresponds to that in the single crystal [9].

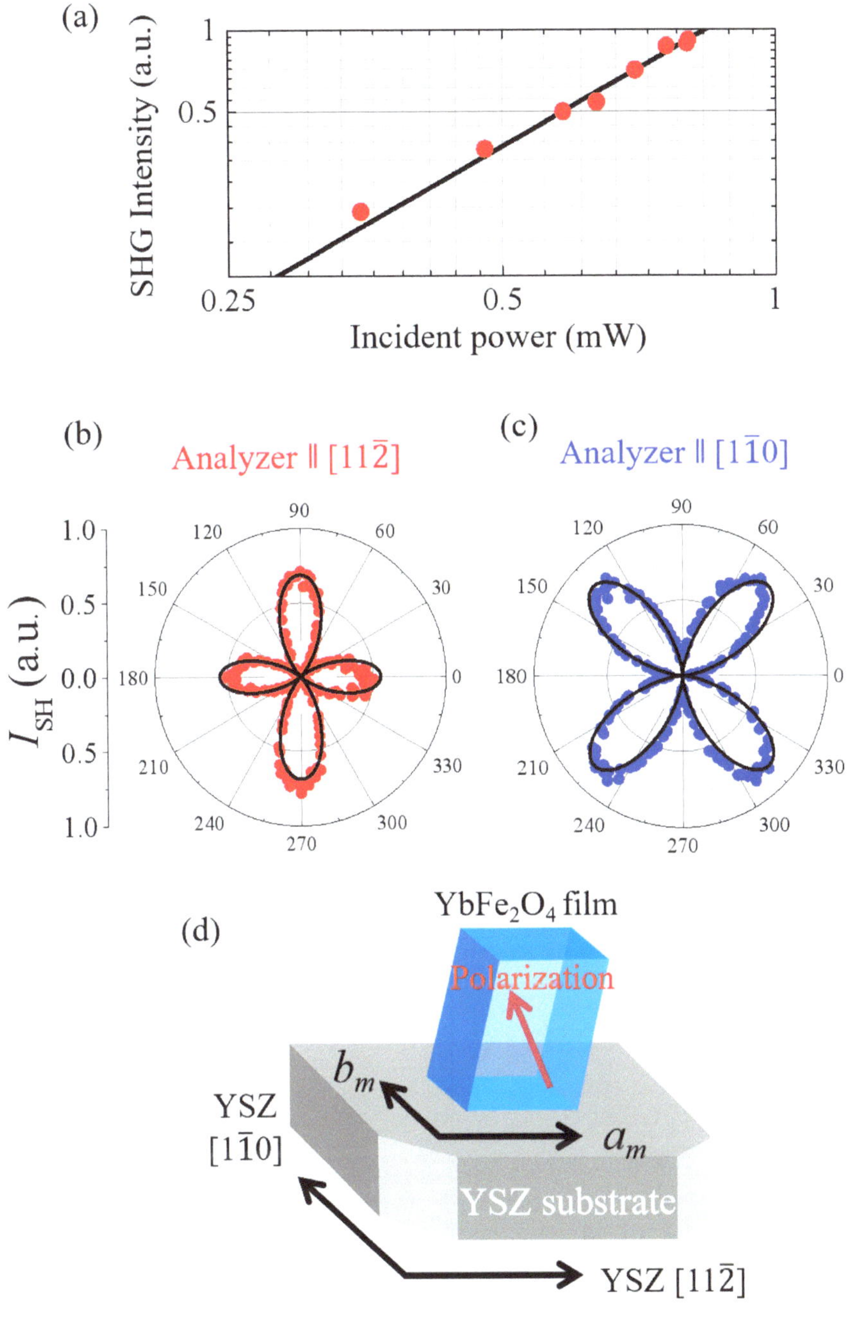

Figure 4. (**a**) Incident light intensity dependence of SHG emitted from $YbFe_2O_4$ thin film on the logarithmic scale. The solid lines are the fitting results (see text). (**b**,**c**) Azimuth angle dependence of SHG parallel (red circles) and perpendicular (blue circles) to the [11$\bar{2}$] direction of YSZ substrate. The solid lines are the fitted results based on the C_m symmetry (see text). (**d**): A schematic of monoclinic unit cell of $YbFe_2O_4$ thin film fabricated on the YSZ substrate determined by SHG polarimetry.

Figure 4d summarizes the relationship between the crystal structure of $YbFe_2O_4$ thin film and the axes of the YSZ substrate. The polarization structure is described based on the C_m point group, which is identical to that in a single crystal. What should be noticed is that the direction of the crystal axes is uniquely determined by that of the YSZ substrate, indicating that the W-layer of the $YbFe_2O_4$ film grows in a hexagonal lattice, reflecting the atomic arrangement of the YSZ substrate.

Finally, terahertz radiation from the film of $YbFe_2O_4$ was investigated. In a crystal with broken inversion symmetry, difference frequency generation (DFG), one of the second-order nonlinear optical effects, is observed in addition to SHG. Since the femtosecond pulses used in the experiment have a finite bandwidth (Full width at half maximum $\approx$ 50 nm), DFG within the incident pulse is expected to occur in a crystal without inversion symmetry, generating terahertz waves [25]. (See Figure 3b for the experimental setup).

In Figure 5c, we plotted waveforms of terahertz pulse polarized along the a_{m} axis, which are generated when the polarization of the fundamental pulse is along the a_{m} axis (${E_{\mathrm{THz}}}^{(a)}$, red circles) and b_{m} axis (${E_{\mathrm{THz}}}^{(b)}$, blue circles). The relationships between the polarization of the incident 800 nm pulse and the generated terahertz pulse in the measurement are schematically depicted in Figure 5a,b. Despite the small thickness of approximately 50 nm, we could observe clear terahertz radiation signals in both polarizations, supporting the polar structure of the $YbFe_2O_4$ film confirmed by the SHG polarimetry. It is important to notice that the shapes of the two profiles of ${E_{\mathrm{THz}}}^{(a)}$ and ${E_{\mathrm{THz}}}^{(b)}$ are symmetrical about the horizontal axis and their signs are opposite to each other at all the delay times. In the C_m symmetry, ${E_{\mathrm{THz}}}^{(a)} \propto d_{11}^{\mathrm{THz}}\varepsilon_0 E^2$ and ${E_{\mathrm{THz}}}^{(b)} \propto d_{12}^{\mathrm{THz}}\varepsilon_0 E^2$ [26] as described in Equations (1) and (2), indicating that the absolute value of d_{11}^{THz} is quite similar to that of d_{12}^{THz} but their signs are opposite. This result is consistent with the conclusion of SHG polarimetry that $\frac{d_{12}}{d_{11}} \approx -1.1$ at the wavelength of 800 nm.

Figure 5d shows the Fourier amplitude spectra calculated from the terahertz waveforms obtained by irradiating the fundamental pulse polarized along the a_m axis (red circles) and the b_m axis (blue circles). The calculated amplitude spectra are almost identical, reflecting the symmetric time profiles in both polarizations. The emitted spectra formed a broad peak at approximately 1 THz and covered the photon energy region up to approximately 2 THz.

To estimate the magnitude of the terahertz electric field emitted from the $YbFe_2O_4$ film, we also measured the terahertz pulse from a ZnTe, a typical nonlinear crystal usually used for terahertz wave generation, using the same experimental configuration for the $YbFe_2O_4$ film, as depicted in Figure 3b. (The thickness of the ZnTe crystal investigated is 1 mm.) The maximum value of the electric field radiated from the bulk ZnTe, whose waveform is shown in Supplementary Figure S1, is approximately 625 times as large as that observed in the $YbFe_2O_4$ film [27]. If we assume that the electric field of the generated pulse is proportional to the thickness of the film and the intensity of the irradiated 800 nm pulse, we can estimate that the magnitude of the terahertz field from $YbFe_2O_4$ corresponds to approximately 9.2% of that in ZnTe. This result suggests that $YbFe_2O_4$ has a large $\chi^{(2)}$ and has a potential application as a future nonlinear optical device.

It is noteworthy that the shape of the generated terahertz waveform is determined by the relative relationship between the polarization of the ferroelectric material and the incident pulse. It is generally difficult to invert the terahertz waveform without rotating the ferroelectric material or nonlinear crystal in real space. However, with $YbFe_2O_4$, the generated terahertz pulse can easily be inverted by rotating an 800 nm incident pulse by 90°, as shown in Figure 5a,b. This result from the asymmetric relationship of d_{11}^{THz} and d_{12}^{THz} implies that $YbFe_2O_4$ can be applied as a terahertz wave generator in which the direction of the generated electric field can be easily switched.

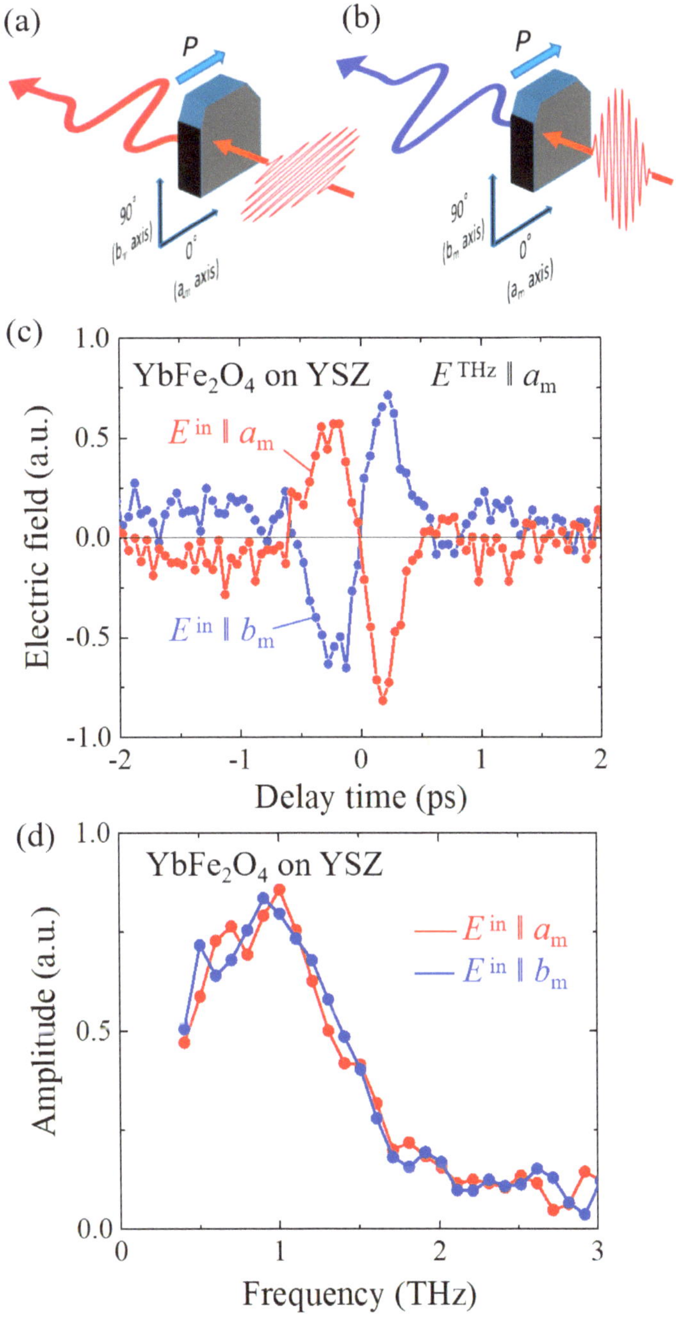

Figure 5. (**a**,**b**) Schematics to illustrate the relationship of the polarization of the 800 nm pulse and that of generated terahertz pulse. (**c**) Waveforms of the terahertz pulse emitted in the direction of a_m axis from $YbFe_2O_4$ thin film. The red (blue) circles denote the terahertz pulse generated when the polarization of the incident fundamental pulse (E_{in}) is set along the a_m (b_m) axis. (**d**) Amplitude spectra calculated by Fourier transform analysis from the waveforms in Figure 5c.

4. Conclusions

In summary, we fabricated $YbFe_2O_4$ thin film on YSZ (111) substrates using a magneto-sputtering technique and investigated two types of second-order nonlinear effects, SHG, and terahertz radiation. The nonlinear phenomena directly reveal a polar structure of the fabricated thin film crystal. The azimuth angle dependence of the SHG profiles for the $YbFe_2O_4$ film clarifies the point group of the thin film crystal (monoclinic C_m) and the relationship between the direction of the polarization in the crystal and the crystal axes of the YSZ substrate. The magnitude of the generated terahertz electric field was found to be as large as 9.2% of that of ZnTe, a typical nonlinear optical device. This study demonstrates the epitaxial growth of $YbFe_2O_4$ thin films with polar symmetry similar to that of bulk crystal and provides the way for their nonlinear optical device applications of electronic ferroelectrics.

Supplementary Materials: The following supporting information can be downloaded at: https://www.mdpi.com/article/10.3390/ma16051989/s1. See supplementary materials for the waveform of the electric field radiated from the bulk ZnTe in our measurement system in Figure 3b.

Author Contributions: Conceptualization, Y.O., N.I. and T.F.; Formal analysis, H.Y. and H.I.; Investigation, H.Y., A.M., S.S., K.T., T.I., S.-y.K., R.M., H.I., S.I., N.I., T.S. and M.N.; Resources, S.-y.K. and S.I.; Writing—original draft, Y.O.; Funding acquisition, Y.O. and T.F.; Supervision, Y.O. and T.F. All authors have read and agreed to the published version of the manuscript.

Funding: This research was supported by Japan Society for the Promotion of Science (JSPS) KAKENHI Grants Nos. 18H05208, 18H02057, 18H01144, 19H01827, 22H01153, 22H01942, 22H01149, 20H05147, 21K03427, and Japan Science and Technology agency (JST) Grants No. JPMJSP2106, CREST (JPMJCR1901), MEXT Q-LEAP (JPMXS0118067426).

Institutional Review Board Statement: Not applicable.

Informed Consent Statement: Not applicable.

Data Availability Statement: Data availble on request.

Acknowledgments: The authors thank R. Seimiya, R. Ota, S. Ogasawara, and K. Fujiwara for their technical assistance and discussion, and Q. J. Wang for his help in the preparation of figures.

Conflicts of Interest: The authors declare no conflict of interest.

References and Notes

1. Miyamoto, T.; Yada, H.; Yamakawa, H.; Okamoto, H. Ultrafast modulation of polarization amplitude by terahertz fields in electronic-type organic ferroelectrics. *Nat. Commun.* **2013**, *4*, 2586–2594. [CrossRef] [PubMed]
2. Hoegen, A.V.; Mankowsky, R.; Fechner, M.; Först, M.; Cavalleri, A. Probing the interatomic potential of solids with strong-field nonlinear phononics. *Nature* **2018**, *555*, 79–82. [CrossRef] [PubMed]
3. Okimoto, Y.; Xia, P.; Itatani, J.; Matsushima, H.; Ishikawa, T.; Koshihara, S.; Horiuchi, S. Ultrafast opto-protonics in a hydrogen-bonded π-molecular ferroelectric crystal. *APL Mater.* **2022**, *10*, 090702. [CrossRef]
4. Ikeda, N.; Ohsumi, H.; Ohwada, K.; Ishii, K.; Inami, T.; Kakurai, K.; Murakami, Y.; Yoshii, K.; Mori, S.; Horibe, Y.; et al. Ferroelectricity from iron valence ordering in the charge-frustrated system LuFe2O4. *Nature* **2005**, *436*, 1136–1138. [CrossRef]
5. Monceau, P.; Nad, F.Y.; Brazovskii, S. Ferroelectric Mott-Hubbard Phase of Organic (TMTTF)2X Conductors. *Phys. Rev. Lett.* **2001**, *86*, 4080–4083. [CrossRef] [PubMed]
6. Yamamoto, K.; Iwai, S.; Boyko, S.; Kashiwazaki, A.; Hiramatsu, F.; Okabe, C.; Nishi, N.; Yakushi, K. Strong Optical Nonlinearity and its Ultrafast Response Associated with Electron Ferroelectricity in an Organic Conductor. *J. Phys. Soc. Jpn.* **2008**, *77*, 074709. [CrossRef]
7. Portengen, T.; Östreich, T.; Sham, L.J. Theory of electronic ferroelectricity. *Phys. Rev. B* **1996**, *54*, 17452–17463. [CrossRef]
8. Ishihara, S. Electronic Ferroelectricity and Frustration. *J. Phys. Soc. Jpn.* **2010**, *79*, 011010. [CrossRef]
9. Fujiwara, K.; Fukada, Y.; Okuda, Y.; Seimiya, R.; Ikeda, N.; Yokoyama, K.; Yu, H.; Koshihara, S.; Okimoto, Y. Direct evidence of electronic ferroelectricity in YbFe2O4 using neutron difraction and nonlinear spectroscopy. *Sci. Rep.* **2021**, *11*, 4277–4283. [CrossRef]
10. Yamada, Y.; Nohdo, S.; Ikeda, N. Incommesurate charge ordering in charge-frustrated LuFe2O4 system. *J. Phys. Soc. Jpn.* **1997**, *66*, 3733–3736. [CrossRef]

11. Angst, M.; Hermann, R.P.; Christianson, A.D.; Lumsden, M.D.; Lee, C.; Whangbo, M.H.; Kim, J.W.; Ryan, P.J.; Nagler, S.E.; Tian, W.; et al. Charge Order in LuFe2O4: Antiferroelectric Ground State and Coupling to Magnetism. *Phys. Rev. Lett.* **2008**, *101*, 227601. [CrossRef]
12. Angst, M. Ferroelectricity from iron valence ordering in rare earth ferrites? *Phys. Stat. Solidi.* **2013**, *7*, 383–400. [CrossRef]
13. Kim, Y.J.; Konishi, S.; Hayasaka, Y.; Ota, R.; Kakeya, I.; Tanaka, K. Magnetic properties of epitaxial TmFe2O4 thin films with an anomalous interfacial structure. *Cryst. Eng. Comm.* **2020**, *22*, 1096–1105. [CrossRef]
14. Sakagami, T.; Ota, R.; Kano, J.; Ikedab, N.; Fujii, T. Single domain growth and charge ordering of epitaxial YbFe2O4 films. *Cryst. Eng. Comm.* **2021**, *23*, 6163–6170. [CrossRef]
15. Shimamoto, K.; Tanaka, J.; Miura, K.; Kiriya, D.; Yoshimura, T.; Fujimura, N. Change in the defect structure of composition controlled single-phase YbFe2O4 epitaxial thin films. *Jpn. J. Appl. Phys.* **2020**, *59*, SPPB07. [CrossRef]
16. Wang, W.; Gai, Z.; Chi, M.; Foelkes, J.D.; Yi, J.; Cheng, X.; Keavney, D.J.; Snijders, P.C.; Ward, T.Z.; Shen, J.; et al. Growth diagram and magnetic properties of hexagonal LuFe2O4 thin films. *Phys. Rev. B* **2012**, *85*, 155411. [CrossRef]
17. Fujii, T.; Numata, T.; Nakahata, H.; Nakanishi, M.; Kano, J.; Ikeda, N. Growth and charge ordering of epitaxial YbFe2O4 films on sapphire using Fe3O4 buffer layer. *Jpn. J. Appl. Phys.* **2018**, *57*, 010305. [CrossRef]
18. Rai, R.C.; Hinz, J.; Pascolini, M.; Sun, F.; Zeng, H. Electronic excitations and optical properties of YbFe2O4 thin films. *Thin Solid Film.* **2019**, *673*, 57–61. [CrossRef]
19. Fujiwara, K.; Karasudani, T.; Fukunaga, M.; Kobayashi, H.; Kano, J.; Janolin, P.-E.; Kiat, J.-M.; Nogami, Y.; Kondo, R.; Ikeda, N. Possible charge order structure of stoichiometric YbFe2O4. *Ferroelectrics* **2017**, *512*, 85–91. [CrossRef]
20. Xiang, H.J.; Whangbo, M.-H. Charge order and origin of giant Megnetoncapacitance in LuFe2O4. *Phys. Rev. Lett.* **2007**, *98*, 246403. [CrossRef]
21. Rai, R.C.; Delmont, A.; Sprow, A.; Cai, B.; Nakarmi, M.L. Spin-charge-orbital coupling in multiferroic LuFe2O4 thin films. *Appl. Phys. Lett.* **2012**, *100*, 212904. [CrossRef]
22. Xu, X.S.; Angst, M.; Brinzari, T.V.; Hermann, R.P.; Musfeldt, J.L.; Christianson, A.D.; Mandrus, D.; Sales, B.C.; McGill, S.; Kim, J.-W.; et al. Charge Order, Dynamics, and Magnetostructural Transition in Multiferroic LuFe2O4. *Phys. Rev. Lett.* **2008**, *101*, 227602. [CrossRef] [PubMed]
23. Yu, H.; (Tokyo Institute of Technology, Tokyo, Japan); Okimoto, Y.; (Tokyo Institute of Technology, Tokyo, Japan); Ikeda, N.; (Okayama university, Okayama, Japan). Unpublished work, 2023.
24. Authier, A. *International Tables for Crystallography D: Physical Properties of Crystals*; International Union of Crystallography: Chester, UK, 2013.
25. Kida, N.; Miyamoto, T.; Okamoto, H. Emission of Terahertz Electromagnetic Waves: A New Spectroscopic Method to Investigate Physical Properties of Solids. *J. Phys. Soc. Jpn.* **2022**, *91*, 112001. [CrossRef]
26. d_{11}^{THz} and d_{12}^{THz} are $\chi^{(2)}$ tensor components at THz frequency region and proportional to $\chi^{(2)}{}_{aaa}(0, -\omega, \omega)$ and $\chi^{(2)}{}_{abb}(0, -\omega, \omega)$, respectively. While d_{11} and $d_{12} \propto \chi^{(2)}{}_{aaa}(-2\omega, \omega, \omega)$ and $\chi^{(2)}{}_{abb}(-2\omega, \omega, \omega)$, respectively.
27. The fluence and the spot area of the irradiated 800 nm pulses to generate the terahertz pulse were 7.7 $\mu J/cm^2$ and 0.79 mm^2 in ZnTe (Supplementary Figure S1), while 150 $\mu J/cm^2$ and 13 mm^2 in $YbFe_2O_4$ (Figure 3b), respectively.

materials

MDPI

Article

Effect of Poling on Multicatalytic Performance of $0.5Ba(Zr_{0.2}Ti_{0.8})O_3$-$0.5(Ba_{0.7}Sr_{0.3})TiO_3$ Ferroelectric Ceramic for Dye Degradation

Akshay Gaur [1,†], Shivam Dubey [1,†], Zainab Mufarreh Elqahtani [2], Samia ben Ahmed [3], Mohammed Sultan Abdulghaffar Al-Buriahi [4], Rahul Vaish [1,*] and Vishal Singh Chauhan [1]

1 School of Mechanical and Materials Engineering, Indian Institute of Technology Mandi Himachal Pradesh, Himachal Pradesh 175005, India
2 Department of Physics, College of Science, Princess Nourah bint Abdulrahman University, P.O. Box 84428, Riyadh 11671, Saudi Arabia
3 Department of Chemistry, College of Sciences, King Khalid University, P.O. Box 9004, Abha 62529, Saudi Arabia
4 Department of Physics, Sakarya University, Sakarya 54050, Turkey
* Correspondence: rahul@iitmandi.ac.in
† These authors contributed equally to this work.

Abstract: Ferroelectric materials with a spontaneous polarization are proven to be potential multicatalysts in water remediation applications. The composition of $0.5Ba(Zr_{0.2}Ti_{0.8})O_3$-$0.5(Ba_{0.7}Sr_{0.3})TiO_3$ (BST-BZT) was examined for photocatalysis, piezocatalysis, and piezo-photocatalysis processes by degrading an azo dye named methylene blue (MB). Generally, dis-aligned dipoles restrict the catalytic activities due to which the BST-BZT powder sample was poled by the corona poling technique. Coupled piezocatalysis and photocatalysis process, i.e., the piezo-photocatalysis process has shown maximum dye degradation. There was a significant improvement in degradation efficiency by using a poled BST-BZT sample compared to the unpoled sample in all processes, thus the results suggest an extensive scope of poled ferroelectric ceramic powder in the catalysis field.

Keywords: ferroelectric material; multicatalytic; dye degradation; corona poling; photocatalysis; piezocatalysis

Citation: Gaur, A.; Dubey, S.; Elqahtani, Z.M.; Ahmed, S.b.; Al-Buriahi, M.S.A.; Vaish, R.; Chauhan, V.S. Effect of Poling on Multicatalytic Performance of $0.5Ba(Zr_{0.2}Ti_{0.8})O_3$-$0.5(Ba_{0.7}Sr_{0.3})TiO_3$ Ferroelectric Ceramic for Dye Degradation. *Materials* **2022**, *15*, 8217. https://doi.org/10.3390/ma15228217

Academic Editor: Dana Popescu

Received: 21 September 2022
Accepted: 11 November 2022
Published: 18 November 2022

1. Introduction

Photocatalysis is one of the most popular advanced oxidation processes for water-cleaning applications. Titanium oxide (TiO2) is a well-studied photocatalyst with a high stability and excellent photochemical properties [1,2]. However, due to low visible light absorption capacity, TiO2 as a photocatalyst is unable to harness the free availability of visible light. Furthermore, the electron-hole pair recombines causing a surface defect, which reduces the efficiency of energy conversion [3–5]. Surface functionalization, heterojunction formation, and tuning defects are some of the strategies that can be used to improve the overall efficiency of the photocatalysis process [6,7]. However, ferroelectric materials with inherent spontaneous polarization are recently being acknowledged as effective photocatalysts in electrochemical processes; $SrTiO_3$, $PbTiO_3$, $BaTiO_3$, and $BiFeO_3$ are examples of ferroelectric materials [8–11]. Ferroelectric materials have shown superior catalytic performances in view of hydrogen generation, dye degradation, and antibacterial properties [12–16]. Moreover, these families of materials have demonstrated multicatalytic processes, which include photocatalysis, piezocatalysis, pyrocatalysis, and a combination of these [17–20]. It means catalysis can be triggered via different input energies such as electromagnetic, mechanical, and thermal vibration. The driving factors for the improved performances are associated with the internal electric field due to non-centrosymmetry. In this regard, the piezocatalysis process is one such process where the non-centrosymmetric structure of the ferroelectric material act as a piezocatalyst for water remediation applications. In the piezocatalysis process, mechanical stress/energy causes charge separation resulting in induced polarization in ferroelectric material, which is used in water splitting, organic dyes, and organic compound synthesis applications [21–24]. $BaTiO_3$, $BiFeO_3$, $Pb(Zr_{0.52}Ti_{0.48})O_3$, $LiNbO_3$, $Bi_4Ti_3O_{12}$, etc. are some of the ferroelectric materials with in-built spontaneous polarization that are utilized in water cleaning applications through the piezocatalysis process [25–29]. Recently, researchers have combined piezocatalysis and photocatalysis processes into one system to enhance the transfer and separation of charges, thereby overall improving the photocatalytic performance [30].

Piezoelectric materials, in particular, generates internal electric fields that separate photogenerated electrons and holes with opposing charges in photocatalysis, thereby improving photocatalytic activity. When a catalyst is allowed to work as a piezocatalyst and photocatalyst, by combining photocatalysis and piezocatalysis processes in one system, charge transfer and separation are accelerated, and high photocatalytic performance is achieved [30].

It is noted that ferroelectric materials can only demonstrate piezoelectric nature after electrical poling. Poling is the process of alignment of dipoles in a unidirectional way. It can be stated that the catalytic performances of ferroelectric material (without poling) will not be very effective. Additionally, in the case of ferroelectric ceramic powder, poling with electrodes is not possible. Thus, for efficient conversion to single-domain grains in ferroelectric powder, the corona poling technique can be applied only. The effect of poling was reported on a few ferroelectric ceramics where significant improvement was observed [31].

Due to the versatile properties of $BaTiO_3$-based ceramics, its various derivatives (Calcium (Ca), Zirconium (Zr), and Strontium (Sr) as dopants) are studied in piezocatalysis, photocatalysis, and piezo-photocatalysis [17,19,20]. As there are promising results from doping Sr and Zr in $BaTiO_3$ solid solution, the multicatalytic properties of $BaTiO_3$ ceramic can be improved by simultaneously introducing Sr and Zr elements. Excellent electrical properties have been reported for Sr and Zr in $BaTiO_3$ solid solutions [32,33]. Thus, the current study focuses on the photocatalytic, piezocatalytic, and piezo-photocatalytic performance of $0.5Ba(Zr_{0.2}Ti_{0.8})O_3$-$0.5(Ba_{0.7}Sr_{0.3})TiO_3$ composition (BST–BZT). BST-BZT has not been examined for any catalytic processes for water remediation application. The study BST-BZT is reported for various catalytic characteristics with the effect of poling on the same.

2. Experimental

2.1. Synthesis of BST-BZT Composition

In the present study, BST-BZT composition was synthesized using the conventional solid oxide reaction route. Barium carbonate ($BaCO_3$) and Titanium dioxide (TiO_2) were purchased from Loba Chemie PVT. LTD, Mumbai, Maharashtra, India (purity ~99%), Zirconium Dioxide (ZrO_2), and Strontium Dioxide ($SrCO_3$) was purchased from Fisher Scientific India PVT. LTD, Mumbai, Maharashtra, India (purity ~99%) were mixed according to their stoichiometric ratio using a mortar and pestle. To attain homogeneity, acetone was used as a mixing medium. Furthermore, the mixed powder was subjected to a calcination temperature of 1350 °C for 6 h in an electric furnace (Nabertherm, Germany).

2.2. Characterization

The phase formation of the BST-BZT sample was revealed using X-ray diffraction data through a Smart Lab X-ray diffractometer (Rigaku Corporation, Tokyo, Japan) in the range of 20–70° at a rate of 0.03°/s. Additionally, Raman spectroscopy was employed for detecting the vibrational modes present in the BST-BZT samples. For this, a 532 nm wavelength green laser was employed on the sample with a power of 25 watts. A field emission scanning electron microscope (FE-SEM) (FEI SEM NOVA Nanosem 450, Hillsboro, OR, USA) was employed for microstructure and surface morphology analysis. The elements detection in BST-BZT composition was performed using X-ray photoelectron spectroscopy (XPS technique) with an X-ray photoelectron spectrophotometer (Thermo scientific, Model: NEXSA, Waltham, MA, USA) with a micro-focused X-ray (400 µm, 72 Watt, and 12,000 V). A monochromatic Al-Kα source with energy, hν = 1486.6 eV with a hemispherical analyzer and a 128-channel plate detector was used for the XPS analysis. The pass energy for the survey scan was 200 eV, whereas 50 eV was used for a core scan. A UV-visible spectrophotometer (SHIMADZU-2600, Tokyo, Japan) aided in the elucidation of absorbed light in samples and for quantification dye degradation.

2.3. Poling of BST-BZT Sample

The BST-BZT powder was poled through an in-house fabricated Corona poling setup. The powder was placed on the ceramic plate (heated ~40 °C) where it was poled at 4 kV/mm for 2 h. Figure 1 shows the schematic representation of the BST-BZT composition preparation using a solid route reaction method and poling of BST-BZT powder by the corona poling technique.

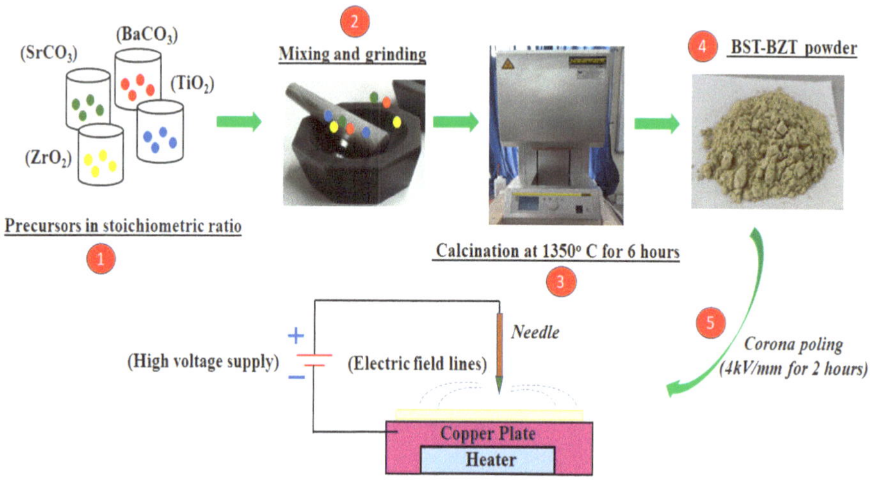

Figure 1. Schematic representation of BST-BZT composition preparation and Corona poling.

2.4. Piezocatalysis, Photocatalysis, and Piezo-Photocatalysis Experiments

The piezocatalysis, photocatalysis, and piezo-photocatalysis experiments were performed on ~5 mg/L concentrated, 10 mL of organic cationic dye named "Methylene Blue" (MB) dye. For piezocatalysis, a conventional ultrasonicator (150 W and 40 kHz) was used, whereas for the photocatalysis process 2 bulbs (Havells) of 15 watts each were used. For piezo-photocatalysis, the two systems, i.e., piezocatalysis and photocatalysis setups were combined for conducting the experiments. Initially, 0.1 g of the BST-BZT powder was immersed in MB dye overnight for adsorption powered by a magnetic stirrer. Then, the adsorbed powder was subjected to visible light, vibrations, and visible light plus vibrations to assess photocatalysis, piezocatalysis, and piezo-photocatalysis process for MB dye degradation. In each experiment after every 30 min, 1 mL of dye was taken out, centrifused and its absorbance was checked using a UV-visible spectrophotometer (SHIMADZU-2600, Tokyo, Japan) for dye degradation quantification.

3. Results and Discussion

Figure 2 shows the XRD pattern for the BST-BZT composition in a range of 20–70°. The obtained peaks were in accordance with tetragonal $BaTiO_3$ ceramics (JCPDS No: 00-005-0626) thus confirming that Zirconium (Zr) and Strontium (Sr) were well doped in $BaTiO_3$ ceramics and no presence of an impurity in the synthesized BST-BZT powder [8]. There is a splitting of the peak at a ~45° angle into two where the peaks corresponding to (002) and (200) Bragg planes as shown in the magnified view of Figure 2, where the intensity of the peak corresponding to the (200) plane was more compared to the (002) plane, which indicates that there is a maximum possibility that BST-BZT crystallized in tetragonal phase. Further information about structural formation in BST-BZT composition was revealed through Raman spectroscopy. Figure 3 shows the Raman spectrum between 150–1250 cm^{-1} wavenumber.

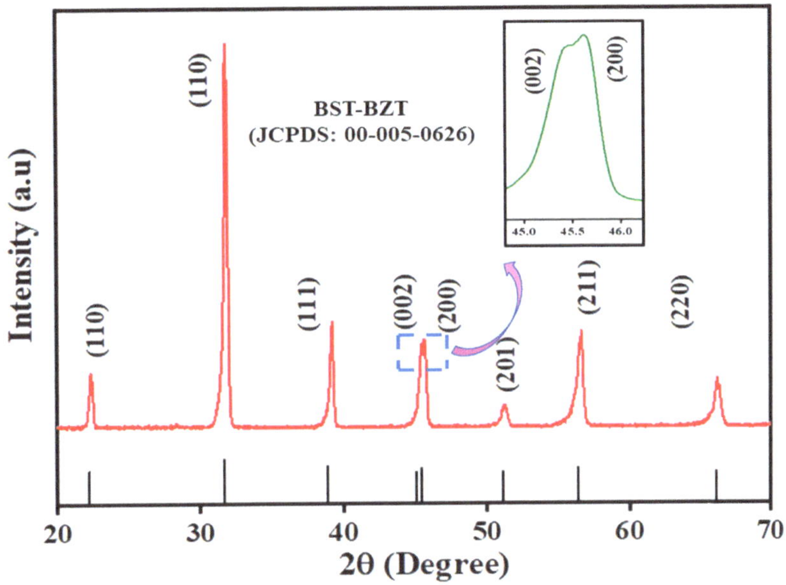

Figure 2. XRD pattern of BST-BZT ceramic powder.

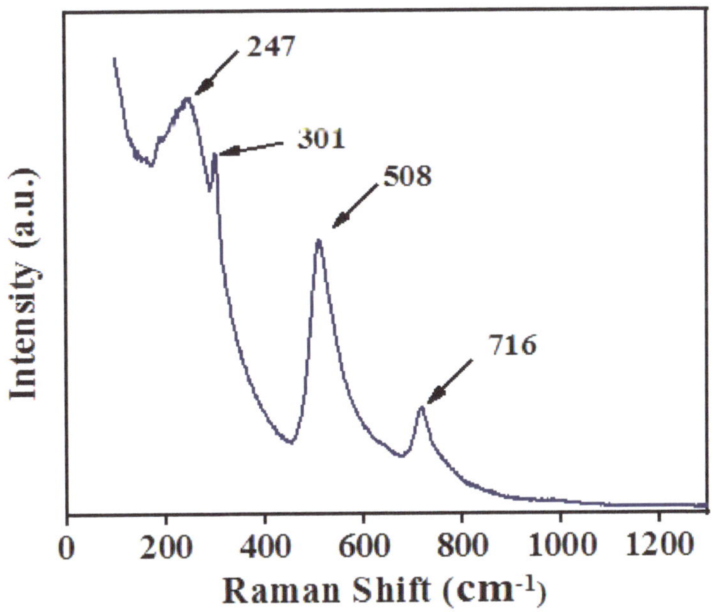

Figure 3. Raman spectrum of BST-BZT powder.

Dominating peaks were observed at ~247, ~304, ~508, and ~716 cm^{-1} wavenumbers. A dip at ~180 wavenumber (cm^{-1}) is observed, which is due to the A1(TO2) anti-symmetry mode. Furthermore, the peaks at ~255 cm^{-1} and ~508 cm^{-1} are due to A1 symmetry caused by transverse optical modes, whereas the peak at ~301 cm^{-1} is due to B1 mode and ~716 cm^{-1} is a consequence of A1(LO3)E(LO) mode, which may indicate the tetragonal phase is formed in a BST-BZT composition [34,35]. Through the XRD and Raman spectroscopy results, there is a high possibility that BST-BZT crystallized in the tetragonal phase. However, there are some works reported in the literature where with an increase in the concentration of dopant in $BaTiO_3$ ceramic, the $BaTiO_3$ crystallized in pseudo tetragonal/cubic phase [36,37]. Figure 4a,b shows the surface morphology of the synthesized BST-BZT composition at a 1 and 3 μm scale obtained from scanning the electron microscope at a certain magnification. The morphology of the BST-BZT composition shows an irregular shape of particles with smooth edges.

Figure 4. SEM images of BST-BZT at certain magnification at scale of (**a**) 1 μm, and (**b**) 3 μm.

The chemical composition of the BST-BZT powder was determined using the XPS technique. Figure 5a shows the XPS survey (0–1000 eV binding energy) of BST-BZT ceramic powder revealing that Ba, Zr, Sr, Ti, and O elements were present in the composition. Furthermore, the XPS spectrum was calibrated according to the binding energy of carbon (~285 eV). Each element was deconvoluted into the peaks using linear as a background. In the C1s spectrum shown in Figure 5b, the obtained binding energy at ~284.6 eV and ~288 eV is due to the C-C/C-H bond and carbonates [38]. The Ba XPS survey in Figure 5c shows that Ba3d3/2 and Ba3d5/2 peaks can be fitted around at a difference of ~1.5 eV. The binding energies ~794.16 and ~778 eV are due to the Ba atoms in BST-BZT ceramics, whereas the peaks around ~779.6 and ~795.18 eV are assigned to Ba atoms obtained during decomposed barium carbonate layers [39]. The O1s spectra can be deconvoluted into three peaks shown in Figure 5d centered at ~528.97 eV, ~531.09 eV, and ~532.83 eV, which are assigned to the oxygen lattice in the sample, oxygen vacancy, and adsorbed oxygen, respectively [40]. Although a peak at ~531.09 eV does not directly indicate the detection of oxygen vacancies, it helps in identifying the adsorbed O^-, O_2^-, and –OH groups trapped in formal oxygen vacancies in synthesized material [41,42], whereas higher binding energy at ~532.83 eV is generally due to the chemisorbed and oxygen-dissociated species. Furthermore, the Sr3d peaks can be deconvoluted at ~132.5 and 134.32 eV binding energies as shown in Figure 5e. The peaks shown in Figure 5f fitted perfectly at binding energy ~458 eV and ~464 eV, which is attributed to the Ti2p3/2 and Ti2p1/2 core level with Ti^{4+} cations [43]. Figure 5g shows profiles representing two-spin orbit peaks of Zr, i.e., the 3d5/2 peak at a binding energy of ~176.4 eV and the 3d3/2 peak at a binding energy of ~178.0 eV. Figure 6 shows the absorbance spectrum of BST-BZT powder sample in a range of 200–800 nm wavelength. The powder sample absorbs much less energy in the range of 500–800 nm wavelength, however there is a gradual increment in the absorbance in the range of 415–550 nm wavelength. After the 415 nm wavelength, there is a sudden rise in the absorption revealing the maximum absorption of the light up to a 200 nm wavelength. Furthermore, the energy band gap of the synthesized BST-BZT composition was calculated through Tauc's relation using Equation (1) [44,45].

$$\alpha(h\nu) = B(h\nu - Eg)^m \quad (1)$$

where, "B" is the energy independent coefficient, "α" is the coefficient of the absorption, "Eg" is the energy band gap of the synthesized sample, "h" is the plank's constant, "ν" is the frequency of the light, and "m" represents the nature of the electronic transition responsible for optical absorption. The energy band gaps (direct and indirect) are determined by using "m" as 1/2 and 2, respectively. The energy band gap of BST-BZT composition was found to be ~3.19 eV using a Tauc's plot as shown in Figure 6. There is not much considerable change in the band gap of the BST-BZT composition compared to the ~3.2 eV band gap of $BaTiO_3$ (tetragonal phase) [8].

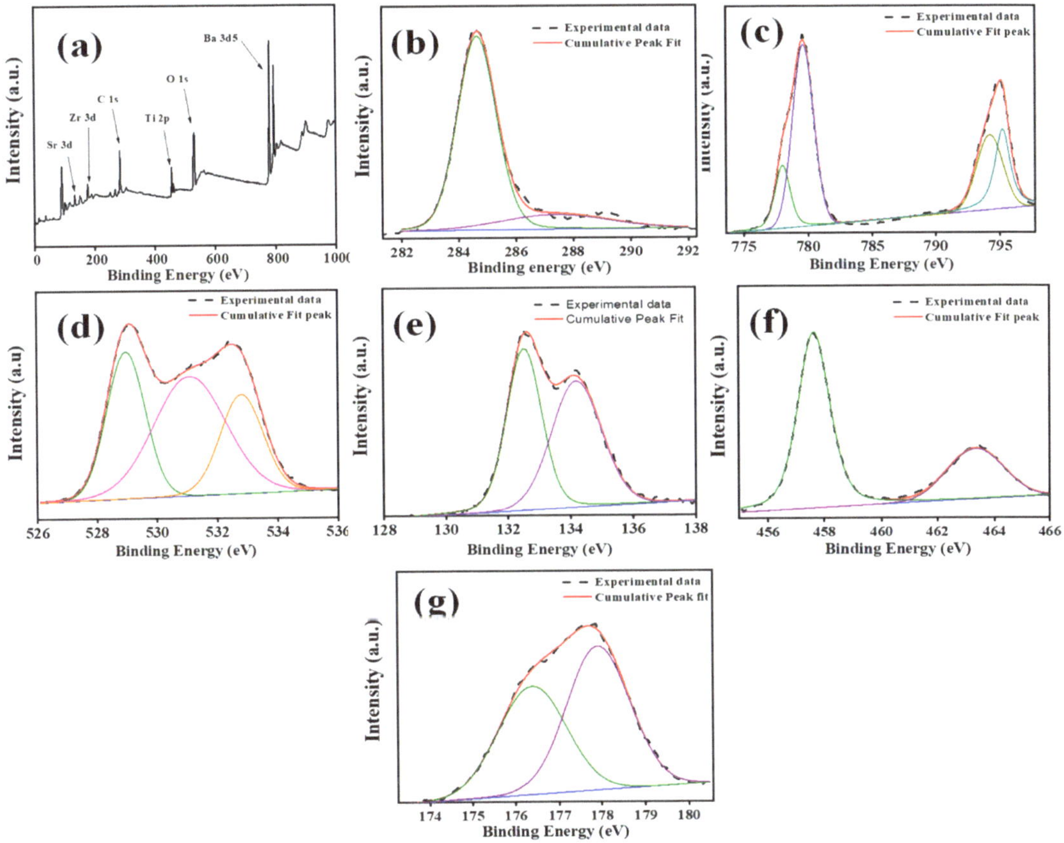

Figure 5. (**a**) XPS survey showing elemental composition in BST-BZT, (**b**) C1s spectrum, (**c**) Ba 3d spectrum, (**d**) O1s spectrum, (**e**) Sr3d spectrum, (**f**) Ti2p spectrum, (**g**) Zr 3d spectrum.

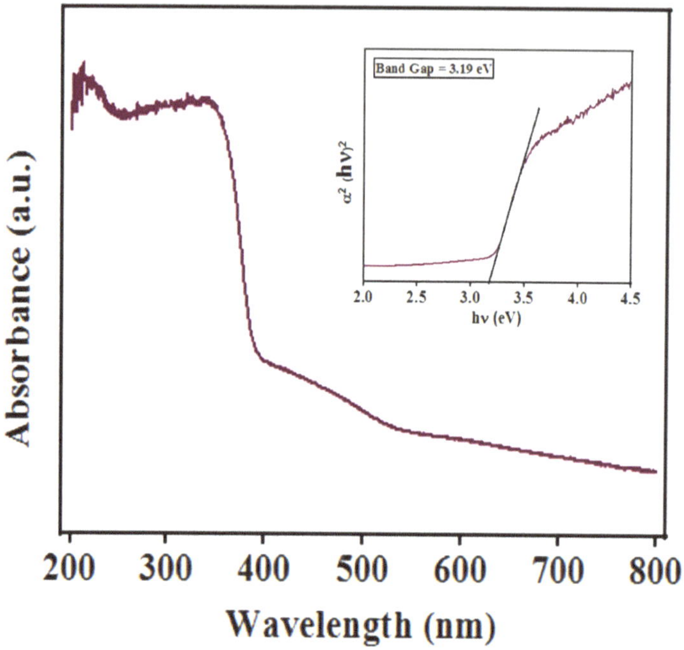

Figure 6. Absorbance vs. wavelength plot showing band gap (Tauc's plot inset).

Under individual and coupled catalytic conditions, the contribution of photocatalytic and piezoelectric properties to degradation was assessed as follows: photocatalysis, piezocatalysis, and piezo-photocatalysis through both unpoled and unpoled BST-BZT powder. In the present study, methylene blue (MB) dye has been selected as an indicative organic pollutant in water. For the assessment of MB dye degradation through photocatalysis, piezocatalysis, and piezo-photocatalysis processes, initially, 0.1 g of BST-BZT powder was immersed in 10 mL of ~5mg/L concentrated MB dye overnight. This was performed to attain an adsorption–desorption equilibrium between the catalyst (BST-BZT composition) and MB dye. Once the adsorption–desorption equilibrium was attained, the BST-BZT powder was subjected to a visible light source, vibrations, and visible light plus vibrations to quantify the weakening of the MB dye. The dye-containing BST-BZT composition was stirred using a magnetic stirrer while placing it in a dark environment for distinguishing adsorption and catalytic processes.

Figure 7a,b show the absorbance spectra of MB dye taken at an interval of 30 min and continuing up to 3 h using 0.1 g of both unpoled and poled BST-BZT samples through the photocatalysis process.

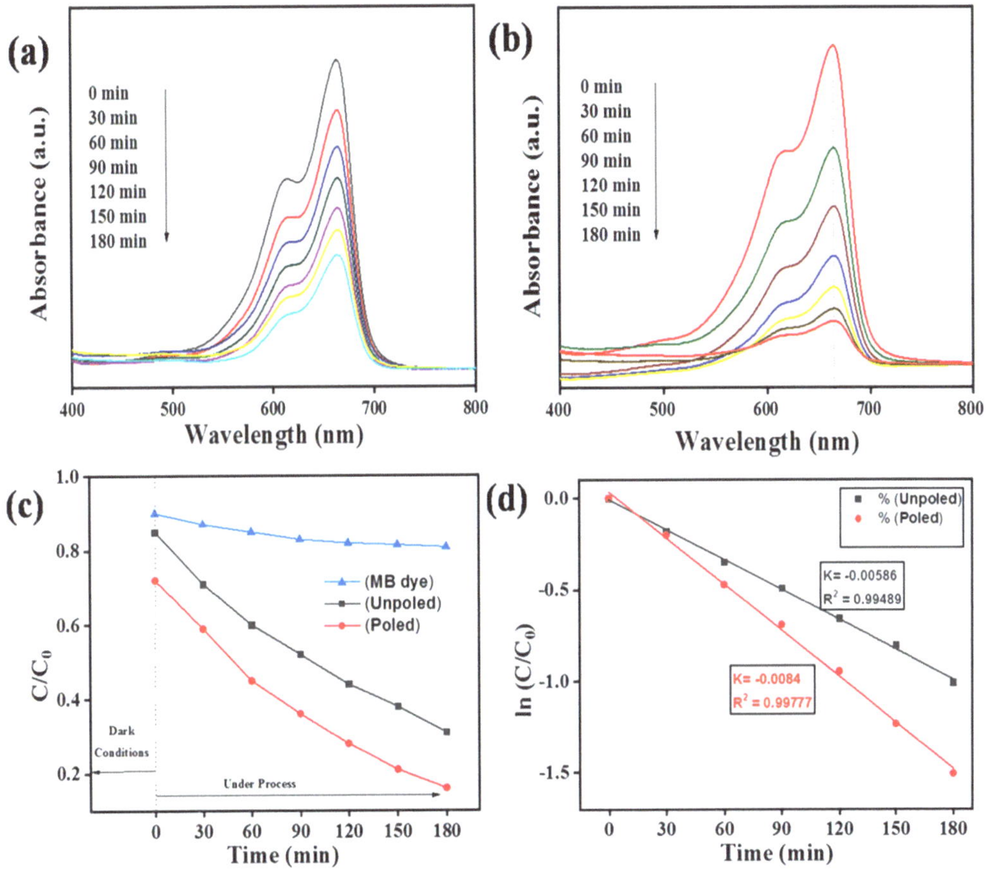

Figure 7. (**a**) Absorbance spectra of MB dye through photocatalysis process under visible light using 0.1 g of unpoled BST-BZT composition, (**b**) absorbance spectra of MB dye through photocatalysis process under visible light using 0.1 g of poled BST-BZT composition, (**c**)% degradation of MB dye after 3 h using 0.1 g of unpoled and poled BST-BZT powder through photocatalysis process, and (**d**) kinetic parameter of MB dye using 0.1 g of unpoled and poled BST-BZT powder under visible light.

The degradation of MB dye was measured by following Equation (2) [46,47].

$$\text{Dye degradation } (\%) = \left(1 - \frac{C}{C_o}\right) \times 100 = \left(1 - \frac{A}{A_o}\right) \times 100 \quad (2)$$

where, "A" and "A_o" are the absorbance peak, and "C" and "C_o" denote the concentration of MB dye at time = "t" mins and at a time "t" = 0 min. The degradation of the MB dye was checked using a UV-Visible spectrophotometer by noting the absorbance peak of the MB dye, which occurs at ~664 nm wavelength [48]. Clearly, there is a decrement in the absorbance spectrum after every 30 min evidencing the weakening of the MB dye. Furthermore, Figure 7c shows the percentage of MB dye degradation achieved after every 30 min using both unpoled and poled BST-BZT samples. The degradation of MB dye was ~10% in 3 h without using any catalyst (control sample) as shown in Figure 7c, whereas Figure 7d shows the kinetic rate constant of MB dye degradation through the photocatalysis process, which is calculated by following Equation (3) [49,50].

$$\ln\left(\frac{C}{Co}\right) = -kt \quad (3)$$

Using unpoled and poled BST-BZT composition, the kinetic rate "*k*" was found to be 0.00586 and 0.0084 min^{-1}, respectively, for the photocatalysis process. Figure 8a,c show the percentage of MB dye degradation using an unpoled and poled BST-BZT sample under piezocatalysis and piezo-photocatalysis process, whereas Figure 8b,d show the kinetic rate for MB dye weakening by using an unpoled and poled BST-BZT sample under a piezocatalysis and piezo-photocatalysis process, respectively. The kinetic rate under the piezocatalysis process was found to be 0.00289 and 0.00698 min^{-1} using an unpoled and poled BST-BZT sample, whereas 0.00738 and 0.01331 min^{-1} was a kinetic rate using unpoled and poled BST-BZT sample under piezo-photocatalysis process, respectively.

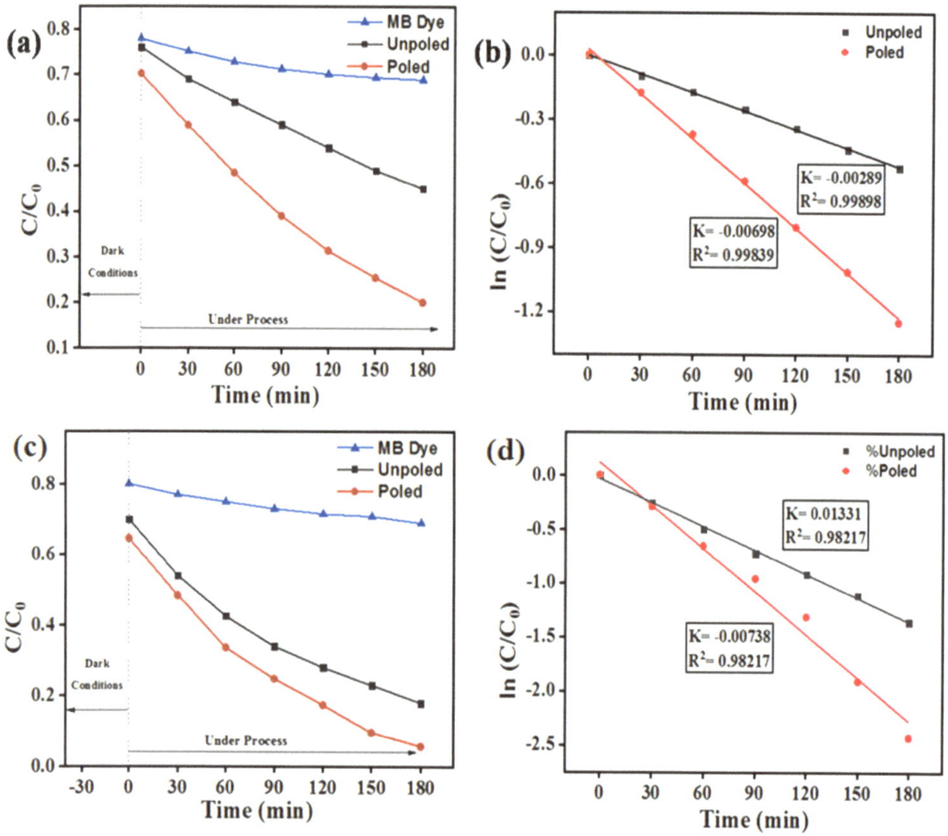

Figure 8. (**a**) % degradation of MB dye after 3 h using 0.1 g of unpoled and poled BST-BZT powder through piezocatalysis process, (**b**) kinetic parameter of MB dye using 0.1 g of unpoled and poled BST-BZT powder in piezocatalysis process, (**c**) % degradation of MB dye after 3 h using 0.1 g of unpoled and poled BST-BZT powder through piezo-photocatalysis process, and (**d**) kinetic parameter of MB dye using 0.1 g of unpoled and poled BST-BZT powder in piezo-photocatalysis process.

There is an effect of poling in the BST-BZT sample for MB dye degradation. The degradation achieved by unpoled and poled BST-BZT samples were ~70 and ~81 through photocatalysis, ~54 and ~79 through piezocatalysis, and 81 and 97% through piezo-photocatalysis processes, respectively. Figure 9 shows a comparison of unpoled and poled BST-BZT samples aiding in MB dye degradation through piezocatalysis, photocatalysis, and piezo-photocatalysis. Coupling piezocatalysis and photocatalysis effect resulted in ~81 and 97% degradation of MB dye from unpoled and poled BST-BZT samples in 3 h. Based upon the first order pseudo kinetics, the kinetic rate of the piezo-photocatalysis process is 1.3 and

1.6 times more compared to the photocatalysis process, while, compared to piezocatalysis, 2.5 and 1.9 times more kinetic rate was achieved through an unpoled and poled BST-BZT sample, respectively.

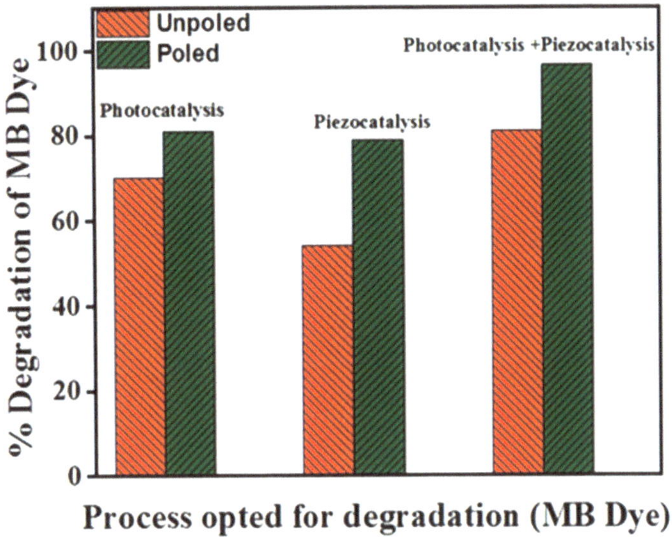

Figure 9. % Degradation achieved by using unpoled and poled BST-BZT sample in photocatalysis, piezocatalysis, and piezo-photocatalysis process.

When the light of a specific wavelength is exposed to a photocatalyst, the electron (e^-) in the valence band (VB) gains enough energy to leave the valence band and enter the conduction band (CB), leaving an equal number of holes (h^+) in the valence band. This formed electron-hole pair reacts with water components in the dye solution, forming reactive species that attack the dye molecules. For this, the conduction band potential should be more negative than the potential of $O_2/{\cdot}O_2$ by which the electron (e^-) can reduce to ${\cdot}O_2{}^-$ from O_2, whereas the valance band potential should be higher than the potential of $H_2O/{\cdot}OH$, thus ·OH can be generated thermodynamically and oxidation may occur, resulting in the weakening of the organic dye [51]. The calculated energy band gap of the BST-BZT composition, i.e., 3.19 eV, should be limited to the ultra-violet region of electromagnetic waves (100–400 nm wavelength). However, in the present study, in the photocatalysis process, significant degradation was achieved from both poled and unpoled BST-BZT samples. The photoactivity could be due to a gradual increment in absorption in the range of 415–550 nm wavelength as shown in Figure 6. The previous literature also showed such results where the energy band gap is greater than 3.0 eV and showed photoactivity in the visible region [52].

However, in the piezocatalysis process, shock waves exert pressure on the catalyst, resulting in a direct piezoelectric effect in the synthesized samples contained within the dye. The charge gets separated, i.e., positive charge (holes) and negative charge (free electron) onto the surface of the BST-BZT sample, which participates in the redox reaction. At the same time, this process would result in a novel property of piezocatalysts, as the piezoelectric potential tilts both CB and VB of the BST-BZT sample. Additionally, in the piezocatalysis process, shock waves create cavities forming bubbles, which grow and ultimately burst during ultrasonication. The formation/growth/collapse of the bubbles lead to an increase in the localized temperature up to 4000–5000 K and generate shock waves up to ~10^8 Pa pressure [53]. This phenomenon is called thermolysis or sonolysis due to which the degradation of the dyes may take place. However, to decrease the thermolysis/sonolysis effect, ice-cold water (<15 °C) was used as a vibration medium.

Figure 8a,c show the degradation of the control sample (without using the BST-BZT sample). Sonocatalysis cannot be avoided under ultrasonication waves. However, the impact of poling supports the intrinsic piezocatalysis processes.

The photogenerated electrons and holes in the photocatalysis process could recombine, resulting in a decrease in charge carriers and thus can obstruct the degradation performance. Furthermore, the piezoelectric effect would separate charges on the surface of the BST-BZT sample when mechanical force was applied. As the piezoelectric potential tilts both CB and VB of the BST-BZT sample, the tilted VB would effectively attract the h^+ to oxidize H_2O to form OH, while the tilted CB would accumulate e^- to reduce O_2 to generate O_2 because the tilted CB and VB after incorporating the piezoelectric effect into photocatalysis photogenerated electrons and holes are more easily excited to participate in the redox reactions for pollutant degradation [54]. The induced piezoelectric field has improved the separation efficiency of photogenerated electron-hole pairs. The induced piezoelectric field separates more photogenerated electrons and holes, which then migrate to the opposite surface of the catalyst and participate in redox reactions, resulting in a significant increase in catalytic degradation activity and as a result enhance catalytic activity in the piezo-photocatalysis process compared to piezocatalysis and photocatalysis processes. Figure 10 shows the plausible mechanism of photocatalysis, piezocatalysis, and piezocatalysis processes separately.

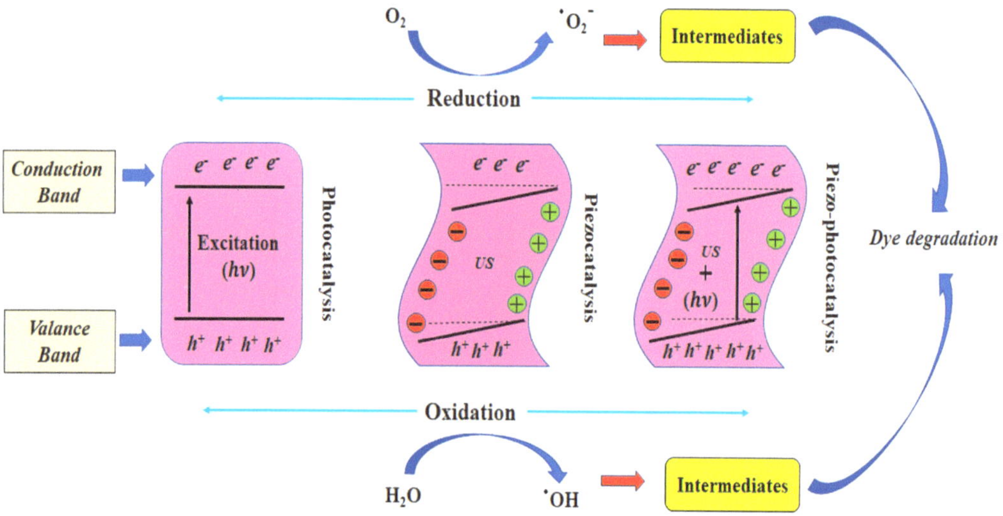

Figure 10. A plausible mechanism of photocatalysis, piezocatalysis, and piezo-photocatalysis process.

In contrast to piezoelectric sensing, actuating, and energy harvesting applications, piezocatalytic can occur in unpoled samples. This is because the piezocatalysis process necessitates the local interaction of pollutant molecules with the surfaces of piezoelectric materials. This enables multiple domains (as individual piezoelectric) in unpoled ferroelectric material particles to participate in catalytic reactions. However, polling, on the other hand, can significantly improve performance by creating a high surface electric potential on the surfaces by aligning the dipoles in a specific direction [31]. This is supported by ~54% and ~79% degradation of MB dye achieved by unpoled and poled BST-BZT samples, respectively, in 3 h of ultrasonication. Additionally, poled ferroelectric ceramics have shown enhanced photocatalytic activities. This is due to the reason that poling accelerates the separation and migration of electron-hole pairs and results in increased photocatalytic activity [55,56]. Additionally, in the case of the piezo-photocatalysis process, enhanced MB dye degradation was achieved using poled BST-BZT compared to an unpoled BST-BZT sample.

The photocatalysis, piezocatalysis, and piezo-photocatalysis processes generate attacking species such as holes (h^+), hydroxyl radicals (OH), electrons (e^-), and superoxide radicals (O_2), which are responsible for the degradation of the MB dye. However, only one of the attacking species has dominant characteristics that cause dye degradation. A specific scavenger traps specific attacking species, which means that if that specific attacking species is trapped, dye degradation will not occur to the extent that it would have without that specific scavenger. The scavenger test was used to identify the reactive species responsible for MB dye degradation using poled the BST-BZT sample. In the current study, a scavenger test was performed in a piezocatalysis process by adding 1000 μL of Ethylene diamine tetra acetic acid (EDTA), Isopropanol (IPA), and Benzoquinone (BQ) scavengers to 10 mL of 5 mg/L concentrated MB dye. Figure 11 depicts the degradation of MB dye by scavengers IPA, BQ, and EDTA at 26%, 36, and 38%, respectively. Scavengers such as EDTA, IPA, and BQ capture reactive species such as OH, h^+, and O_2 [17]. Figure 11 shows that among all scavengers, MB dye containing IPA has shown the least degradation, indicating the least participation of OH radical in 3 h of ultrasonication. This indicates that scavenger IPA has trapped the OH radical, which was the primary attacking species in the piezocatalysis process.

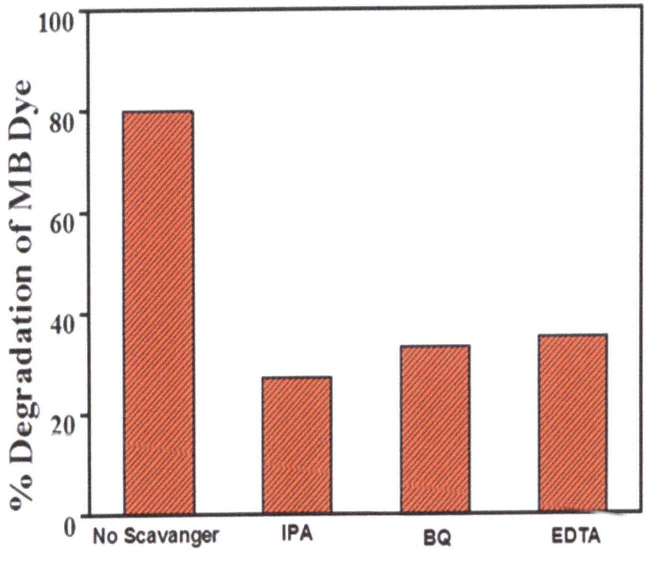

Figure 11. % Degradation achieved by using EDTA, IPA, and BQ scavengers in piezocatalysis process using poled BST-BZT sample in 3 h.

4. Conclusions

The ferroelectric ceramic $0.5Ba(Zr_{0.2}Ti_{0.8})O_3$-$0.5(Ba_{0.7}Sr_{0.3})TiO_3$ composition (BST–BZT) was synthesized via the solid-state reaction route. The synthesized BST-BZT ceramic was studied for multicatalytic activity, i.e., photocatalysis, piezocatalysis, and piezo-photocatalysis for degrading MB dye. The degradation was found to be ~70, ~54 and ~81% for photocatalysis, piezocatalysis, and piezo-photocatalysis processes, respectively. Clearly, the performance was improved by the synergistic effect of photocatalysis and the piezocatalysis process. Further in the study, the BST-BZT sample was poled using the corona poling technique where significant enhancement was observed compared to the unpoled BST-BZT powder sample. The kinetic rate was 1.4, 2.4, and 1.9 times more when poled the BST-BZT sample was used compared to the unpoled BST-BZT sample in photocatalysis, piezocatalysis, and piezo-photocatalysis processes, respectively. Thus, the present study brings out a novel study by combining photocatalysis and piezocatalysis processes for organic pollutant removal, which can be further accelerated by the electric poling of ferroelectric catalysts.

Author Contributions: Methodology, Validation, writing-original draft preparation: A.G. and S.D.; Project administration, funding acquisition, formal analysis, writing—review and editing: Z.M.E., S.b.A. and M.S.A.A.-B.; Conceptualization, Visualization, Supervison, investigation, writing—review and editing: R.V. and V.S.C.; All authors have read and agreed to the published version of the manuscript.

Funding: Princess Nourah bint Abdulrahman University Researchers Supporting Project (Grant No. PNURSP2022R124), Princess Nourah bint Abdulrahman University, Riyadh, Saudi Arabia. Deanship of Scientific Research at King Khalid University, Saudi Arabia for funding this work through Large Research Groups Program under grant number L.R.G.P2/171/43.

Institutional Review Board Statement: Not applicable.

Informed Consent Statement: Not applicable.

Data Availability Statement: The data that support the findings of this study are available from the corresponding author upon reasonable request.

Acknowledgments: The authors express their gratitude to Princess Nourah bint Abdulrahman University Researchers Supporting Project (Grant No. PNURSP2022R124), Princess Nourah bint Abdulrahman University, Riyadh, Saudi Arabia. The authors extend their appreciation to the Deanship of Scientific Research at King Khalid University, Saudi Arabia for funding this work through Large Research Groups Program under grant number L.R.G.P2/171/43.

Conflicts of Interest: The authors declare no conflict of interest.

References

1. Ni, M.; Leung, M.K.H.; Leung, D.Y.C.; Sumathy, K. A review and recent developments in photocatalytic water-splitting using TiO_2 for hydrogen production. *Renew. Sustain. Energy Rev.* **2007**, *11*, 401–425. [CrossRef]
2. Xu, H.; Ouyang, S.; Liu, L.; Reunchan, P.; Umezawa, N.; Ye, J. Recent advances in TiO_2-based photocatalysis. *J. Mater. Chem. A* **2014**, *2*, 12642–12661. [CrossRef]
3. Qureshy, A.M.M.I.; Ahmed, M.; Dincer, I. Performance assessment study of photo-electro-chemical water-splitting reactor designs for hydrogen production. *Int. J. Hydrogen Energy* **2019**, *44*, 9237–9247. [CrossRef]
4. Licht, S.; Wang, B.; Mukerji, S.; Soga, T.; Umeno, M.; Tributsch, H. Efficient solar water splitting, exemplified by RuO_2-catalyzed AlGaAs/Si photoelectrolysis. *J. Phys. Chem. B* **2000**, *104*, 8920–8924. [CrossRef]
5. Khan, S.U.M.; Al-Shahry, M.; Ingler, W.B., Jr. Efficient photochemical water splitting by a chemically modified n-TiO_2. *Science* **2002**, *297*, 2243–2245. [CrossRef] [PubMed]
6. Yu, Y.; Zhang, Z.; Yin, X.; Kvit, A.; Liao, Q.; Kang, Z.; Yan, X.; Zhang, Y.; Wang, X. Enhanced photoelectrochemical efficiency and stability using a conformal TiO_2 film on a black silicon photoanode. *Nat. Energy* **2017**, *2*, 17045. [CrossRef]
7. Wang, X.; Li, Z.; Shi, J.; Yu, Y. One-dimensional titanium dioxide nanomaterials: Nanowires, nanorods, and nanobelts. *Chem. Rev.* **2014**, *114*, 9346–9384. [CrossRef]
8. Gaur, A.; Sharma, M.; Chauhan, V.S.; Vaish, R. Solar/visible light photocatalytic dye degradation using $BaTi_1$-xFexO_3 ceramics. *J. Am. Ceram. Soc.* **2022**, *105*, 5140–5150. [CrossRef]
9. Niu, F.; Chen, D.; Qin, L.; Gao, T.; Zhang, N.; Wang, S.; Chen, Z.; Wang, J.; Sun, X.; Huang, Y. Synthesis of Pt/$BiFeO_3$ heterostructured photocatalysts for highly efficient visible-light photocatalytic performances. *Sol. Energy Mater. Sol. Cells* **2015**, *143*, 386–396. [CrossRef]
10. Huang, X.; Lei, R.; Yuan, J.; Gao, F.; Jiang, C.; Feng, W.; Zhuang, J.; Liu, P. Insight into the piezo-photo coupling effect of $PbTiO_3$/CdS composites for piezo-photocatalytic hydrogen production. *Appl. Catal. B Environ.* **2021**, *282*, 119586. [CrossRef]
11. Konta, R.; Ishii, T.; Kato, H.; Kudo, A. Photocatalytic activities of noble metal ion doped $SrTiO_3$ under visible light irradiation. *J. Phys. Chem. B* **2004**, *108*, 8992–8995. [CrossRef]
12. Kumar, S.; Sharma, M.; Frömling, T.; Vaish, R. Antibacterial ferroelectric materials: Advancements and future directions. *J. Ind. Eng. Chem.* **2021**, *97*, 95–110. [CrossRef]
13. Kim, S.; Nguyen, N.T.; Bark, C.W. Ferroelectric materials: A novel pathway for efficient solar water splitting. *Appl. Sci.* **2018**, *8*, 1526. [CrossRef]
14. Khan, M.A.; Nadeem, M.A.; Idriss, H. Ferroelectric polarization effect on surface chemistry and photo-catalytic activity: A review. *Surf. Sci. Rep.* **2016**, *71*, 1–31. [CrossRef]
15. Liang, Z.; Yan, C.-F.; Rtimi, S.; Bandara, J. Piezoelectric materials for catalytic/photocatalytic removal of pollutants: Recent advances and outlook. *Appl. Catal. B Environ.* **2019**, *241*, 256–269. [CrossRef]
16. Singh, S.; Faraz, M.; Khare, N. Recent advances in semiconductor-graphene and semiconductor-ferroelectric/ferromagnetic nanoheterostructures for efficient hydrogen generation and environmental remediation. *ACS Omega* **2020**, *5*, 11874–11882. [CrossRef]
17. Sharma, M.; Singh, G.; Vaish, R. Dye degradation and bacterial disinfection using multicatalytic $BaZr_{0.02}Ti_{0.98}O_3$ ceramics. *J. Am. Ceram. Soc.* **2020**, *103*, 4774–4784. [CrossRef]

18. Sharma, M.; Singh, V.P.; Kumar, S.; Vaish, R. Multicatalytic behavior of $Ba_{0.85}Ca_{0.15}Ti_{0.9}$ $Zr_{0.1}O_3$ ceramics for pharmaceutical/dye/bacterial treatments. *J. Appl. Phys.* **2020**, *127*, 135103. [CrossRef]
19. Qifeng, L.; Jingjun, M.; Sharma, M.; Vaish, R. Photocatalytic, piezocatalytic, and piezo-photocatalytic effects in ferroelectric $(Ba_{0.875}Ca_{0.125})(Ti_{0.95}Sn_{0.05})$ O_3 ceramics. *J. Am. Ceram. Soc.* **2019**, *102*, 5807–5817. [CrossRef]
20. Sharma, M.; Vaish, R. Piezo/pyro/photo-catalysis activities in $Ba_{0.85}Ca_{0.15}(Ti_{0.9}Zr_{0.1})$1-xFex$O_3$ ceramics. *J. Am. Ceram. Soc.* **2021**, *104*, 45–56. [CrossRef]
21. Starr, M.B.; Shi, J.; Wang, X. Piezopotential-driven redox reactions at the surface of piezoelectric materials. *Angew. Chem. Int. Ed.* **2012**, *51*, 5962–5966. [CrossRef] [PubMed]
22. Hong, K.-S.; Xu, H.; Konishi, H.; Li, X. Piezoelectrochemical effect: A new mechanism for azo dye decolorization in aqueous solution through vibrating piezoelectric microfibers. *J. Phys. Chem. C* **2012**, *116*, 13045–13051. [CrossRef]
23. Feng, Y.; Ling, L.; Wang, Y.; Xu, Z.; Cao, F.; Li, H.; Bian, Z. Engineering spherical lead zirconate titanate to explore the essence of piezo-catalysis. *Nano Energy* **2017**, *40*, 481–486. [CrossRef]
24. Hong, K.-S.; Xu, H.; Konishi, H.; Li, X. Direct water splitting through vibrating piezoelectric microfibers in water. *J. Phys. Chem. Lett.* **2010**, *1*, 997–1002. [CrossRef]
25. Lin, H.; Wu, Z.; Jia, Y.; Li, W.; Zheng, R.-K.; Luo, H. Piezoelectrically induced mechano-catalytic effect for degradation of dye wastewater through vibrating Pb $(Zr_{0.52}Ti_{0.48})$ O_3 fibers. *Appl. Phys. Lett.* **2014**, *104*, 162907. [CrossRef]
26. Singh, G.; Sharma, M.; Vaish, R. Exploring the piezocatalytic dye degradation capability of lithium niobate. *Adv. Powder Technol.* **2020**, *31*, 1771–1775. [CrossRef]
27. Lv, W.; Kong, L.; Lan, S.; Feng, J.; Xiong, Y.; Tian, S. Enhancement effect in the piezoelectric degradation of organic pollutants by piezo-Fenton process. *J. Chem. Technol. Biotechnol.* **2017**, *92*, 152–156. [CrossRef]
28. Tu, S.; Huang, H.; Zhang, T.; Zhang, Y. Controllable synthesis of multi-responsive ferroelectric layered perovskite-like $Bi_4Ti_3O_{12}$: Photocatalysis and piezoelectric-catalysis and mechanism insight. *Appl. Catal. B Environ.* **2017**, *219*, 550–562. [CrossRef]
29. You, H.; Jia, Y.; Wu, Z.; Xu, X.; Qian, W.; Xia, Y.; Ismail, M. Strong piezo-electrochemical effect of multiferroic $BiFeO_3$ square micro-sheets for mechanocatalysis. *Electrochem. Commun.* **2017**, *79*, 55–58. [CrossRef]
30. Jiang, Z.; Tan, X.; Huang, Y. Piezoelectric effect enhanced photocatalysis in environmental remediation: State-of-the-art techniques and future scenarios. *Sci. Total Environ.* **2022**, *806*, 150924. [CrossRef]
31. Xu, J.; Zang, T.; Du, D.; Kumar, S.; Sharma, M.; Vaish, R. Effect of poling on piezocatalytic removal of muti-pollutants using $BaTiO_3$. *J. Am. Ceram. Soc.* **2021**, *104*, 1661–1668. [CrossRef]
32. Patel, S.; Srikanth, K.S.; Steiner, S.; Vaish, R.; Froemling, T. Pyroelectric and impedance studies of the 0.5 Ba $(Zr_{0.2}Ti_{0.8})$ O_3-0.5 $(Ba_{0.7}Sr_{0.3})$ TiO_3 ceramics. *Ceram. Int.* **2018**, *44*, 21976–21981. [CrossRef]
33. Patel, S.; Yadav, H. Electrical conduction properties of the BZT–BST ceramics. *J. Adv. Dielectr.* **2020**, *10*, 2050026. [CrossRef]
34. Hayashi, H.; Nakamura, T.; Ebina, T. In-situ Raman spectroscopy of $BaTiO_3$ particles for tetragonal-cubic transformation. *J. Phys. Chem. Solids* **2013**, *74*, 957–962. [CrossRef]
35. Mondal, T.; Majee, B.P.; Das, S.; Sinha, T.P.; Middya, T.R.; Badapanda, T.; Sarun, P.M. A comparative study on electrical conduction properties of Sr-substituted Ba1−x Sr x $Zr_{0.1}Ti_{0.9}O_3$ (x = 0.00–0.15) ceramics. *Ionics* **2017**, *23*, 2405–2416. [CrossRef]
36. Muhammad, R.; Khalil, M.A.; Castro, M.S. Structure and dielectric characteristics of $Ba_{1-x}CaxTi_{1-x}CaxO_{3-\delta}$ ceramics. *Ceram. Int.* **2020**, *46*, 1059–1064. [CrossRef]
37. Buscaglia, M.T.; Buscaglia, V.; Viviani, M.; Nanni, P.; Hanuskova, M. Influence of foreign ions on the crystal structure of $BaTiO_3$. *J. Eur. Ceram. Soc.* **2000**, *20*, 1997–2007. [CrossRef]
38. Viviani, M.; Buscaglia, M.T.; Nanni, P.; Parodi, R.; Gemme, G.; Dacca, A. XPS investigation of surface properties of Ba (1−x) SrxTiO_3 powders prepared by low temperature aqueous synthesis. *J. Eur. Ceram. Soc.* **1999**, *19*, 1047–1051. [CrossRef]
39. Craciun, V.; Singh, R.K. Characteristics of the surface layer of barium strontium titanate thin films deposited by laser ablation. *Appl. Phys. Lett.* **2000**, *76*, 1932–1934. [CrossRef]
40. Kushwaha, A.; Aslam, M. Hydrogen-incorporated ZnO nanowire films: Stable and high electrical conductivity. *J. Phys. D Appl. Phys.* **2013**, *46*, 485104. [CrossRef]
41. Dupin, J.-C.; Gonbeau, D.; Vinatier, P.; Levasseur, A. Systematic XPS studies of metal oxides, hydroxides and peroxides. *Phys. Chem. Chem. Phys.* **2000**, *2*, 1319–1324. [CrossRef]
42. Tabata, K.; Hirano, Y.; Suzuki, E. XPS studies on the oxygen species of $LaMn_{1-x}CuxO_{3+\lambda}$. *Appl. Catal. A Gen.* **1998**, *170*, 245–254. [CrossRef]
43. Nasser, S.A. X-ray photoelectron spectroscopy study on the composition and structure of $BaTiO_3$ thin films deposited on silicon. *Appl. Surf. Sci.* **2000**, *157*, 14–22. [CrossRef]
44. Singh, G.; Sharma, M.; Vaish, R. Influence of $LiNbO_3$ crystallization on the optical, dielectric and nanoindentation properties of the $30SiO_2$–$35Li_2O$–$35Nb_2O_5$ glass. *J. Appl. Phys.* **2019**, *126*, 214101. [CrossRef]
45. George, A.; Solomon, S.; Thomas, J.K.; John, A. Characterizations and electrical properties of $ZrTiO_4$ ceramic. *Mater. Res. Bull.* **2012**, *47*, 3141–3147. [CrossRef]
46. Wang, S.; Wu, Z.; Chen, J.; Ma, J.; Ying, J.; Cui, S.; Yu, S.; Hu, Y.; Zhao, J.; Jia, Y. Lead-free sodium niobate nanowires with strong piezo-catalysis for dye wastewater degradation. *Ceram. Int.* **2019**, *45*, 11703–11708. [CrossRef]
47. Ma, J.; Ren, J.; Jia, Y.; Wu, Z.; Chen, L.; Haugen, N.O.; Huang, H.; Liu, Y. High efficiency bi-harvesting light/vibration energy using piezoelectric zinc oxide nanorods for dye decomposition. *Nano Energy* **2019**, *62*, 376–383. [CrossRef]

48. Singh, G.; Sharma, M.; Vaish, R. Flexible Ag@ $LiNbO_3$/PVDF composite film for piezocatalytic dye/pharmaceutical degradation and bacterial disinfection. *ACS Appl. Mater. Interfaces* **2021**, *13*, 22914–22925. [CrossRef]
49. Wu, J.; Qin, N.; Bao, D. Effective enhancement of piezocatalytic activity of $BaTiO_3$ nanowires under ultrasonic vibration. *Nano Energy* **2018**, *45*, 44–51. [CrossRef]
50. Ma, W.; Yao, B.; Zhang, W.; He, Y.; Yu, Y.; Niu, J.; Wang, C. A novel multi-flaw MoS_2 nanosheet piezocatalyst with superhigh degradation efficiency for ciprofloxacin. *Environ. Sci Nano* **2018**, *5*, 2876–2887. [CrossRef]
51. Jiang, Z.; Wan, W.; Li, H.; Yuan, S.; Zhao, H.; Wong, P.K. A hierarchical Z-Scheme α-Fe_2O_3/g-C_3N_4 hybrid for enhanced photocatalytic CO_2 reduction. *Adv. Mater.* **2018**, *30*, 1706108. [CrossRef] [PubMed]
52. Srisuvetha, V.T.; Rayar, S.L.; Shanthi, G. Role of cerium (Ce) dopant on structural, optical and photocatalytic properties of MgO nanoparticles by wet chemical route. *J. Mater. Sci. Mater. Electron.* **2020**, *31*, 2799–2808. [CrossRef]
53. Singh, G.; Sharma, M.; Vaish, R. Transparent ferroelectric glass-ceramics for wastewater treatment by piezocatalysis. *Commun. Mater.* **2020**, *1*, 100. [CrossRef]
54. Lin, E.; Wu, J.; Qin, N.; Yuan, B.; Kang, Z.; Bao, D. Enhanced piezocatalytic, photocatalytic and piezo-/photocatalytic performance of diphasic Ba_{1-x} Ca_x TiO_3 nanowires near a solubility limit. *Catal. Sci. Technol.* **2019**, *9*, 6863–6874. [CrossRef]
55. Inoue, Y.; Hayashi, O.; Sato, K. Photocatalytic activities of potassium-doped lead niobates and the effect of poling. *J. Chem. Soc. Faraday Trans.* **1990**, *86*, 2277–2282. [CrossRef]
56. Lan, Y.; Sun, Z.; Yuan, C.; Xue, X.; Chen, J.; Miao, L.; Guo, Y.; Zhou, C.; Xu, J.; Zhou, J.; et al. Enhanced Visible Photocatalytic Hydrogen Evolution of KN-Based Semiconducting Ferroelectrics via Band-Gap Engineering and High-Field Poling. *ACS Appl. Mater. Interfaces* **2022**, *14*, 8916–8930. [CrossRef]

materials

MDPI

Article

Antiferromagnet–Ferromagnet Transition in $Fe_{1-x}Cu_xNbO_4$

Diego S. Evaristo [1], Raí F. Jucá [1], João M. Soares [2], Rodolfo B. Silva [2], Gilberto D. Saraiva [3], Robert S. Matos [4], Nilson S. Ferreira [1,*], Marco Salerno [5] and Marcelo A. Macêdo [1,*]

1 Department of Physics, Federal University of Sergipe, São Cristóvão 49100-000, Brazil
2 Department of Physics, State University of Rio Grande do Norte, Mossoró 59610-210, Brazil
3 Faculty of Education Sciences and Letters of Sertão Central, State University of Ceará, Quixadá 63900-000, Brazil
4 Postgraduate Program in Materials Science and Engineering (P2CEM), Federal University of Sergipe, São Cristovão 49100-000, Brazil
5 Chair of Materials Science and Technology, Technische Universität Dresden, Budapester Str. 27, 010169 Dresden, Germany
* Correspondence: nilson@academico.ufs.br (N.S.F.); mmacedo@academico.ufs.br (M.A.M.)

Abstract: Iron niobates, pure and substituted with copper ($Fe_{1-x}Cu_xNbO_4$ with x = 0–0.15), were prepared by the solid-state method and characterized by X-ray diffraction, Raman spectroscopy, and magnetic measurements. The results of the structural characterizations revealed the high solubility of Cu ions in the structure and better structural stability compared to the pure sample. The analysis of the magnetic properties showed that the antiferromagnetic–ferromagnetic transition was caused by the insertion of Cu^{2+} ions into the $FeNbO_4$ structure. The pure $FeNbO_4$ structure presented an antiferromagnetic ordering state, with a Néel temperature of approximately 36.81K. The increase in substitution promoted a change in the magnetic ordering, with the state passing to a weak ferromagnetic order with a transition temperature (T_c) higher than the ambient temperature. The origin of the ferromagnetic ordering could be attributed to the increase in super-exchange interactions between Fe/Cu ions in the Cu^{2+}-O-Fe^{3+} chains and the formation of bound magnetic polarons in the oxygen vacancies.

Keywords: structure properties; magnetic transition; antiferromagnetic–ferromagnetic

Citation: Evaristo, D.S.; Jucá, R.F.; Soares, J.M.; Silva, R.B.; Saraiva, G.D.; Matos, R.S.; Ferreira, N.S.; Salerno, M.; Macêdo, M.A. Antiferromagnet–Ferromagnet Transition in $Fe_{1-x}Cu_xNbO_4$. *Materials* **2022**, *15*, 7424. https://doi.org/10.3390/ma15217424

Academic Editor: Laurent Daniel

Received: 27 September 2022
Accepted: 18 October 2022
Published: 22 October 2022

1. Introduction

$FeNbO_4$ is a polymorphic compound that crystallizes in three different crystalline phases depending on the annealing temperature [1].Considered an n-type semiconductor, it has a narrow bandgap of 1.81–2.25 eV [2,3], making it attractive for multipleapplications, e.g., as a photocatalyst [4,5], due to its excellent visible light activity attributed to the higher energy levels of the Nb 4d orbital;in gas sensors [3], capacitors [6], and lithium-ion batteries [7,8]; and as an anode material in solid oxide fuel cells (SOFCs) [9].The most stable phase, under ambient conditions, has monoclinic symmetry (m-$FeNbO_4$, space group P2/c) and is obtained at temperatures below 1085 °C [10]. Its structure has ordered cations, with both Fe^{3+} and Nb^{5+} forming regular octahedra, coordinated by six oxygen ions, forming zig-zag chains of FeO_6 and NbO_6 (see Figure 1a) [10,11]. A crystal structure with orthorhombic symmetry (o-$FeNbO_4$, Pbcn space group) is formed in the temperature range of 1085 to 1380 °C [12].Unlike the m-$FeNbO_4$ phase, above 1100 °C, the distribution of cations in this structure becomes disordered. Above 1380 °C, it crystallizes in the tetragonal phase (t-$FeNbO_4$, space group P42/mnm) [12,13].

The order/disorder of the cations and the super-exchange interactions via oxygen are significant for the magnetic properties of these compounds, and the degree of the order depends on the specific conditions of synthesis [14]. m-$FeNbO_4$ is isostructural to $FeWO_4$, with the sharing of the edges of the Fe-O-Fe octahedral chains resulting in magnetic ordering, forming infinite ferromagnetic (FM) sheets along the [1 0 0] direction [1]. Cross-chains of Fe-O-Nb-O-Fe result in antiferromagnetic (AFM) arrangements and, therefore, in net antiferromagnetism [15]. Recently, Wang et al., through theoretical calculations of functional density (DFT), hypothesized the existence of three distinct and stable magnetic ordering configurations, which can be explained as the possible formation of magnetic domains in the material [10].

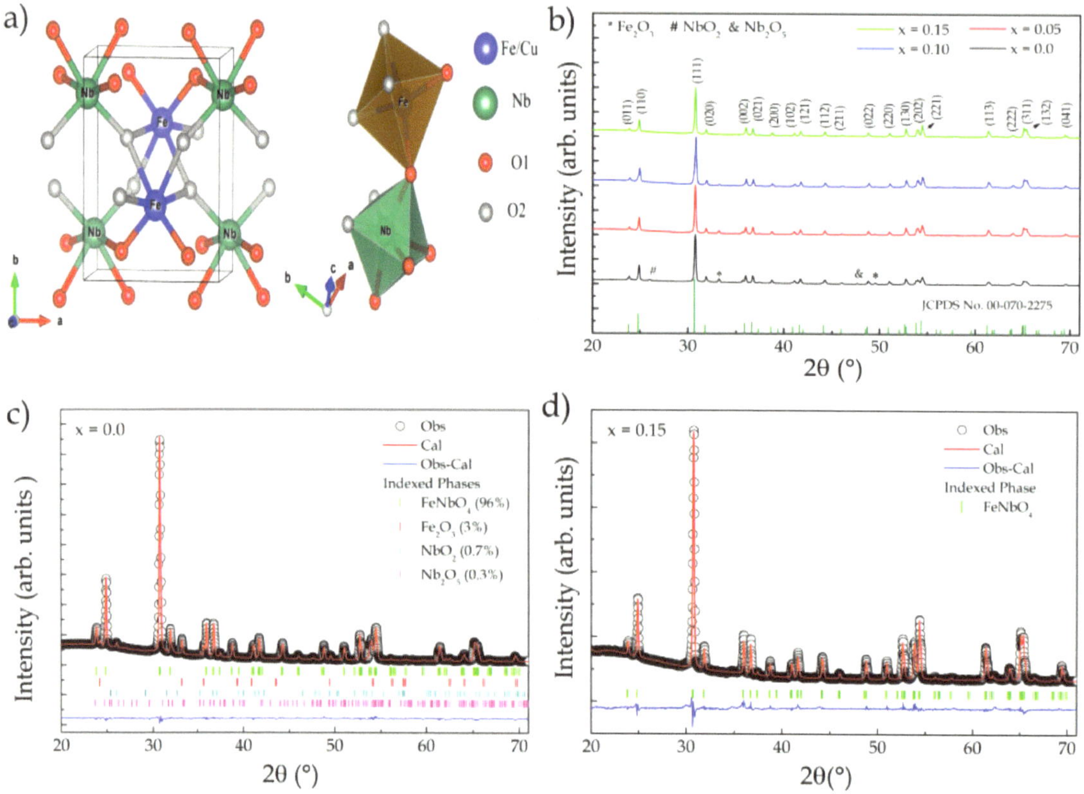

Figure 1. (**a**) Structural representation of $FeNbO_4$; (**b**) XRD standard for $Cu_xFe_{1-x}NbO_4$ samples (x = 0–0.15); (**c**,**d**) adjustment profiles for x = 0 and 0.15, respectively (Rietveld refinement).

The o-$FeNbO_4$ phase presents different magnetic behavior to the monoclinic phase. Cationic disorder disrupts the super-exchange, intra-chain, and inter-chain pathways, resulting in the absence of any magnetic ordering up to 4.2 K [1]. Lakshminarasimhan et al. compared the magnetic properties of both phases (m- and o-$FeNbO_4$) and observed that cationic disorder in the orthorhombic structure could lead to magnetic frustration and thus induce spin-glass behavior, as well as spin-glass characteristics, such as memory effects [15].

The antiferromagnetic–ferromagnetic transition is a phenomenon that has been reported for several structures, such as FeRh [16], $CeFe_2$ [17], ZnO [18,19], $LaFeO_3$ perovskite-like structures [20], $LaMnO_3$ [21,22], $LaMn_{0.5}Ni_{0.5}O_3$ [23], $YTiO_3$ [24], and $LaCoO_3$ [25]. Several mechanisms have been suggested to explain the nature of the ferromagnetic state of these compounds: super-exchange, based on localized electronic interactions through an oxygen ion; double exchange via charge transfer; and itinerant electronic ferromagnetism [25].

Replacement at the lattice site by ions with different ionic radii is an effective way to modulate the crystal structure and magnetic ordering, which results in a reduction in the AFM and an increase in the FM of $La_{1-x}A_xFeO_3$ (A = Bi, Al, Zn, Ce, etc.) [20]. Another way to modulate the magnetic ordering is substitution with non-magnetic acceptor ions, whereby the magnetic transition arises from super-exchange interactions between mixed-valence magnetic ions, as reported for the compound $La_{1-x}Sr_xMn_{0.5}Ni_{0.5}O_3$ [23]. Oxygen-mediated super-exchange interactions also play an essential role in magnetic ordering properties. Oxygen deficiency can change the magnetic ordering in structures such as $LaCoO_3$ [25] and $LaMnO_3$ [22]. One of the most efficient methods for changing the magnetic ordering is magnetic ion doping. Co-doping with the magnetic ions Cu and Fe is reported to be the main cause of the emergence of ferromagnetism in ZnO [18,19].

Ferromagnetic semiconductor materials have attracted much interest due to their potential application as spin-polarized carriers and their easy integration in semiconductor devices [19].

In this work, we evaluated the influence of the substitution of Fe^{3+} ions with Cu^{2+} ions on the structural and magnetic properties of ordered-structure $FeNbO_4$. Our results showed that copper insertion induced an antiferromagnetic–ferromagnetic transition, which may have been associated with the increase in ferromagnetic interactions caused by the rise in super/exchange interactions between Cu^{2+}-O-Fe^{3+} ions and the contribution of defects such as oxygen vacancies.

2. Materials and Methods

$Fe_{1-x}Cu_xNbO_4$ samples (x = 0–0.15) were prepared by the modified solid-state reaction method. The compounds iron III nitrate ($Fe(NO_3)_3 \cdot 9H_2O$), copper oxide (CuO), and niobium pentoxide (Nb_2O_5) were calculated stoichiometrically and dissolved in ethanol. The product was dried at 80 °C for 1 h in a mini-muffle furnace and sintered at 1050 °C for 16 h in an ambient atmosphere.

Structural characterization was carried out by X-ray diffraction (XRD) performed on a Miniflex instrument (Rigaku, Tokyo, Japan) using Cu-Kα radiation (λ = 1.54 Å) in the range of 10–80° with a 0.01° step and a counting time step of 2 s. The lattice parameters of the crystal structures were refined using the Rietveld method with GSAS software.

Raman spectroscopy was carried out with a T64000 spectrometer (HORIBA-Jobin Yvon, Kyoto, Japan) using a 100 nW laser (λ = 514.5 nm) as an excitation source, with 2 cm^{-1} of spectral resolution. The temperature-dependent magnetization (field-cooled (FC) and zero-field-cooled (ZFC)) and magnetization (M) versus applied magnetic field (H) curves were obtained at a low temperature and ambient temperature using a physical property measurement system (PPMS) (Quantum Design, Les Ulis, France).

3. Results and Discussion

3.1. Structural Analysis

The samples were first characterized by XRD at room temperature. The resulting patterns are shown in Figure 1b. The analysis revealed that the samples presented a monoclinic phase, belonging to the P2/c space group (No. 13), indexed through the ICDD standard (00-070-2275), with the reflection planes (001) and (110) observed at 19.30° and 23.6° being characteristic of this type of structure [15]. Impurities were also observed, which were indexed with the following phases: Nb_2O_5 (ICSD-017027), NbO_2 (ISCD-035181), and Fe_2O_3 (ICSD-082903). In Figure 1c,d, the patterns fitted by Rietveld refinement to x = 0 and 0.15 are presented, respectively.

The values obtained for the refinement quality factors R_{wp}, R_p, and χ^2 indicated that the refined results were reliable. The results obtained for all samples are listed in Table 1. The insertion of Cu^{2+} ions did not promote significant changes in the lattice parameters of the structure (see Table 1). However, it did cause local changes in the octahedral sites FeO_6 and NbO_6 (see Figure 1a) regarding the bond length and angle at each site.

Table 1. Structural parameters and quality factors obtained by refinement.

	Samples	0.00	0.05	0.10	0.15
Lattice parameters	a (Å)	4.650(3)	4.651(6)	4.651(5)	4.652(9)
	b (Å)	5.616(8)	5.619(0)	5.620(2)	5.623(7)
	c (Å)	4.996(3)	4.99(0)	4.997(5)	4.999(5)
	$\alpha(^0)$	90	90	90	90
	$\beta(^0)$	90.181(4)	90.130(4)	90.120(9)	90.079(4)
	$\gamma(^0)$	90	90	90	90
	V(Å^3)	130.5	130.42	130.64	130.81
Bond length	Fe-O1 × 2(Å)	1.93	1.95	2.03	1.9
	Fe-O2 × 2(Å)	2.04	2.12	2.18	2.1
	Fe-O2 × 2(Å)	2.07	2.02	2.04	2.11
	Nb-O1 × 2(Å)	2.02	1.87	1.96	1.84
	Nb-O1 × 2(Å)	2.2	2.16	2.05	1.98
	Nb-O2 × 2(Å)	1.86	1.87	1.83	1.84
Bond angle	Fe-O-Fe(^{0})	98.84	99.09	97.5	96.94
	Nb-O-Fe(^{0})	125.19	127.95	123.43	129.03
Quality factors	χ^2	7.034	2.61	6.95	4.50
	R_{wp} (%)	3.21	2.29	3.82	3.13
	R_p (%)	2.20	5.5	3.13	8.00

The term χ^2 in Table 1 indicates that this bond was repeated in the octahedron (see Figure 2b). A reduction in the Nb-O bond length was observed, while the Fe-O bond length increased because of the substitution. This was expected due to the small difference in the ionic radius of the substituent (Fe^{3+} = 0.064 nm and Cu^{2+} = 0.073 nm), since the lattice parameters did not undergo significant changes with doping, providing overall structural stability. The absence of the CuO phase may indicate that Cu^{2+} ions replaced Fe^{3+} ions at site A, which was confirmed by Rietveld refinement with the respective occupancy fractions 4, 8.5 and 14% (x = 0.05, 0.10, and 0.15). In addition, the majority phase of m-$FeNbO_4$ was quantified as 96, 97, 98, and 100% for x = 0, 0.05, 0.10, and 0.15, respectively.

3.2. Raman Spectroscopy

According to group theory, the calculations for the space group P2/c (No. 13), with point symmetry C_{2h}, showed that the 3N degrees of freedom for the 12 atoms in each primitive cell were divided into 36 modes of vibrations, as follows:

$$\Gamma_{vibration} = 8Ag + 10Bg + 8Au + 10Bu, \quad (1)$$

where all even vibrations (g) are Raman active and all odd vibrations (u) are IR active [26–29]. Of the 18 possible active Raman vibration modes, at least 16 distinct vibration bands were identified in the spectra shown in Figure 2a. Prominent vibration frequencies were located at 110, 135, 150, 175, 213, 226, 274, 303, 321, 362, 392, 416, 467, 498, 597, and 818 cm^{-1}. The 622 cm^{-1} band observed for x = 0 was attributed to the vibration modes associated with the secondary phases of NbO_2 and Nb_2O_5 [30], previously confirmed by XRD (Figure 1b). Based on the calculations of structures belonging to the same point group, it was possible to identify the attributions of each vibration mode [27,31] (see Table 2).

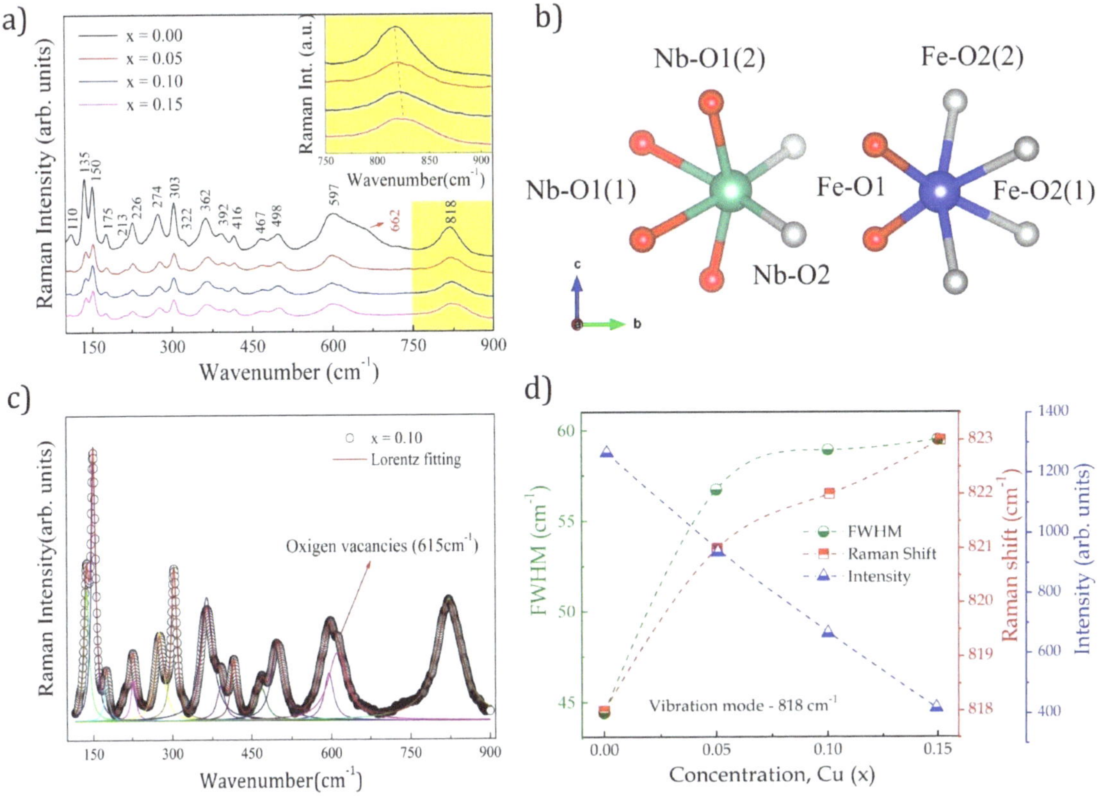

Figure 2. Raman $Fe_{1-x}Cu_xNbO_4$ spectroscopy (x = 0.0-0.15). (**a**) inset shows the peak magnification at 818 cm^{-1} evidencing the Raman shift as a function of the Cu insertion; (**b**) schematic representation of NbO_6 and FeO_6 octahedral symmetry and their respective connections; (**c**) fit profile calculated for x = 0.10, the colored lines show the deconvolution of peaks; (**d**) FWHM, intensity, and Raman shift due to Cu substitution.

All 16 bands detected were attributed to the vibrations of the $FeNbO_4$ monoclinic phase base units [4,8,32]. Low-frequency modes below 303 cm^{-1} are vibration modes that originate from network vibrations [32]. The most intense peaks identified in the regions of higher frequencies, 362, 597, and 818 cm^{-1}, were associated with the frequencies of the internal vibration modes of the NbO_6 octahedrons corresponding to the balance symmetries and stretching of the Nb-O bonds [32].

The strong interactions between ions caused sharper and more intense Raman modes [33]. Therefore, the most intense modes at 818, 597, and 362 cm^{-1} suggested the existence of strong interactions in the NbO_6 octahedron.

The Raman spectra were fitted using the peak profile of the Lorentz distribution to confirm the effect of Cu^{2+} substitution on the vibrational characteristics of the Nb-O bonds; the results are shown in Figure 2c. The broad peak near 600 cm^{-1} was indicative of overlapping vibration modes, so we set it to three modes. The mode at 597 cm^{-1} was attributed to the m-$FeNbO_4$ phase, that at 615 cm^{-1} was associated with the contribution of oxygen vacancies (Vo) [34], and that at 662 cm^{-1} corresponded with the vibration modes of Nb_2O_5 impurities. The Raman shift, width at half height, and the intensity of the Nb-O symmetric stretching modes at 818 cm^{-1} are shown in Figure 2d.

Table 2. Active Raman modes and assignments.

Band No.	Symmetry	ω_{obs} (cm^{-1})				Assignments
		0	0.05	0.10	0.15	
1	Bg	110				Rotational mode of NbO_6 around an axis perpendicular to the b-axis
2	Ag	135	137	137	137	Rotational mode of NbO_4^{3-} around the b-axis
3	Bg	150	151	150	150	Rotational mode of NbO_4^{3-} around an axis perpendicular to the b-axis
4	Bg	175	176	175	175	Translational mode parallel to the ac-plane
5	Ag	213	212	212	212	Translational mode along the b-axis
6	Bg	226	227	226	225	Translational mode parallel to the ac-plane
7	Ag	274	277	276	276	Translational mode along the b-axis
8	Bg	303	304	303	303	Translational mode parallel to the ac-plane
9	Ag	322	322	321	321	Scissors mode of NbO_4^{3-}
10	Ag	362	366	364	364	Scissors mode of NbO_4^{3-}
11	Bg	392	394	391	392	Translational mode parallel to the ac-plane
12	Bg	416	415	415	415	Rocking mode of NbO_4^{3-}
13	Bg	467	468	468	468	Rocking mode of NbO_4^{3-}
14	Ag	498	500	498	500	Twist mode of NbO_4^{3-} around the b-axis
15	Bg	597	598	597	601	Stretching mode of NbO_4^{3-}
16	Ag	818	821	822	823	Stretching mode of NbO_4^{3-}

The increase in the Raman shift due to the rise in doping was attributed to the shrinkage of the Nb-O bonds [4]. The reduction in the NbO_6 octahedron could be attributed to the insertion of ions with a smaller ionic radius in the A site, resulting in a smaller bond length (see Table 1). Additionally, a decrease in intensity and an increase in FWHM values were observed in the main vibration modes at 362, 597, and 818 cm^{-1}, which implied that the degradation of the ordered structure and the disordered arrangement of the A-site ions were the causes of the Raman spectrum broadening [33]. Thus, the increase in the substitution of Fe^{3+} ions by Cu^{2+}ions in the ordered structure increased the cationic disorder, which was directly reflected in the magnetic ordering of the structure.

3.3. Magnetic Properties

The compound $FeNbO_4$, in the monoclinic phase, is reported in the literature to have a structure with a liquid antiferromagnetic order due to the FM and AFM competitions associated with the interactions between the octahedral Fe-O-Fe chains and the crossed Fe-O-chains. Nb-O-Fe [1,14,15]. The results obtained through magnetic characterizations, the magnetization measurements as a function of temperature (FC and ZFC), and the field-dependent magnetization are shown in Figure 3a, which confirmed the antiferromagnetic nature with a Néel temperature of approximately 36.81K. The magnetic transition temperature (TN) was defined by the derivative of the product of susceptibility with temperature ($d(\chi T)/dT$) and the derivative of susceptibility concerning temperature ($d\chi/dT$), see Figure 3b. Generally, a peak in $d\chi/dT$ represents a PM (paramagnetic) to FM (ferromagnetic) transition region. For PM to AFM transitions, it has been shown theoretically and experimentally that the peak accurately determines TN in $d(\chi T)/dT$ since χT is proportional to the magnetic energy near TN in an antiferromagnet. Thus, the peak in $d(\chi T)/dT$ corresponds to the peak in specific heat [35]. In our case, the peaks of both derivatives were in the same region (36.81K), confirming the magnetic transition region.

The ZFC and FC curves showed similar behavior, with a smooth drop below TN, which may have been associated with weak AFM interactions in the structure [15] (Figure 3a). We also observed an increase in magnetization below 29K, which could be attributed to short-range spin interactions with a small fraction of Fe^{3+} in the niobium chains [15], as well as the contribution of α-Fe_2O_3 secondary-phase ferromagnetic interactions. Figure 3c provides the $M \times H$ curves measured for $x = 0$ at 5 and 300K, respectively. In the insert of Figure 3c, one can observe a small gap in the hysteresis curve for the M vs. H curve measured at 300K, with respective coercive field (Hc) and remanent magnetization (Mr) values of

Hc = 201.5 Oe and Mr = 0.156 emu/g, which could be attributed to the contributions of the impurities present in the sample.

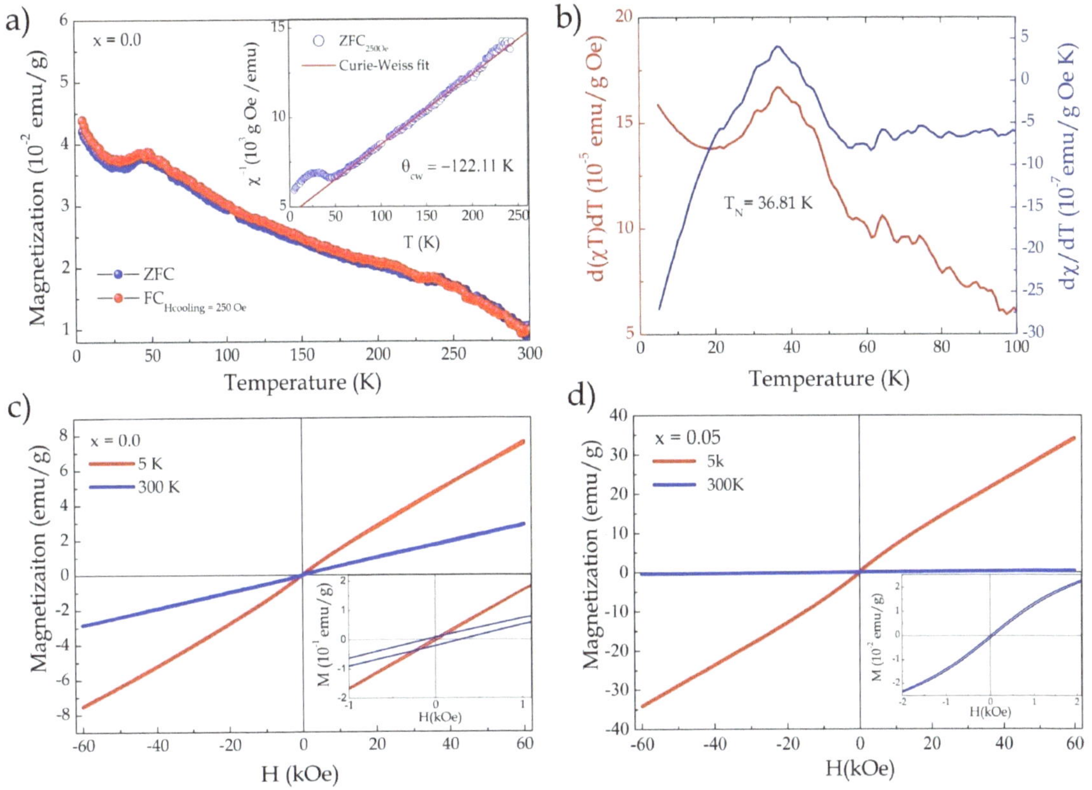

Figure 3. (**a**) Magnetization as a function of temperature (FC and ZFC) with an applied field of 250 Oe (FC)—inset shows the inverse of susceptibility as a function of temperature (χ^{-1}/T) adjusted by the Curie–Weiss law; (**b**) derivative of the product of susceptibility with temperature ($d(\chi T)/dT$) and derivative of magnetic susceptibility ($d\chi/dT$) as a function of temperature for x = 0.0, showing the Néel transition temperature; (**c**) magnetization curves as a function of applied field (M–H) measured at 5 and 300 K for x = 0.0; (**d**) M–H curves measured at 5 and 300 K for x = 0.05.

The inverse magnetic susceptibility data, obtained from the curve in ZFC mode, were fitted, as shown in the inset of Figure 3a. Above 60K, they displayed paramagnetic behavior according to the Curie–Weiss law ($\chi = C/(T - \theta)$), where C is the Curie–Weiss constant, θ_{cw} is the Curie–Weiss temperature, and T is the absolute temperature. The values obtained were C = 2.56 emu·K/g Oe and $\theta_{cw} = -122$ K, respectively. The negative θ_{cw} value confirmed the antiferromagnetic nature. The effective magnetic moment was calculated using the Curie constant according tothe following relationship:

$$\mu_{eff}^{exp} = \sqrt{8C}\mu_B, \tag{2}$$

where μ_B is the Bohr magneton [15]. The effective magnetic moment obtained was 4.53 μ_B for Fe^{3+}, which agreed with the values reported in previous works [11,15].

The influence of Cu^{2+} insertion on the magnetic properties of $FeNbO_4$ was evaluated as follows. Figure 3d shows the magnetization versus temperature curves (M–H) measured at 5 and 300K for x = 0.05. The insertion of 5% Cu^{2+} promoted a significant increase in magnetization and changed the behavior of the curves for low fields (see the inset in

Figure 3d). This result indicated that the change in magnetic ordering in the structure began as a result of replacement. Figure 4a shows the M–H curves as a function of the increase in Cu^{2+} insertion, where one can observe a significant change in the curve's behavior as a result of substitution, i.e., the magnetic transition induced by the insertion of Cu^{2+} ions into the $FeNbO_4$ structure. The curves for x = 0.10 and 0.15 showed weak ferromagnetic behavior, with respective values of Hc = 85 Oe, Mr = 0.25 emu/g and Hc = 41 Oe, Mr = 0.19 emu/g (inset in Figure 4a). The M–H curves measured at 5K are shown in Figure 4b, where an increase in magnetization and the coercive field and remanent magnetization values can be observed: Hc = 140 Oe, Mr = 0.54 emu/g and Hc = 73.3 Oe, Mr = 0.39emu/g for x = 0.10 and 0.15, respectively (inset in Figure 4b).

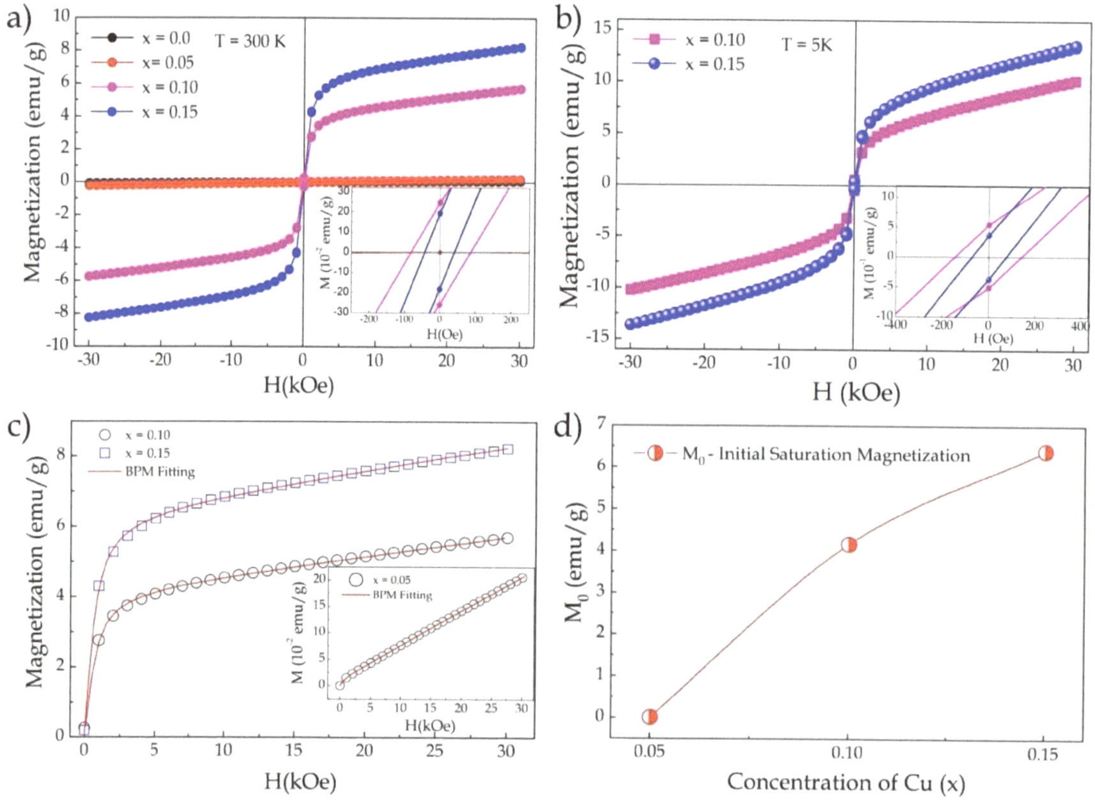

Figure 4. (**a**) Hysteresis curves (M–H) at room temperature for x = 0–0.15; (**b**) M–H curves measured at 5 K for x = 0.10 and 0.15; (**c**) adjustments of the M–H curves at 300K for the BMP model for x = 0.05, 0.10, and 0.15; (**d**) initial saturation magnetization (M_0) as a function of Cu^{2+} substitution extracted from the fitted model.

To try to understand the main mechanisms responsible for the origin of magnetism in the doped $FeNbO_4$ structure, the bound magnetic polaron (BMP) model was proposed:

$$M = M_0 L(x) + \chi_m H, \quad (3)$$

where the first term represents the contribution of the BMP, and the second term represents the contribution of the paramagnetic matrix. The value of M_0 is equal to $N_p x m_s$, where N_p is the density number of BMPs involved per cm^3, and m_s is the effective spontaneous magnetic moment per BMP. L(x) is the Langevin function defined as L(x) = coth(x)−1/x, where $x = \mu_{eff}H/k_BT$. μ_{eff} is the true spontaneous moment per BMP at room temperature approximated by μ_{eff}~m_s [36]. μ_{eff} can be expressed in Bohr magneton units, see Table 3. Figure 4c shows the fit of the M–H curves for the BMP model, where one can see that the experimental data perfectly fit the proposed model, and the theoretical parameters extracted can be seen in Table 3.

Table 3. Parameters obtained by fitting to the BMP model.

Samples (x)	M_0 (emu/g)	χ_m (emu/g Oe)	m_s (emu)	N_p (cm^{-3})	μ_{eff} (μB)
0.05	0.01139 (±0.00005)	0.00204 (±0.00008)	1.32×10^{-23}	8.62×10^{20}	1.42
0.10	4.166 (±0.03)	0.0196 (±0.0009)	1.11×10^{-23}	3.74×10^{23}	1.20
0.15	6.40 (±0.05)	0.0233 (±0.001)	1.14×10^{-23}	5.61×10^{23}	1.23

The theoretical BMP model affirmed that the presence of magnetic carriers and defects such as oxygen vacancies were the main factors responsible for the magnetism of defects induced by carriers, which formed magnetic polarons. Therefore, substitution with a lower oxidation state could increase the generation of oxygen vacancies that behave like polarons, altering the material's magnetic properties [36].

The increased insertion of Cu^{2+} ions caused a magnetic transition from antiferromagnetic to weak ferromagnetic. The increase in the initial saturation magnetization as a result of substitution can be seen in Figure 4d. Our results showed that the origin of magnetism in the $Fe_{1-x}Cu_xNbO_4$ may have been associated with the contribution of super-exchange interactions between Cu^{2+}-O-Fe^{3+} as well as induced defects such as oxygen vacancies that behaved as bound magnetic polarons. Figure 5 shows an illustrative scheme that helps elucidate the change in magnetic ordering and the contribution of each interaction. Figure 5a shows a net AFM configuration in the structure as already reported in the literature for m-$FeNbO_4$ [10]. This configuration contains FM interactions associated with Fe-O-Fe interactions and AFM interactions associated with Fe chains mediated by Nb, resulting in a liquid AFM order. In Figure 5b, an FM configuration for x = 0.10 is provided as an example. The yellow underlined circle shows a possible FM interaction for Cu^{2+}-O-Nb-O-Fe^{3+}. The theoretical magnetic moment of copper (μ_{Cu} = 1.73μ_B) is much smaller than that of iron (μ_{Fe} = 5.9μ_B), so the substitution could promote this configuration state. However, these interactions alone may not be enough to encourage the transition or explain our results. In this way, the polaron model was proposed and helped us to understand the contribution of defects to the magnetic properties of our samples.

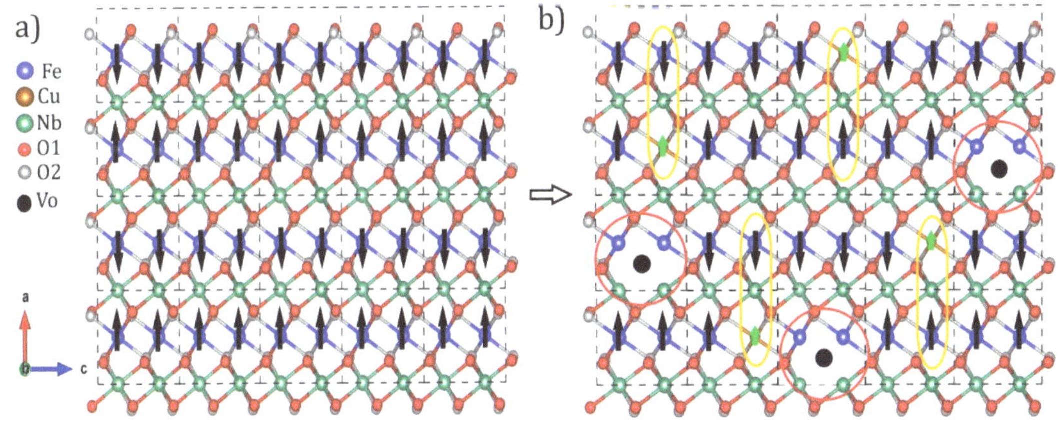

Figure 5. Representative illustration of the magnetic ordering configurations: (**a**) AFM and (**b**) FM.

In Figure 5b, the area highlighted in red shows the regions of possible magnetic polaron formation due to oxygen vacancies. The center of a polaron is occupied by a defect, such as an oxygen vacancy, which would be responsible for aligning the spins of the closest elements. Atoms that are not close enough to an oxygen vacancy do not align and may even form an AFM configuration [37].

The fit data can show us the density of polarons (Np) formed, which was in the order of 10^{20}–10^{23} and increased as more polarons formed (see Table 3); this order of magnitude was sufficient to make significant contributions to the magnetic properties [38]. Each polaron formed could contribute an effective polaron magnetic moment (μ_p) of approximately 1.2 μ_B. Thus, we could determine that the AFM/FM transition was the origin of FM interactions between Cu^{2+}-O-Fe^{3+} ions, with a contribution from defects.

4. Conclusions

We prepared samples of $Fe_{1-x}Cu_xNbO_4$ (x = 0–0.15) by the conventional solid-state reaction method. Structure analysis by XRD and Raman spectroscopy confirmed the monoclinic phase of $FeNbO_4$ and the presence of impurities (Fe_2O_3, NbO_2, and Nb_2O_5). The refinement experiments confirmed the substitution of Cu^{2+} in the Fe site and quantified the obtained phase fractions. The increase in the insertion of Cu^{2+} did not cause significant changes in the network parameters. However, it promoted local changes in each network site. The Raman spectroscopy results complemented and confirmed the XRD results. The results confirmed the monoclinic phase and the local changes promoted by the insertion of Cu^{2+} (oxygen vacancies). Studies of the magnetic properties confirmed the AFM nature of the pure sample, as well as evidence of a magnetic transition caused by the insertion of Cu^{2+} into the structure. The insertion of ions in the lower oxidation state induced the generation of oxygen vacancies that behaved like magnetic polarons and contributed to the change in magnetic ordering. Adjustments to the bound magnetic polaron model confirmed the contribution of structural defects to the magnetic properties. The AFM to FM transition could be associated with exchange interactions between Cu^{2+}-O-Fe^{3+} ions and the contribution of defects such as oxygen vacancies.

Author Contributions: D.S.E. and M.A.M.—conceptualization, methodology, original draft preparation, and data analysis; R.F.J., J.M.S., R.B.S. and G.D.S.—data collection; N.S.F., R.S.M. and M.S.—writing (review and editing). All authors have read and agreed to the published version of the manuscript.

Funding: This research was partially funded by the Brazilian funding agencies Conselho Nacional de Desenvolvimento Científico e Tecnológico (CNPq), grants number 147782/2019-8, 309054/2019-2, 309054/2019-2 and 141486/2021-0, and Coordenacão de Aperfeiçoamento de Pessoal de Nível Superior (CAPES), finance code 001.

Institutional Review Board Statement: Not applicable.

Informed Consent Statement: Not applicable.

Data Availability Statement: Not applicable.

Acknowledgments: G.D. Saraiva acknowledges the support from FUNCAP/Edital N 03/2019 (process DEP-0164- 00350.01.00/19); MCTI/CNPQ/ Universal 28/2018 (grant number 426995/2018-0); and MCTI/CNPQ PQ—09/2020 (grant number 311898/2020-3).

Conflicts of Interest: The authors declare no conflict of interest.

References

1. Harrison, W.; Cheetham, A. Structural and Magnetic Properties of $FeNbO_4$-II. *Mater. Res. Bull.* **1989**, *24*, 523–527. [CrossRef]
2. Ahmed, S.H.; Bakiro, M.; Alzamly, A. Effect of PH and Surfactant on Band Gap Properties of Monoclinic $FeNbO_4$ Prepared via Different Routes. *Nano-Struct. Nano-Objects* **2019**, *20*, 100400. [CrossRef]
3. Balamurugan, C.; Maheswari, A.R.; Lee, D.W.; Subramania, A. Selective Ethanol Gas Sensing Behavior of Mesoporous N-Type Semiconducting $FeNbO_4$ Nanopowder Obtained by Niobium–Citrate Process. *Curr. Appl. Phys.* **2014**, *14*, 439–446. [CrossRef]
4. Cho, I.-S.; Lee, S.; Noh, J.H.; Choi, G.K.; Jung, H.S.; Kim, D.W.; Hong, K.S. Visible-Light-Induced Photocatalytic Activity in $FeNbO_4$ Nanoparticles. *J. Phys. Chem. C* **2008**, *112*, 18393–18398. [CrossRef]

5. Jones, B.M.F.; Maruthamani, D.; Muthuraj, V. Construction of Novel N-Type Semiconductor Anchor on 2D Honey Comb like $FeNbO_4$/RGO for Visible Light Drive Photocatalytic Degradation of Norfloxacin. *J. Photochem. Photobiol. A Chem.* **2020**, *400*, 112712. [CrossRef]
6. Devesa, S.; Graça, M.P.; Henry, F.; Costa, L.C. Dielectric Properties of $FeNbO_4$ Ceramics Prepared by the Sol-Gel Method. *Solid State Sci.* **2016**, *61*, 44–50. [CrossRef]
7. Wang, T.; Shi, S.; Kong, F.; Yang, G.; Qian, B.; Yin, F. The Role of Stable Interface in Nano-Sized $FeNbO_4$ as Anode Electrode for Lithium-Ion Batteries. *Electrochim. Acta.* **2016**, *203*, 206–212. [CrossRef]
8. Sri Devi Kumari, T.; Vinith Gandhi, R.; Rahul, G.; Kamalanathan, G.; Prem Kumar, T.; Jeyakumar, D.; Lakshminarasimhan, N. Electrochemical Lithium Insertion Behavior of $FeNbO_4$: Structural Relations and in Situ Conversion into $FeNb_2O_6$ during Carbon Coating. *Mater. Chem. Phys.* **2014**, *145*, 425–433. [CrossRef]
9. Liu, X.; Xie, D.; Irvine, J.T.S.; Ni, J.; Ni, C. An $FeNbO_4$-Based Oxide Anode for a Solid Oxide Fuel Cell (SOFC). *Electrochim. Acta.* **2020**, *335*, 135692. [CrossRef]
10. Wang, X.; Santos-Carballal, D.; de Leeuw, N.H. Density Functional Theory Study of Monoclinic $FeNbO_4$: Bulk Properties and Water Dissociation at the (010), (011), (110), and (111) Surfaces. *J. Phys. Chem. C.* **2021**, *125*, 27566–27577. [CrossRef]
11. Liu, C.B.; Chen, R.; Yue, X.Y.; Liu, Y.J.; Shi, M.M.; Zhu, H.P.; Dong, C.; Liu, Y.; Han, Y.B.; Wang, J.F.; et al. Crystal Growth, Magnetic Property and Phase Transition of the Zigzag-Chain Antiferromagnet $FeNbO_4$. *J. Magn. Magn. Mater.* **2018**, *464*, 108–111. [CrossRef]
12. Roth, R.S.; Waring, J.L. Ixiolite and Other Polymorphic Types of $FeNbO_4$. *Am. Mineral.* **1964**, *49*, 242–246.
13. Theissmann, R.; Ehrenberg, H.; Weitzel, H.; Fuess, H. Domain Structure and Lattice Strains in $FeNbO_4$. *Solid State Sci.* **2005**, *7*, 791–795. [CrossRef]
14. Ehrenberg, H.; Wltschek, G.; Theissmann, R.; Weitzel, H.; Fuess, H.; Trouw, F. The Magnetic Structure of $FeNbO_4$. *J. Magn. Magn. Mater.* **2000**, *218*, 261–265. [CrossRef]
15. Lakshminarasimhan, N.; Kumar, A.K.N.; Chandrasekaran, S.S.; Murugan, P. Structure-Magnetic Property Relations in $FeNbO_4$ Polymorphs: A Spin Glass Perspective. *Prog. Solid State Chem.* **2019**, *54*, 20–30. [CrossRef]
16. Li, G.; Medapalli, R.; Mentink, J.H.; Mikhaylovskiy, R.V.; Blank, T.G.H.; Patel, S.K.K.; Zvezdin, A.K.; Rasing, T.; Fullerton, E.E.; Kimel, A.V. Ultrafast Kinetics of the Antiferromagnetic-Ferromagnetic Phase Transition in FeRh. *Nat. Commun.* **2022**, *13*, 2998. [CrossRef]
17. Sokhey, K.J.S.; Chattopadhyay, M.K.; Nigam, A.K.; Roy, S.B.; Chaddah, P. Signatures of Phase Separation across the Disorder Broadened First Order Ferromagnetic to Antiferromagnetic Transition in Doped-$CeFe_2$ Alloys. *Solid State Commun.* **2004**, *129*, 19–23. [CrossRef]
18. Han, S.-J.; Song, J.W.; Yang, C.-H.; Park, S.H.; Park, J.-H.; Jeong, Y.H.; Rhie, K.W. A Key to Room-Temperature Ferromagnetism in Fe-Doped ZnO: Cu. *Appl. Phys. Lett.* **2002**, *81*, 4212–4214. [CrossRef]
19. Shim, J.H.; Hwang, T.; Lee, S.; Park, J.H.; Han, S.-J.; Jeong, Y.H. Origin of Ferromagnetism in Fe- and Cu-Codoped ZnO. *Appl. Phys. Lett.* **2005**, *86*, 082503. [CrossRef]
20. Yao, Q.; Tian, C.; Lu, Z.; Wang, J.; Zhou, H.; Rao, G. Antiferromagnetic-Ferromagnetic Transition in Bi-Doped LaFeO3 Nanocrystalline Ceramics. *Ceram. Int.* **2020**, *46*, 20472–20476. [CrossRef]
21. Pavlov, V.I.; Bychkov, G.L.; Bogush, A.K. Concentration Magnetic Phase Transitions in the ($La_{1-x}CaxMnO_3$ System. *Cryst. Res. Technol.* **1986**, *21*, 487–494. [CrossRef]
22. Trukhanov, S.V.; Lobanovski, L.S.; Bushinsky, M.V.; Troyanchuk, I.O.; Szymczak, H. Magnetic Phase Transitions in the Anion-Deficient $La_{1-x}Ba_xMnO_{3-x/2}$ ($0 \leq x \leq 0.50$) Manganites. *J. Phys. Condens. Matter.* **2003**, *15*, 1783–1795. [CrossRef]
23. Troyanchuk, I.O.; Karpinsky, D.V.; Bushinsky, M.V.; Sirenko, V.A.; Sikolenko, V.V.; Franz, A. Antiferromagnet-Ferromagnet Transition in $La_{1-x}Sr_xMn_{0.5}Ni_{0.5}O_3$ ($0 \leq x \leq 0.2$) Ceramics. *Low Temp. Phys.* **2017**, *43*, 982–985. [CrossRef]
24. Hameed, S.; El-Khatib, S.; Olson, K.P.; Yu, B.; Williams, T.J.; Hong, T.; Sheng, Q.; Yamakawa, K.; Zang, J.; Uemura, Y.J.; et al. Nature of the Ferromagnetic-Antiferromagnetic Transition in $Y_{1-x}La_xTiO_3$. *Phys. Rev. B.* **2021**, *104*, 024410. [CrossRef]
25. Troyanchuk, I.O.; Bushinsky, M.V.; Karpinsky, D.V.; Sirenko, V.A. Antiferromagnet-Ferromagnet Transitions in Cobaltites. *Low Temp. Phys.* **2012**, *38*, 662–668. [CrossRef]
26. Liu, Y.; Wang, H.; Chen, G.; Zhou, Y.D.; Gu, B.Y.; Hu, B.Q. Analysis of Raman Spectra of $ZnWO_4$ Single Crystals. *J. Appl. Phys.* **1988**, *64*, 4651–4653. [CrossRef]
27. Ishii, K.; Morita, N.; Nakayama, H.; Tsunekawa, S.; Fukuda, T. Raman Spectra of $LaNbO_4$ in the Ferroelastic Phase and the Relaxation after the State Shift. *Phys. Status Solidi.* **1989**, *112*, 207–214. [CrossRef]
28. Basiev, T.T.; Karasik, A.Y.; Sobol, A.A.; Chunaev, D.S.; Shukshin, V.E. Spontaneous and Stimulated Raman Scattering in $ZnWO_4$ Crystals. *Quantum Electron.* **2011**, *41*, 370–372. [CrossRef]
29. Ruiz-Fuertes, J.; Errandonea, D.; López-Moreno, S.; González, J.; Gomis, O.; Vilaplana, R.; Manjón, F.J.; Muñoz, A.; Rodríguez-Hernández, P.; Friedrich, A.; et al. High-Pressure Raman Spectroscopy and Lattice-Dynamics Calculations on Scintillating $MgWO_4$: Comparison with Isomorphic Compounds. *Phys. Rev. B.* **2011**, *83*, 214112. [CrossRef]
30. Carvalho, R.C.; Mendonça, M.E.V.; Tavares, M.S.; Moreira, E.; Azevedo, D.L. Optoelectronic and Thermodynamic Properties, Infrared and Raman Spectra of NbO_2 and Nb_2O_5 from DFT Formalism. *J. Phys. Chem. Solids.* **2022**, *163*, 110549. [CrossRef]
31. Blasse, G. Vibrational Spectra of Yttrium Niobate and Tantalate. *J. Solid State Chem.* **1973**, *7*, 169–171. [CrossRef]

32. Babu, R.; Kelkar, S.; Kashid, V.; Achary, S.N.; Salunke, H.G.; Gupta, N.M. Photophysical, Bandstructural, and Textural Properties of o-$FeNbO_4$ in Relation to Its Cocatalyst-Assisted Photoactivity for Water Oxidation. *RSC Adv.* **2014**, *4*, 33435–33445. [CrossRef]
33. Zhang, Q.; Xu, L.; Tang, X.; Huang, F.; Wu, X.; Li, Y.; Jing, Y.; Han, Z.-K.; Su, H. Electronic Structure, Raman Spectra, and Microwave Dielectric Properties of Co-Substituted $ZnWO_4$ Ceramics. *J. Alloys Compd.* **2021**, *874*, 159928. [CrossRef]
34. Góis, M.M.; Valença, E.; Machado, R.; Lopez, E.O.; Mello, A.; Rodrigues, C.L.; Macêdo, M.A. Study of Filament Resistive Switching in New $Pt/Co_{0.2}TiO_{3.2}/ITO$ Devices for Application in Non-Volatile Memory. *Phys. Status Solidi* **2018**, *215*, 1800369. [CrossRef]
35. Thota, S.; Ghosh, S.; Maruthi, R.; Joshi, D.C.; Medwal, R.; Rajdeep, S.; Rawat, R.S.; Seehra, M.S. Magnetic ground state and exchange interactions in the Ising chain ferromagnet $CoNb_2O_6$. *Phys. Rev. B.* **2021**, *103*, 064415. [CrossRef]
36. Bhardwaj, P.; Singh, J.; Kumar, R.; Kumar, R.; Verma, V. Structural, Optical and Magnetic Characterization of Ni^{2+} Ions Doped Chromium Oxide (Cr_2O_3) Nanoparticles. *Solid State Sci.* **2021**, *115*, 106581. [CrossRef]
37. Wolf, P.A.; Bhatt, R.N.; Durst, A. Polaron-polaron interactions in diluted magnetic semiconductors. *J. Appl. Phys.* **1966**, *79*, 5196–5198. [CrossRef]
38. Franco, A.; Pessoni, H.V.S.; Soares, M.P. Room Temperature Ferromagnetism in Eu-Doped ZnO Nanoparticulate Powders Prepared by Combustion Reaction Method. *J. Magn. Magn. Mater.* **2014**, *355*, 325–330. [CrossRef]

materials

Article

Stress Evaluation of Welded Joints with Metal Magnetic Memory Testing Based on Tension–Compression Fatigue Test

Huipeng Wang [1,2,*], Zhiwei Xu [1], Dongwei Cai [1], Lihong Dong [3,*], Guozheng Ma [3], Haidou Wang [3,4] and Bin Liu [5]

1 School of Mechanical and Electrical Engineering, Jiangxi University of Science and Technology, Ganzhou 341000, China; 6720210731@mail.jxust.edu.cn (Z.X.); 6720210730@mail.jxust.edu.cn (D.C.)
2 National United Engineering Laboratory for Advanced Bearing Tribology, Henan University of Science and Technology, Luoyang 471023, China
3 National Key Laboratory for Remanufacturing, Army Academy of Armored Forces, Beijing 100072, China; magz0929@163.com (G.M.); wanghaidou@tsinghua.org.cn (H.W.)
4 National Engineering Research Center for Remanufacturing, Army Academy of Armored Forces, Beijing 100072, China
5 School of Material Science and Engineering, Jiangsu University of Science and Technology, Zhenjiang 212003, China; liubindely@163.com
* Correspondence: wanghuipeng1983@126.com (H.W.); lihong.dong@126.com (L.D.)

Abstract: Metal magnetic memory testing (MMMT) is an effective nondestructive technique for fatigue damage monitoring of weldments because of its capacity for stress evaluation. An experimental investigation of the effect of the applied fatigue stress on MMMT signals, including the tangential component B_x and the normal component B_z, during tension–compression fatigue tests in welded joints was carried out systematically. The B_x and B_z signals at different fatigue cycles and fatigue stresses were collected and analyzed, and the results showed that there was a peak of B_x and abnormal peaks of B_z that existed at the welded joint before loading. After loading, the peak of B_x and the abnormal peaks of B_z reversed, and the B_x signals moved upward and the B_z signals rotated anticlockwise dramatically in the first few fatigue cycles. After the fatigue cycle number was larger than 1000, B_x and B_z were stable, with very little fluctuation. In addition, the characteristics of B_x signals, the mean value, and the peak value of the average of B_x had an extremely significant linear relationship with the applied fatigue stress during the stable stage of the fatigue test, which indicates that MMMT is a feasible method for fatigue stress evaluation and even residual fatigue life estimation for weldments in service.

Keywords: welded joint; tension-compression fatigue test; metal magnetic memory testing; characteristic extraction; stress evaluation

Citation: Wang, H.; Xu, Z.; Cai, D.; Dong, L.; Ma, G.; Wang, H.; Liu, B. Stress Evaluation of Welded Joints with Metal Magnetic Memory Testing Based on Tension–Compression Fatigue Test. *Materials* **2022**, *15*, 3103. https://doi.org/10.3390/ma15093103

Academic Editor: Dana Popescu

Received: 18 March 2022
Accepted: 21 April 2022
Published: 25 April 2022

1. Introduction

Metal magnetic memory testing (MMMT) can not only detect already-developed defects but can also evaluate stress concentration zones (SCZs), which can evolve into macroscopic defects under the effect of an external load on the basis of the magneto-mechanical effect [1,2]. As a result of the interaction between external stress and the geomagnetic field, magnetic domains in the SCZs are reoriented, a spontaneous magnetic field near the surface can form even when the load is released, and the characteristics of the spontaneous magnetic field can be used to identify the SCZs. Because of its prominent advantage in early damage diagnosis, along with passive magnetic field measurement, simple operation, high sensitivity, time-saving, etc., MMMT has attracted extensive attention since it was proposed by Doubov in the late 1990s [3,4], and many experimental studies focusing on the relationship between the MMMT signals of ferromagnetic materials and the stress and damage status have been carried out. Static tensile tests show that characteristics of MMMT signals can indicate plastic deformation [5,6], the stress concentration degree [7,8], and the length of backside cracks [9]. Furthermore, tension–tension fatigue tests [10,11], three-point bending fatigue tests [12,13], and four-point bending fatigue tests [14] show that the characteristics of MMMT signals increase linearly with increases in crack length. Moreover, many theories have also been presented to explore the foundation of MMMT. Magnetic dipole models [15–17], magneto-elastoplastic coupling models [18,19], the thermo-magneto-elastoplastic coupling model [20], density functional theory [21], and the modified Jiles–Atherton model [22] have been introduced to reveal the quantitative relationship between the MMMT signals and the stress and damage of ferromagnetic materials.

Welding is commonly used as an important mechanical structure connection method for equipment manufacturing in the aerospace, pressure vessel, and petroleum pipeline industries, among others [23,24]. Additionally, for safety assurance, welded structures are usually inspected by conventional nondestructive testing methods (visual testing, radiography testing, ultrasonic testing, magnetic particle testing, magnetic flux leakage testing, etc.) to detect the already-developed defects resulting from the welding process [25]. However, stress concentration introduced during the welding process, which cannot be detected by conventional NDT methods, plays a much more dangerous role in the service life of welded structures. During service, fatigue cracks usually initiate from the SCZs of weldment under the effect of an applied alternative load, ultimately causing the failure of the weld structure. Therefore, it is vital to inspect and monitor the fatigue stress of weldments.

MMMT has drawn much interest in the inspection and evaluation of weldment for its great advantages in SCZ detection. Additionally, it has been widely applicated in the engineering field, especially in weldment detection and evaluation [26–28], and qualitative and quantitative research has also been carried out by many researchers. Tensile and bending tests of butt welds show that MMMT can predict weld failure with early warnings [29], and three-point bending fatigue tests of weld joints show that the maximum gradient of the normal component of MMMT signals increases exponentially with an increase in crack length [30]. Experimental investigations also show that MMMT can detect buried welding cracks effectively [31].

However, it should be noted that most MMMT research and applications in weldment evaluation have focused on macroscopic defect measurement or crack propagation detection, and far fewer works have been concerned with early damage or stress evaluation. For these interests, the work presented here focused on the magnetic behavior of welded joints during fatigue tests, analyzing variations in both the tangential and the normal component of MMMT signals during the fatigue tests; the relationship between the characteristics of MMMT signals and fatigue stress is discussed, and an MMMT stress evaluation method for welded joints is proposed.

2. Experimental Procedure

2.1. Specimen Preparation

Butt welded joints with a base metal of Optim 900QC steel, welding wire of wire CARBOFIL NiMoCr, and shielding gas of CO_2 were investigated in this study. Optim 900QC steel, a Swiss high-strength structural steel with excellent weldability and bendability, is a low-alloy quenched and tempered steel that can be welded directly, and no additional tempering treatment is required after welding. Wire CARBOFIL NiMoCr is a kind of copper-plated solid wire that is suitable for gas-shielded welding of high-strength steel. The chemical compositions and mechanical properties of the base metal and welding wire are shown in Tables 1 and 2, respectively.

Table 1. Chemical compositions of the base metal and welding wire (in wt.%).

Brand	C	Si	Mn	P	S	Cr	Ni	Mo	Cu	Ti
Optim 900QC steel	0.091	0.27	1.8	0.009	0.002	1.11	0.08	0.152	-	-
CARBOFIL NiMoCr	0.096	0.41	1.34	0.0078	0.014	0.23	1.0	0.2	0.1	0.0021

Table 2. Mechanical properties of the base metal and welding wire.

Brand	Yield Strength/MPa	Ultimate Strength/MPa	Elongation/%
Optim 900QC steel	960	1107	11
CARBOFIL NiMoCr	≥690	900	≥17

Sheets of 900QC steel with a nominal thickness of 5 mm were investigated, and they were machined into a size of 150 mm × 180 mm parallel to the welding direction. The welded joints were fabricated by the CO_2 gas-shielded arc welding technique automatically, and the welding parameters are shown in Table 3. After welding, the fatigue specimens of the welded joints of 900QC steel were machined according to the Chinese industrial standard NB/T 47014-2011 welding procedure qualification for pressure equipment, and the sampling location and specimen configuration are shown in Figure 1.

Table 3. Welding parameters for 900QC steel.

Weld Seam	Current/A	Voltage/V	Gas Flow/L·min^{-1}
Backing weld	100~110	19~20	15
Capping weld	180~200	22~24	15

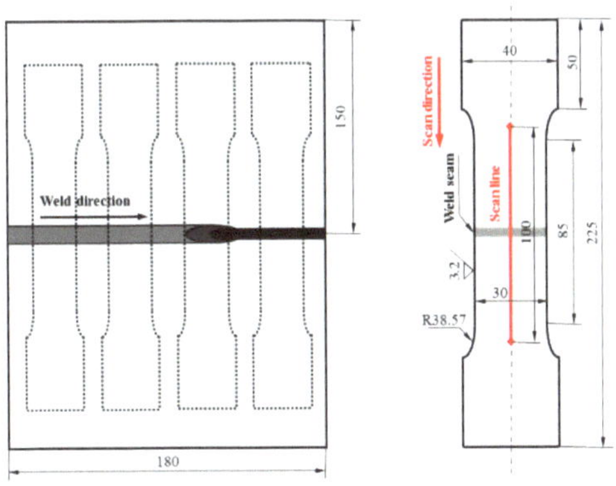

(**a**) (**b**)

Figure 1. Sketch of specimen preparation. (**a**) Welding direction and sampling location, and (**b**) geometry and dimensions of the fatigue specimen. All dimensions are in mm.

Non-destructive testing methods, both visual testing and radiographic testing, were performed to examine the quality of all the specimens, and only specimens with no examined defects were selected for the fatigue tests. The selected specimens were induction-demagnetized by a handheld induction demagnetizer, WLM-TB60, with an effective demagnetization rate of 1 μT to reduce the magnetic field interference induced by joint fabricating and specimen machining before the tests.

2.2. Experimental Instruments

Tension–compression fatigue tests of welded joints were performed at room temperature on a servo-hydraulic MTS 809 testing system, whose dynamic load error is ±1% and fatigue loading frequency is 0~50 Hz.

The MMM signals, both the tangential component B_x and the normal component B_z, of the specimens were collected by a 3D gaussmeter, the CH-3600 Gaussmeter, with a resolution of 0.1 μT. B_x is parallel to the surface of the specimen and along the scanning direction of the probe, while B_z is perpendicular to the surface of the specimen. A non-magnetic three-dimensional electrical control displacement instrument was employed to fix the probe firmly and move it automatically with a constant speed and lift-off value to avoid interference from external magnetic fields and manual operation errors.

2.3. Experimental Arrangement

The specimen was first placed on the electrical control displacement instrument along the south–north direction, and the initial MMMT signals of the scan line were measured with a probe scanning speed of 10 mm/s and a lift-off value of 1 mm. After that, the specimen was clamped between the gripers vertically, and the fatigue test was performed. When loaded to a preset fatigue cycle number, the specimen was removed from the gripers and placed on the displacement instrument, and the MMMT signals were collected in the same way as the initial MMMT signals were collected. Then, the specimen was loaded to another predetermined fatigue cycle number, and the above procedure was repeated until the specimen failed.

3. Results and Discussion

Fatigue tests of specimens with maximum fatigue stress of 150, 200, 300, and 400 MPa, respectively, were carried out according to the mechanical strength of welded joints, and the corresponding fatigue cycle numbers of each specimen were 194,014, 80,792, 19,262, and 10,026 in sequence. The stress ratio and frequency of all fatigue tests were the same, −1 and 10 Hz, respectively, to facilitate comparisons. For convenience, the specimens were named specimens 1, 2, 3, and 4.

3.1. MMMT Signals Variation during Fatigue Test

The variation law of MMT signals of specimens with different maximum fatigue stress showed the same tendency. Therefore, only the results of specimen 3 are presented in this section.

Figure 2 shows the MMMT signals before and after induction–demagnetization. It can be seen in Figure 2a that the MMMT signals were highly fluctuant and varied from −1400 μT to 0 μT before being demagnetized. There were abnormal peaks around the welded joint, but one cannot judge the existence of a welded joint because of the great fluctuation of the magnetic field.

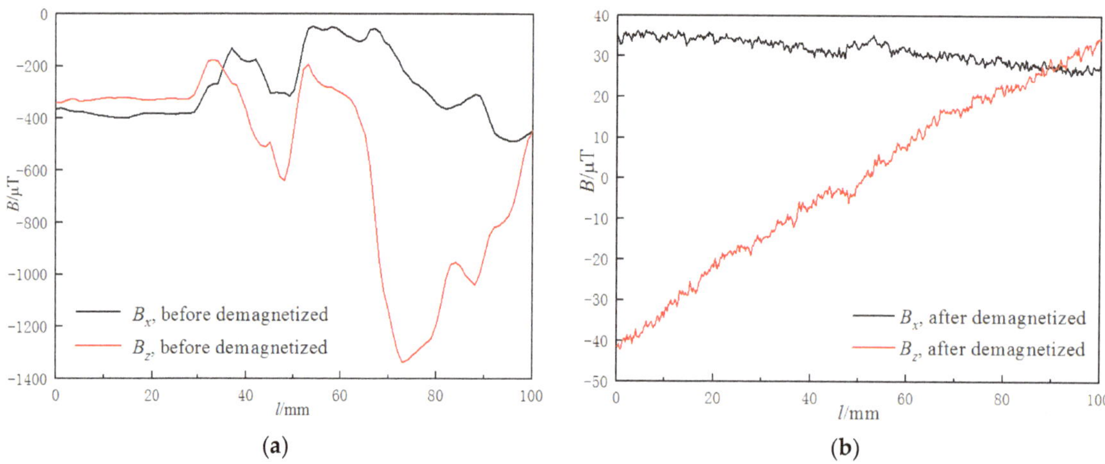

Figure 2. MMMT signals of the specimen (**a**) before and (**b**) after demagnetization.

The MMMT signals became much weaker and more regular after induction–demagneti zation, as shown in Figure 2b. The B_x signal is almost horizontal with an upward peak at the center of the scan line, which represents the right part of the welded joint. The mean value of B_x was 31.47 μT, and the peak value was 35.1 μT. The B_z signal is an inclined line ranging from −40.0 to 32.0 μT with a slope value of −0.7065 μT/mm, and there are two weak abnormal peaks around the welded joint, with a negative peak on the left and a positive one on the right.

The strong magnetic field relates to the welding, machining, and transporting process. During welding and machining, the specimens were fixed by an electromagnet, which introduced an additional magnetic field to the specimen. In addition, the specimen could also be magnetized by other magnetic materials during transportation.

The magnetic field interference could be reduced significantly by induction demagnetization. After demagnetization, magnetic moments became aligned with the direction of the geomagnetic field, and magnetic flux leakage took place around the welded joint as a result of magnetic permeability differences between the base metal and the welding wire. Hence, there was an upward peak of the B_x signal, while there were two weak abnormal peaks, with a negative one on the left and a positive one on the right of the B_z signal around the welded joint.

After that, a tension–compression fatigue test was carried out, and the MMM signals of the specimen were collected during the test. The specimen fractured at 19,262 cycles, and the MMM signals of the specimen during the fatigue test are shown in Figure 3.

The signal curves changed obviously after loading, as shown in Figure 3a. The B_x signal elevated upward, and there was a downward peak with a peak value of 69.2 μT at the welded joint. Both halves of the B_x signal were horizontal, but the amplitudes were not the same, and the mean value of the left half was 137.62 μT, while that of the right half was 116.61 μT. The B_z signal rotated anticlockwise with a positive abnormal peak on the left of the welded joint and a negative one on the right. The B_z signal ranged from −170.3 to 119.8 μT, and both halves were inclined, but they were not on the same inclined line, with a slope value of 4.641 μT/mm on the left and 2.900 μT/mm on the right, and the slope of the B_z signal between the abnormal peaks was −0.707 μT/mm.

The B_x signal continued elevating upward, and the B_z signal continued rotating anticlockwise as the fatigue cycle number increased for the first few fatigue cycles, as shown in Figure 3b,c. After that, when the fatigue cycle number became greater than 100, the MMMT signals were stable, with a little fluctuation as the fatigue cycle number increased, as shown in Figure 3d–f.

When loaded, the specimen was magnetized by the stress-induced magnetic field, and the amplitude of the MMMT signals increased. When the applied stress is greater than 70 MPa, the polarity of the stress-induced magnetic field is opposite to the geomagnetic field [32]. As a result, the MMMT signal around the welded joint reversed even one fatigue cycle was executed, with a downward peak of B_x, a positive abnormal peak on the left, and a negative one on the right of B_z.

Under the influence of the fatigue stress-induced magnetic field, magnetic domains tend to align themselves in the direction of the stress-induced magnetic field; the 180° domains, whose directions of moment are parallel to the direction of the magnetic field, grow, while the 90° domains, whose moment directions are perpendicular to the magnetic field, are reduced [33]. As a result, the magnetization of the specimen changes, and the magnetic flux leakage increases. Hence, the B_x signal elevated upward significantly, and the B_z signal rotated anticlockwise dramatically during the first few fatigue cycles, as shown in Figure 3b,c.

The fatigue stress-induced magnetic field reached the saturation limit, and no more irreversible magnetic domain movement occurred after 100 fatigue cycles. While there was still a reversible domain change, domain wall rotation [34] occurred during the following fatigue test. Hence, when the fatigue test proceeded, the MMMT signals were stable with a little fluctuation, as shown in Figure 3d–f.

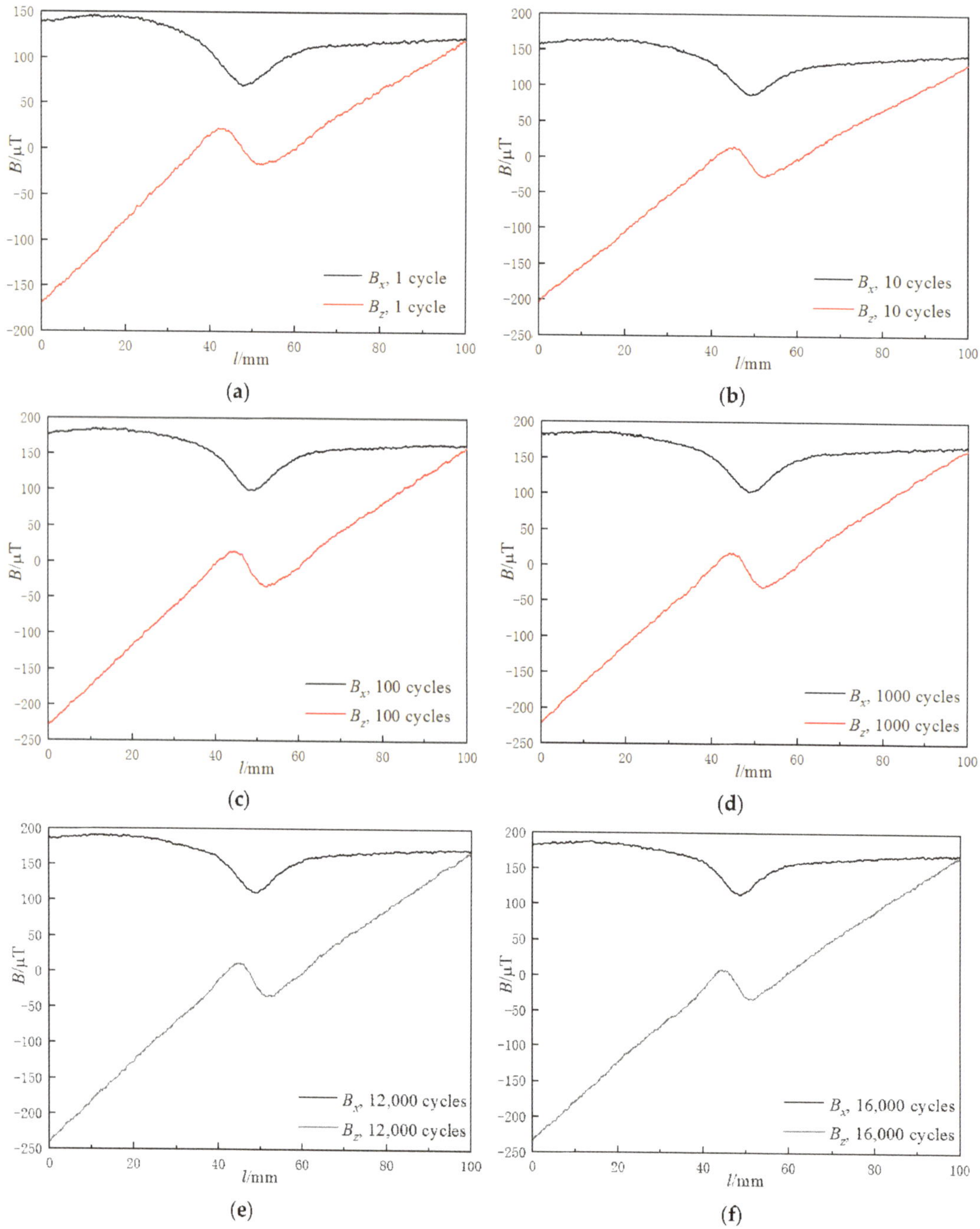

Figure 3. MMMT signals of welded joint under different fatigue cycle numbers: (**a**) 1 cycle; (**b**) 10 cycles; (**c**) 100 cycles; (**d**) 1000 cycles; (**e**) 8000 cycles; (**f**) 16,000 cycles.

It can be seen in Figure 3 that the distribution of the MMMT signals was regular during the fatigue test. There was a peak at the welded joint of the measured line for the B_x signal, and both halves beyond the welded joint were horizontal, while the amplitudes were different. There were two abnormal peaks around the welded joint for the B_z signal, and both halves were inclined, while the slopes were not the same. Thus, it was feasible to extract the characteristics of the MMMT signals for further investigation. According to the features of the MMMT signals, the characteristics were defined as follows: B_{xpeak} is the peak value of the B_x signal; $B_{x_{ave}-L}$ and $B_{x_{ave}-R}$ are the mean values of the left and right side of the B_x signal, respectively; K_G is the gradient between the abnormal peaks of the B_z signal, K_{B_z-L} and K_{B_z-R} are the slopes of the left and right sides of the B_z signal, respectively. The characteristic variations of the MMMT signals during the fatigue test are shown in Figure 4.

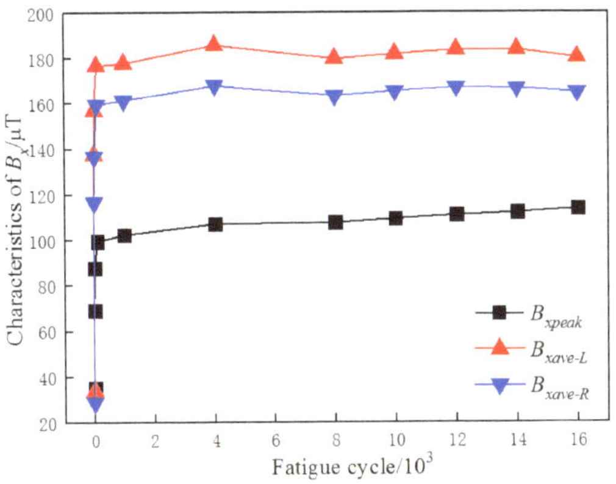

(**a**)

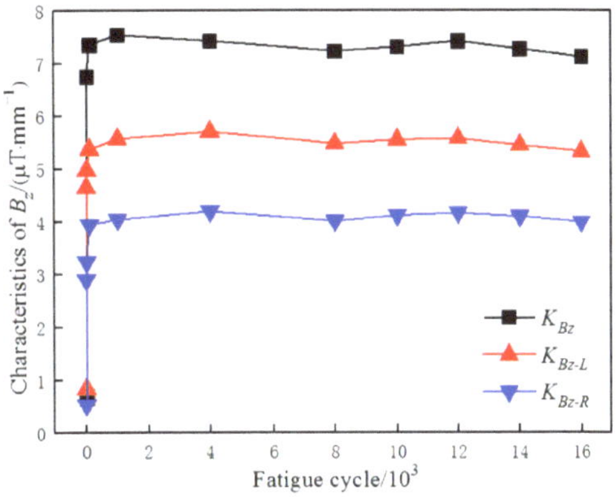

(**b**)

Figure 4. Characteristics variation of (**a**) B_x signals and (**b**) B_z signals during the fatigue test.

It can be seen in Figure 4 that all characteristics of the B_x and B_z signals show the same variation tendency. In the first few fatigue cycles, they increased dramatically, and when the fatigue cycle number was more than 100, they were stable with a little fluctuation. It should also be noted that $B_{x_{ave}-L}$ and $B_{x_{ave}-R}$ had almost the same variation except for the value during the fatigue test, and the same was true for K_{B_z-L} and K_{B_z-R}.

3.2. Characteristics of MMMT Signals under Different Fatigue Stress

The MMMT signals of specimens 1, 2, and 4 during the fatigue tests were also collected, and the variation laws were similar to those of specimen 3. Therefore, the same characteristics, including B_{xpeak}, $B_{x_{ave}-L}$, $B_{x_{ave}-R}$, K_G, K_{B_z-L}, and K_{B_z-R}, were extracted. As shown in Section 3.1, $B_{x_{ave}-L}$ and $B_{x_{ave}-R}$ and K_{B_z-L} and K_{B_z-R} of specimen 3 showed the same variation during the fatigue test except for the value, so it was important to find out whether the situations were the same for other specimens; the variations of the mentioned characteristics of all specimens are shown in Figure 5.

It can be seen from Figure 5 that $B_{x_{ave}-L}$ and $B_{x_{ave}-R}$ and K_{B_z-L} and K_{B_z-R} shared the same variation law for each specimen. For quantitative description, the cross-correlation coefficient ρ was introduced to evaluate the relationship between $B_{x_{ave}-L}$ and $B_{x_{ave}-R}$ and between K_{B_z-L} and K_{B_z-R}. ρ is defined according to statistics:

$$\rho_{x,y} = \frac{\sum_1^n (X_i - \overline{X})(Y_i - \overline{Y})}{\sqrt{\sum_1^n (X_i - \overline{X})^2 \sum_1^n (Y_i - \overline{Y})^2}}, \quad (1)$$

where $\overline{X}$ and $\overline{Y}$ are the arithmetic mean values of X and Y, respectively.

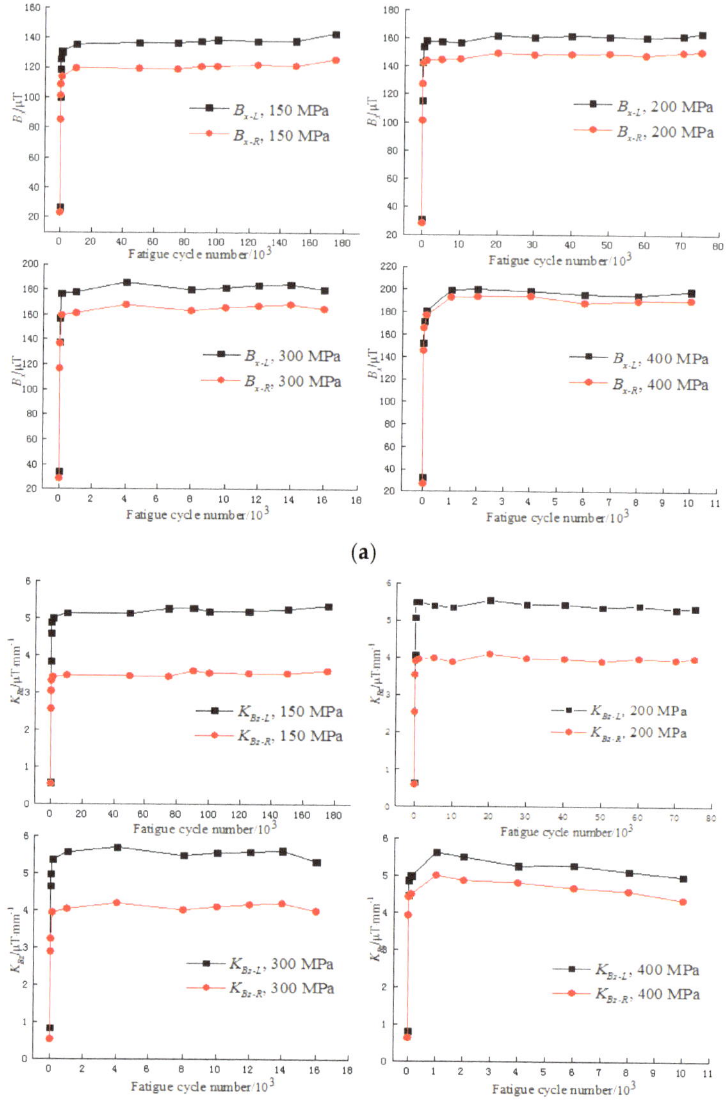

Figure 5. Characteristic variation of (**a**) B_x signals and (**b**) B_z signals of different specimens during the fatigue test.

As only the variation of the MMMT signals during the stable stage was concerned, data collected when the fatigue cycle number was less than 100 were ignored, and the cross-correlation coefficients of $B_{x_{ave}-L}$ and $B_{x_{ave}-R}$ and K_{B_z-L} and K_{B_z-R} of each specimen are shown in Table 4.

Table 4. Cross-correlation coefficients of the characteristics of all the specimens.

	Specimen 1	Specimen 2	Specimen 3	Specimen 4
$\rho_{B_{x_{ave}-L},B_{x_{ave}-R}}$	0.9919	0.9728	0.9684	0.9757
$\rho_{K_{B_z-L},K_{B_z-R}}$	0.8846	0.8289	0.8828	0.9132

It can be seen in Table 4 that, for each specimen, $B_{x_{ave}-L}$ and $B_{x_{ave}-R}$ and K_{B_z-L} and K_{B_z-R} were highly correlated with each other during the fatigue tests. Therefore, the new characteristics $B_{x_{ave}}$ and K_{B_z} were introduced for simplification. $B_{x_{ave}}$ is the mean value of $B_{x_{ave}-L}$ and $B_{x_{ave}-R}$, while K_{B_z} is the slope of the angular bisector of line $y = K_{B_z-L}\cdot x$ and $y = K_{B_z-R}\cdot x$, as shown in Equation (2).

$$B_{x_{ave}} = \frac{B_{x_{ave-L}} + B_{x_{ave-R}}}{2} K_{B_z} = \frac{K_{B_z-L}\cdot K_{B_z-R} - 1 + \sqrt{(1 - K_{B_z-L}\cdot K_{B_z-R})^2 + (K_{B_z-L} + K_{B_z-R})^2}}{K_{B_z-L} + K_{B_z-R}}, \tag{2}$$

Therefore, the number of characteristics can be reduced to four, i.e., B_{xpeak}, B_{xave}, K_{B_z} and K_G, and their variations during the fatigue tests are shown in Figure 6.

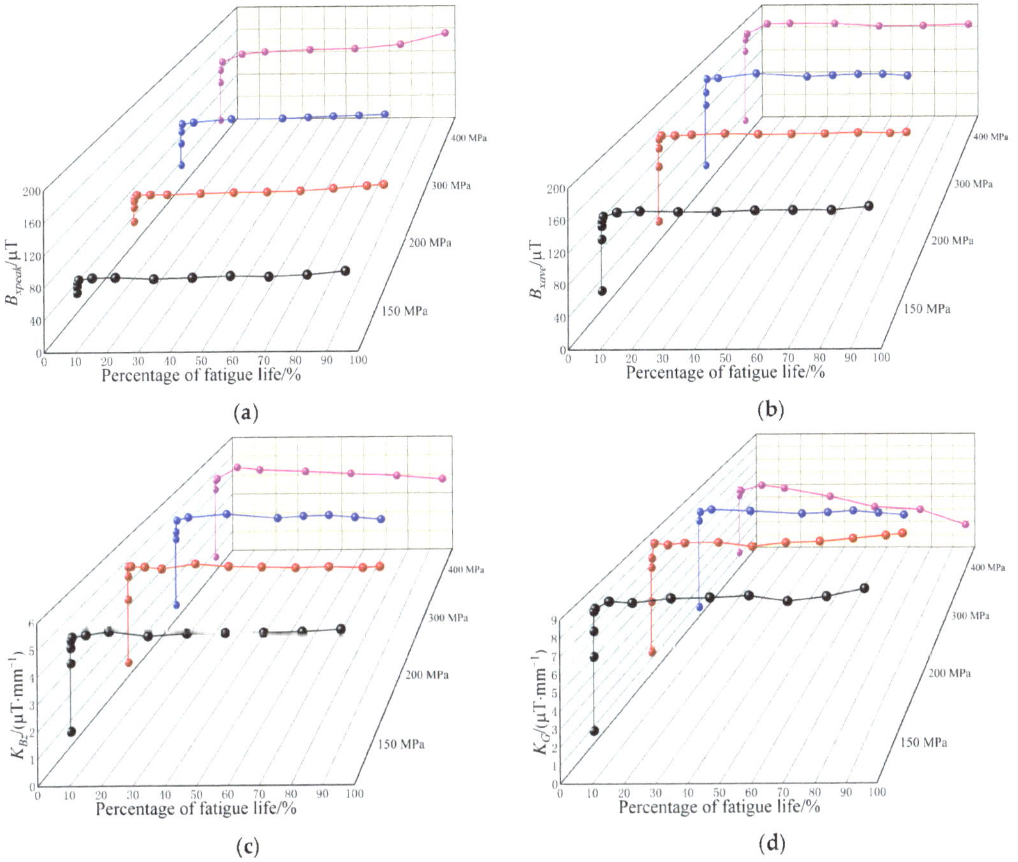

Figure 6. Characteristic variations of MMMT signals during the fatigue tests (**a**) B_{xpeak}; (**b**) B_{xave}; (**c**) K_{B_z}; (**d**) K_G.

Figure 6a shows the variation in B_{xpeak} for all the specimens during the fatigue tests. B_{xpeak} of the specimens increased significantly with the fatigue cycle number at first, and then became stable and fluctuated a little as the fatigue test proceeded. It also can be seen that when the specimens were about to fracture, the amplitudes of B_{xpeak} of the last measurement of specimen 1 (175,000 cycles, 90.20% of fatigue life), specimen 2 (75,000 cycles, 92.83% of fatigue life), and specimen 4 (10,000 cycles, 99.74% of fatigue life) were much larger than those measured before, while that of specimen 3 (16,000 cycles, 83.07% of fatigue life) was not.

It can be seen in Figure 6b that B_{xave} of all the specimens during the fatigue tests showed almost the same variation: they increased dramatically in the first few fatigue cycle numbers, and after that, they were stable with a little fluctuation.

It can be concluded from Figure 6a,b that when the specimen is about to fracture (above 90% of fatigue life), B_{xave} barely changes, while B_{xpeak} changes obviously. This phenomenon could be a significant criterion for failure pre-warning of welded joints.

As shown in Figure 6c, K_{B_z} of all the specimens increased as the fatigue cycle number increased in the first few fatigue cycle numbers, and then became stable with a little fluctuation until the specimens failed. By contrast, the variations in K_G showed the same tendency, as shown in Figure 6d, except that of specimen 4. K_G of specimen 4 increased with the fatigue cycle at the beginning of the fatigue test and then fluctuated much more obviously until the specimen fractured.

To explore the relationship between the characteristics of MMMT signals and the fatigue stress, the mean values of the characteristics of the MMMT signals, $\overline{B}_{xave}$, $\overline{B}_{xpeak}$, $\overline{K}_{B_z}$, and $\overline{K}_G$, during the stable stage of the fatigue tests (when fatigue cycle number was larger than 100) were analyzed, and their variations with maximum fatigue stress are shown in Figure 7. The values of $\overline{B}_{xpeak}$ of the last measurements of specimens 1, 2, and 4 were neglected, as they were much larger than those measured before (in other words, they were about to fracture).

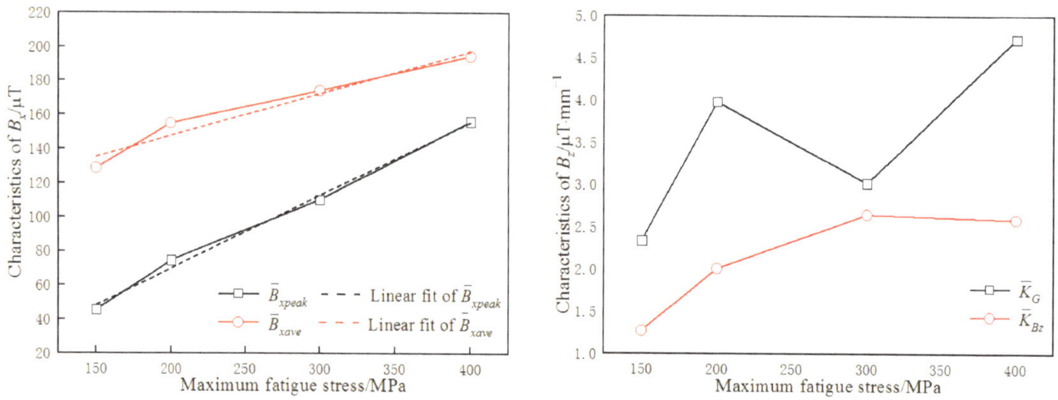

(**a**) (**b**)

Figure 7. Characteristic variation of MMMT signals with the maximum fatigue stress (**a**) B_x signals, (**b**) B_z signals.

It can be seen in Figure 7 that there is a linear relationship between $\overline{B}_{xave}$ and the applied maximum fatigue stress σ and between $\overline{B}_{xpeak}$ and σ. The functions can be expressed as follows:

$$\overline{B}_{xpeak} = 0.4282 \cdot \sigma - 15.71 \overline{B}_{xave} = 0.2458 \cdot \sigma + 98.52, \quad (3)$$

Unfortunately, there is no obvious relationship between $\overline{K}_{B_z}$ or $\overline{K}_G$ and σ.

It can be seen from Equation (3) that there was a linear relationship between $\overline{B}_{xave}$ or $\overline{B}_{xpeak}$ and the applied maximum fatigue stress σ of the welded joints in this work. Therefore, it is feasible to evaluate σ by the measurement of the B_x signal, as shown in Equation (4):

$$\sigma = 37.927 + 2.323 \cdot \overline{B}_{xpeak} \sigma = -370.397 + 3.882 \cdot \overline{B}_{xave}, \quad (4)$$

It should be noted that the premise of the σ calculation by Equation (4) is that the fatigue stress ratio is −1. As a matter of fact, fatigue with a stress ratio that is not −1 is commonly experienced in both experimental research and engineering application. In this case, the Goodman formula can be introduced for equivalent transformation. The Goodman formula is shown in Equation (5):

$$.\sigma_a = \sigma_{-1}\left(1 - \frac{\sigma_m}{\sigma_b}\right), \quad (5)$$

where $\sigma_a = (\sigma_{max} - \sigma_{min})/2$, is the stress amplitude, $\sigma_m = (\sigma_{max} + \sigma_{min})/2$, is the mean stress, and σ_b is the tensile strength of the material.

When measurement of B_x is executed, the characteristics $\overline{B}_{xave}$ and $\overline{B}_{xpeak}$ can be extracted. Then, the applied maximum fatigue stress σ_{-1} when the stress ratio is -1 can be calculated. The actual stress amplitude σ_a can be calculated according to Equation (5).

4. Conclusions

The systematic investigation of MMMT signal variation of welded joints under different fatigue cycle numbers during the whole tension–compression fatigue tests was presented in this work. There was a peak at the welded joint for the B_x signal, and both halves beyond the welded joint were horizontal, while the amplitudes were different. There were two abnormal peaks around the welded joint for the B_z signal, and both halves were inclined, while the slopes were not the same. The following conclusions can be drawn:

(1) The B_x signal elevated upward, and the B_z signal rotated counterclockwise dramatically with an increase in the fatigue cycle number in the first few fatigue cycles; after that, both signals were stable with a little fluctuation.

(2) When the welded joint is about to fracture (about 90% of the fatigue life in this work), the mean value of the B_x signal barely changes, while the peak value increases obviously, which can be used as a service status pre-warning for weld structures.

(3) Fatigue stress of welded joints can be estimated and monitored by the mean value and peak value, i.e., $\overline{B}_{xave}$ and $\overline{B}_{xpeak}$, of the B_x signal.

Author Contributions: Conceptualization, H.W. (Huipeng Wang) and L.D.; methodology, L.D. and H.W. (Haidou Wang); validation, Z.X. and B.L.; formal analysis, D.C.; investigation, G.M.; resources, Z.X.; data curation, G.M.; writing—original draft preparation, H.W. (Huipeng Wang); writing—review and editing, L.D.; visualization, B.L.; supervision, H.W. (Haidou Wang); project administration, H.W. (Huipeng Wang); funding acquisition, H.W. (Huipeng Wang). All authors have read and agreed to the published version of the manuscript.

Funding: This research was funded by the National Natural Science Foundation of China, grant number 52065026, and the National United Engineering Laboratory for Advanced Bearing Tribology, Henan University of Science and Technology, grant number 202110.

Institutional Review Board Statement: Not applicable.

Informed Consent Statement: Not applicable.

Data Availability Statement: Not applicable.

Conflicts of Interest: The authors declare no conflict of interest.

References

1. Dubov, A. A study of metal properties using the method of magnetic memory. *Met. Sci. Heat Treat.* **1997**, *39*, 401–405. [CrossRef]
2. Dubov, A. Screening of weld quality using the magnetic metal memory effect. *Weld. World* **1998**, *41*, 196–199.
3. Shi, P.P.; Su, S.Q.; Chen, Z.M. Overview of Researches on the Nondestructive Testing Method of Metal Magnetic Memory: Status and Challenges. *J. Nondestruct. Eval.* **2020**, *39*, 43. [CrossRef]
4. Bao, S.; Jin, P.; Zhao, Z.; Fu, M. A Review of the Metal Magnetic Memory Method. *J. Nondestruct. Eval.* **2020**, *39*, 11. [CrossRef]
5. LI, Y.F.; Zeng, X.G.; Wei, L.M.; Wan, Q. Characterizations of damage-induced magnetization for X80 pipeline steel by metal magnetic memory testing. *Int. J. Appl. Electromagn. Mech.* **2017**, *54*, 23–35. [CrossRef]
6. Singh, W.S.; Stegemann, R.; Kreutzbruck, M.; Mukhopadhyay, C.K.; Rao, B.P. Mapping of deformation-induced magnetic fields in carbon steels using a GMR sensor based metal magnetic memory technique. *J. Nondestruct. Eval.* **2018**, *37*, 352–358. [CrossRef]
7. Fu, M.; Bao, S.; Lou, H. Characterization of Stress Concentration by Tangential Component $H_p(x)$ of Metal Magnetic Memory Signals. *J. Wuhan Univ. Technol.* **2018**, *33*, 1486–1490. [CrossRef]

8. Moonesan, M.; Kashefi, M. Effect of sample initial magnetic field on the metal magnetic memory NDT result. *J. Magn. Magn. Mater.* **2018**, *460*, 285–291. [CrossRef]
9. Jung, G.J.; Seo, J.S.; Park, Y.W. Comparison between Magnetic Field Distribution Analysis and Metal Magnetic Memory (MMM) Testing Results Around Artificial Cracks under Loads. *J. Magn.* **2019**, *24*, 281–285. [CrossRef]
10. Li, C.C.; Dong, L.H.; Wang, H.D.; Li, G.L.; Xu, B.S. Metal magnetic memory technique used to predict the fatigue crack propagation behavior of 0.45%C steel. *J. Magn. Magn. Mater.* **2016**, *405*, 150–157.
11. Venkatachalapathi, N.; Jameel Basha, S.M.; Janardhan Raju, G.; Raghavulu, P. Charact-erization of fatigued steel states with metal magnetic memory metho. *Mater. Today Proc.* **2018**, *5*, 8645–8654. [CrossRef]
12. Ni, C.; Hua, L.; Wang, X.K. Crack propagation analysis and fatigue life prediction for structural alloy steel based on metal magnetic memory testing. *J. Magn. Magn. Mater.* **2018**, *462*, 144–152. [CrossRef]
13. Zhou, W.; Fan, J.C.; Ni, J.L.; Liu, S.J. Variation of magnetic memory signals in fatigue crack initiation and propagation behavior. *Metals* **2019**, *9*, 89. [CrossRef]
14. Zhao, X.R.; Su, S.Q.; Wang, W.; Zhang, X.H. Metal magnetic memory inspection of Q345B steel beam in four point bending fatigue test. *J. Magn. Magn. Mater.* **2020**, *514*, 167155. [CrossRef]
15. Xu, K.S.; Qiu, X.Q.; Tian, X.S. Theoretical investigation of metal magnetic memory testing technique for detection of magnetic flux leakage signals from buried defect. *Nondestruct. Test. Eval.* **2018**, *33*, 45–55. [CrossRef]
16. Shi, P.P.; Hao, S. Analytical solution of magneto-mechanical magnetic dipole model f-or metal magnetic memory method. *Acta Phys. Sin.* **2021**, *7*, 034101. [CrossRef]
17. Han, G.; Huang, H.H. A dual-dipole model for stress concentration evaluation basedon magnetic scalar potential analysis. *NDT&E Int.* **2021**, *118*, 102394. [CrossRef]
18. Shi, P.P. Magneto-elastoplastic coupling model of ferromagnetic material with plastic deformation under applied stress and magnetic fields. *J. Magn. Magn. Mater.* **2020**, *512*, 166980. [CrossRef]
19. Zhang, P.C.; Shi, P.P.; Jin, K.; Zheng, X.J. An anisotropic magneto-mechanical model of ferromagnetic materials for the magnetic memory testing method. *J. Appl. Phys.* **2019**, *125*, 233901. [CrossRef]
20. Shi, P.P.; Zhang, P.C.; Jin, K.; Chen, Z.M.; Zheng, X.J. Thermo-magneto-elastoplastic coupling model of metal magnetic memory testing method for ferromagnetic materials. *J. Appl. Phys.* **2018**, *123*, 145102. [CrossRef]
21. Liu, B.; Fu, Y.; Jian, R. Modeling and analysis of magnetic memory testing method based on the density functional theory. *Nondestruct. Test. Eval.* **2015**, *30*, 13–25. [CrossRef]
22. Luo, X.; Zhu, H.Y.; Ding, Y.P. A modified model of magneto-mechanical effect on magnetization in ferromagnetic materials. *Acta Phys. Sin.* **2019**, *68*, 187501. [CrossRef]
23. Yokoyama, K.; Miki, C. Participatory database of repair cases on fatigue damaged welded structures. *Int. J. Fatigue* **2009**, *101 Pt 2*, 385–396. [CrossRef]
24. Pei, X.J.; Dong, P.S.; Xing, S.Z. A structural strain parameter for a unified treatment of fatigue behaviors of welded components. *Int. J. Fatigue* **2019**, *124*, 444–460. [CrossRef]
25. Smith, A.T.; Schneider, C.R.; Bird, C.R.; Wall, M. Use of non-destructive testing for engineering critical assessment: Background to the advice given in BS 7910:2013. *Int. J. Press. Vessel. Pip.* **2019**, *169*, 153–159. [CrossRef]
26. Dubov, A.; Kolokolnikov, S. Detection of local stress concentration zones in engineering products-the lacking link in the non-destructive testing system. *Weld. Word* **2018**, *62*, 301–309. [CrossRef]
27. Kolokolnikov, S.; Dubov, A.; Medvedev, A.; Boriskin, D. Comprehensive inspection of refrigerated ammonia storage tank welded joints by the metal magnetic memory technique and conventional NDT methods. *Weld. World* **2020**, *64*, 1659–1670. [CrossRef]
28. Kolokolnikov, S.M.; Steklov, O.I. Evaluation of the heterogeneity of the stress state of welded joints prior to and after heat treatment using the method of the magnetic memory of metal. *Weld. Int.* **2017**, *31*, 487–492. [CrossRef]
29. Su, S.Q.; Zhao, X.R.; Wang, W.; Zhang, X.H. Metal Magnetic Memory Inspection of Q345 Steel Specimens with Butt Weld in Tensile and Bending Test. *J. Nondestruct. Eval.* **2019**, *38*, 64. [CrossRef]
30. Huang, H.H.; Qian, Z.C.; Yang, C.; Han, G.; Liu, Z.F. Magnetic memory signals of ferr-omagnetic weldment induced by dynamic bending load. *Nondestruct. Eval.* **2017**, *32*, 166–184. [CrossRef]
31. Xu, K.S.; Yang, K.; Liu, J.; Chen, X.Q.; Wang, Y. Investigation of Magnetic Memory Signal of Propagation of Buried Crack under Applied Load. *Res. Nondestruct. Eval.* **2021**, *32*, 1–9. [CrossRef]
32. Dong, L.H.; Xu, B.S.; Dong, S.Y.; Li, S.; Chen, Q.Z. Stress dependence of the spontaneous stray field signals of ferromagnetic steel. *NDT E Int.* **2009**, *42*, 323–327. [CrossRef]
33. Cullity, E.D. *Introduction to Magnetic Materials*; Addison-Wesley: New York, NY, USA, 1972.
34. Jiles, D.C. Theory of the magnetomechanical effect. *J. Phys. D Appl. Phys.* **1999**, *28*, 1537–1546. [CrossRef]

materials

MDPI

Article

Switching Diagram of Core-Shell FePt/Fe Nanocomposites for Bit Patterned Media

Yuhui Wang [1,2], Ying Zheng [1,2], Ziyi Zhong [1,2], Zijun Wang [3], Yongfeng Liang [4] and Pingping Wu [1,2,*]

1 Department of Materials Science and Engineering, Xiamen Institute of Technology, Xiamen 361021, Fujian Province, China; yuhuiwang2175@163.com (Y.W.); zhengying_w@126.com (Y.Z.); zzy17321619268@gmail.com (Z.Z.)
2 The Higher Educational Key Laboratory for Flexible Manufacturing Equipment Integration of Fujian Province, Xiamen Institute of Technology, Xiamen 361021, Fujian Province, China
3 Department of Nuclear Physics, China Institute of Atomic Energy (CIAE), Beijing 102413, Beijing, China; wangzijun@ciae.ac.cn
4 State Key Laboratory for Advanced Metals and Materials, University of Science and Technology Beijing, Beijing 100083, Beijing, China; liangyf@skl.ustb.edu.cn
* Correspondence: pingpingwu@xit.edu.cn

Abstract: In the current work, a core-shell type exchange coupled composite structure was constructed by micromagnetic simulation with a phase FePt core and an iron shell. Four types of switching loops with magnetic domain structure evolution were demonstrated. Based on the simulation results, a switching type diagram was constructed, which displays various hysteresis loops as a function of core radius and shell thickness. Furthermore, the effects of switching type and composite structure on the coercivity and remanent magnetization were predicted and discussed. This finding indicates that core-shell type FePt/Fe composite structure film has a large advantage in designing exchange-coupled bit patterned media to realize high-density storage devices at the nanoscale.

Keywords: magnetization switching; micromagnetic simulation; core-shell; bit patterned media

Citation: Wang, Y.; Zheng, Y.; Zhong, Z.; Wang, Z.; Liang, Y.; Wu, P. Switching Diagram of Core-Shell FePt/Fe Nanocomposites for Bit Patterned Media. *Materials* **2022**, *15*, 2581. https://doi.org/10.3390/ma15072581

Academic Editor: Ovidiu Oprea

Received: 10 March 2022
Accepted: 28 March 2022
Published: 31 March 2022

1. Introduction

Exchange coupled bit-patterned media are a very promising concept to delay the superparamagnetic size limit for next-generation ultra-high-density magnetic recording storage applications [1,2]. For massive data storage, $L1_0$ FePt-based thin film/nanodot structure has many advantages, including very large density, low cost, magnetical stability, and chemical stability [3,4]. Therefore, it was considered the most promising candidate for future magnetic recording media material. Beyond this, a composite structure combined with a soft magnetic layer (e.g., an iron phase) creates an exchange coupling structure and helps reverse the magnetization of the hard phase during the writing process with a reduced writing field [5–8]. This exchange coupling composite (ECC) structure with a sharp/graded interface can further decrease the coercive field and enhance the storage density of the bit patterned media [5,9–11].

For traditional double-layered ECC, as a thick, soft layer is introduced, such a composite structure requires more fabrication challenges and large head keeper spacing. Furthermore, it introduces large demagnetization fields and strong dependence on the exchange-coupling strength. Goll et al. proposed several types of ledge-type isolated nanocomposite based on FePt/Fe structure [12]. The investigations show that the coercive field can be significantly decreased compared to the traditional ECC structure [12], which can be experimentally realized by Lomakin et al. [13], Goll et al. [14], and Speliotis et al. recently [15]. In order to further reduce the thickness of composite and keep the advantage of the switching field [16], a structure of enclosed composite patterned media was developed by Goh et al. [17]. Different from the ledge-type structure, the soft layer of the composite is covered all over the surface of the hardcore and creates a core-shell structure in Goh's work.

Their simulation results suggested that this structure provides a lower switching field with lower exchange coupling strength. Following experimental works by Ma et al. [18,19] demonstrate that the switching field, thermal stability, and gain factor of the core-shell structure composite have a large advantage over traditional exchange coupling composite. Some experimental works can present the evolution of the magnetic pattern of composite structure [20,21]; however, the size of the composite is relatively large such as around several hundred nanometers.

It was shown that micromagnetic simulations could predict the magnetic properties for FePt based exchange-coupled bit patterned media [22–27]; however, only a few works for core-shell shaped island composite were reported. It should be noted that the hysteresis loop of the core-shell island is very sensitive to the stacking structure of the composite. Our prior works [16,22] on double-layered exchange spring nanoislands demonstrated that the micromagnetic simulations could predict the magnetization hysteresis loop with domain structure evolution at a switching external field. Therefore, the main objective of this letter is to simulate various core-shell type ECC FePt/Fe structures with different core radii and shell thickness by micromagnetic simulation. Four types of switching processes with domain structure evolution were demonstrated in this simulation. The switching diagram was expected to contribute to the design of the core-shell type ECC for bit patterned media by choosing an appropriate core size and shell thickness. The magnetic properties of the core-shelled composite structure were also analyzed and discussed in this work.

2. Simulation Method

A core-shell type exchange coupled composite structure was constructed by computer simulation. A hemispherical FePt core island of radius r nm is enveloped by an iron outer layer with a thickness of d nm, as shown in the schematic Figure 1. In our study, a set of FePt/Fe core-shell composite structures was simulated with different core sizes and shell thicknesses. The interfaces between FePt/Fe layers are coherent, and the composite structure is grown epitaxially on a substrate. It should be noted that the interface between the hard and soft magnetic phase was assumed to be a sharp interface in this work; a controlled anisotropy graded interface is not discussed in our current work.

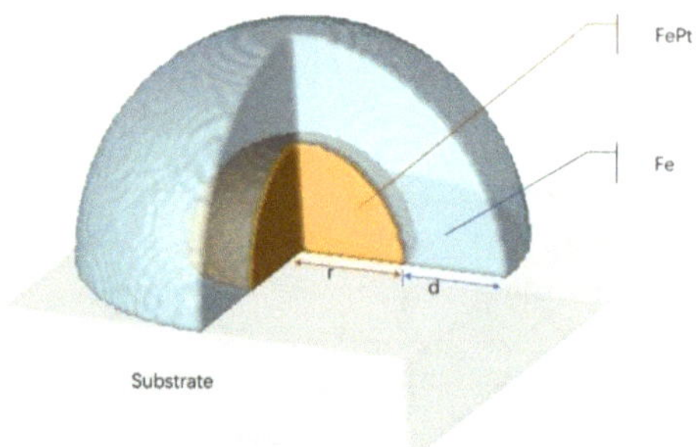

Figure 1. A 3D schematic figure of simulated core-shell structure of FePt/Fe nanocomposite. The FePt core is enveloped by an iron layer.

In order to model a proper composite magnetic domain structure by the micromagnetic method, the spontaneous magnetization $\mathbf{M}_i(i = 1, 2, 3)$ were chosen as the order parameter. In micromagnetic simulations, the equilibrium ferromagnetic domain structure can be achieved by solving the Landau–Lifshitz–Gilbert equation, i.e.,

$$\left(1+\alpha^2\right)\frac{\partial \mathbf{M}}{\partial t} = -\gamma_0 \mathbf{M} \times \mathbf{H}_{eff} - \frac{\gamma_0 \alpha}{M_s}\mathbf{M} \times (\mathbf{M} \times \mathbf{H}_{eff}), \quad (1)$$

where α is the damping factor, γ_0 is the gyromagnetic ratio, M_s is saturation magnetization, and $\mathbf{H}_{eff}$ is the external magnetic field, i.e.,

$$\mathbf{H}_{eff} = -\frac{1}{\mu_0}\frac{\partial E}{\partial \mathbf{M}}, \tag{2}$$

where μ_0 is the vacuum magnetic permeability, E is the total free energy of magnets; in this paper, the total free energy, E, of a FePt/Fe composite is expressed by

$$E = E_{anis} + E_{exch} + E_{ms} + E_{Zeeman}, \tag{3}$$

where E_{anis}, E_{exch}, E_{ms}, and E_{Zeeman} are magnetocrystalline anisotropy, exchange coupling, magnetostatic, and Zeeman energies, respectively.

The magnetocrystalline anisotropy energy of a cubic magnetic crystal is given by

$$E_{anis} = \int \left[K_1\left(m_1^2m_2^2 + m_1^2m_3^2 + m_2^2m_3^2\right) + K_2m_1^2m_2^2m_3^2\right]dV, \tag{4}$$

where K_1 and K_2 are anisotropy constants, $\mathbf{m}_i = \mathbf{M}_i/M_s$ is unit magnetization, and V is nanostructure volume.

The exchange coupling energy in Equation (3) is determined solely by the spatial variation in the magnetization orientation and can be written as:

$$E_{exch} = \int (\nabla \mathbf{m})^2 dV, \tag{5}$$

where A is the exchange coupling constant. In order to calculate the magnetostatics energy of the composite, convolution of magnetization with the three-dimensional LaBonte interaction matrix, the analytic formula developed by Schabes and Aharoni [28] is written in the form

$$E_{ms} = \frac{1}{2}\sum\nolimits'_{ijk,i'j'k'} E_{ijk,i'j'k'}, \tag{6}$$

where $E_{ijk,i'j'k'}$ represents the interaction energy between the grids at the site (i, j, k), and the site (i′, j′, k′), (i, j, k) and (i′, j′, k′) are lattice indices in our 3D grids.

The Zeeman energy can be calculated by considering the interactions between magnetization and external magnetic fields $\mathbf{H}_{ex}$:

$$E_{Zeeman} = -\mu_0 M_s \int \mathbf{H}_{ex} \cdot \mathrm{m}dV, \tag{7}$$

In this work, the evolution of the magnetic domain structure and the magnetization distributions are obtained by solving Equations (1)–(7) with the Gauss–Seidel projection method in Ref. [29].

The simulation box of a FePt/Fe composite unit cell is discretized as $32\Delta x_1 \times 32\Delta x_2 \times 16\Delta x_3$, where Δx_1, Δx_2, and Δx_3 are grid spacing, and $\Delta x_1 = \Delta x_2 = \Delta x_3 = 1$ nm for real space. In this work, the radius of the FePt core changes from 3 nm to 10 nm, the nominal thickness of the iron layer changes from 3 nm to 7 nm, and the shell thickness are similar to that of experimental works reported by Ma et al. [18]. All the corresponding magnetic coefficients, including magnetization saturation, anisotropy constant, exchange coupling constant, were taken from experimental data from Ref. [30] and listed as follows: for the FePt core: $M_s = 1.050 \times 10^6$ A/m, $K_1 = 1.7 \times 10^6$ J/m^3, $K_2 = 0$ J/m^3, $A = 10^{-11}$ J/m; for the Fe shell: $M_s = 1.575 \times 10^6$ A/m, $K_1 = 4.7 \times 10^4$ J/m^3, $K_2 = 0$ J/m^3, $A = 2 \times 10^{-11}$ J/m. In this work, the micromagnetic simulation is performed using GNU Octave, and the code was programmed by the author.

3. Results and Discussion

In order to simulate a switching process of the composite island, the initial domain structure was magnetized to saturation alongx direction, and then we decreased the external field by a step of 10 kA/m (~125 Oe) and finally reversed the field direction to saturate magnetization in the opposite direction. In our simulation, we found that different simulation loops could be obtained according to the different core radii and shell thickness; at this point, we found four representative switching loops. Examples of the switching loop with the type of A–D and the magnetic domain evolution are shown in Figure 2a–d, respectively. For type A, a constricted hysteresis loop was obtained in FePt(5 nm)/Fe(5 nm) core-shell type composites. The magnetization sharply decreased to zero at a negative field, then slowly switched to the opposite direction. For type B, as the hard material FePt increases to 7 nm while Fe decreases to 3 nm, the simulation results of the composite material show a square hysteresis loop; however, while the soft layer increases to 7 nm, a type C switching loop shows a narrow hysteresis with small amounts of residual magnetism and coercive field. For type D, a typical exchange-coupled switching loop for a FePt(8 nm)/Fe(5 nm) composite was observed, resulting in a smaller coercivity and saturation field.

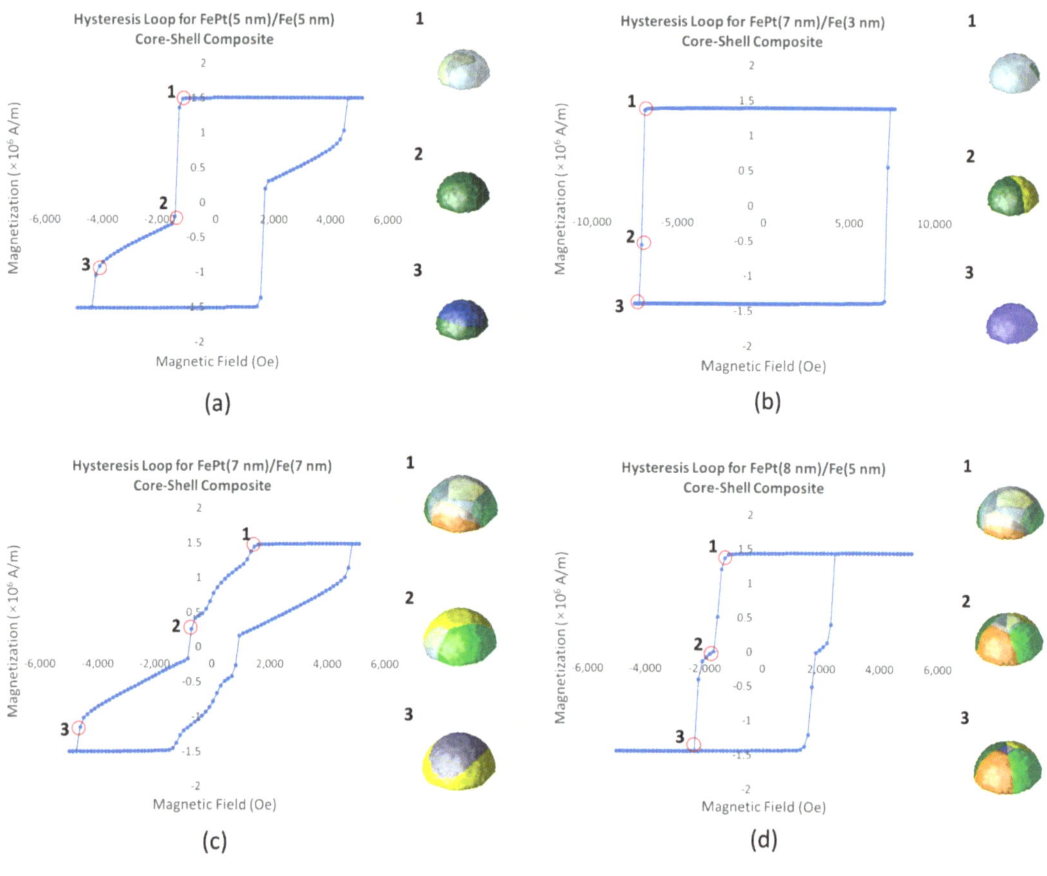

Figure 2. (**a**–**d**) Magnetic hysteresis loops of FePt/Fe core-shelled structure measured in the out-of-plane directions for the switching type of A–D, respectively. The simulated corresponding magnetic domain structure evolution during the switching process at the position of three circles 1, 2, 3 are shown at the right of the loop. Each color represents a type of ferromagnetic domain: yellow/orange represents +x/−x domains, green/dark green represents +y/−y domains, and blue/light blue represents +z/−z domains.

In order to explore the influence of the ferromagnetic domain structures on the switching loops, we analyzed the effect of ferromagnetic domain structures evolution for different switching types, which are illustrated in the right column of Figure 2. The light blue/blue domains represent the domains with magnetization towards +z/−z directions, i.e., external field directions. The yellow/orange/green/dark green domains represent the in-plane domains with magnetizations along with +x/−x/+y/−y directions, respectively.

According to classical magnetic theory [31], one can estimate the size of the particles by balancing the energy needed to create a domain wall spanning a spherical particle and the magnetostatic energy saved by reducing the single domain state to a multi-domain state; thus, this method gives the critical radius of the particle r_c(Fe) ~3 nm and r_c(FePt) ~27 nm. According to Kikuchi et al., experimental report [32], the critical diameter of the single domain FePt particles is around 55 nm, which is consistent with the theory. In this simulation work, the FePt core is always observed to stay in a single domain state because the radius of the FePt core is much less than the critical radius; on the other hand, the thickness of the iron shell is larger than 3 nm, which exceed the predicted critical value; therefore, the iron shell always exhibits a more stable multi-domain state in our simulation.

In the FePt(5 nm)/Fe(5 nm) core-shell composites, the magnetization quickly decreased to below zero at −1500 Oe. In the domain structures, one can see both the soft shell and hardcore are switched to in-plane direction. In the next stage, the magnetization in the soft magnetic phase was gradually switched to −z-direction and resulted in a slight decrease in magnetization. Finally, the whole structure was switched to the −z-direction with a jump of magnetization led by the switching of the hard FePt core, and, as a result, the magnetization reached full saturation in the opposite direction, and a constricted loop (type A) was constructed.

A FePt(7 nm)/Fe(3 nm) composite exhibited a square-like hysteresis loop (type B), which is analogous to the loops of hard magnets. A sharp decrease in magnetization was observed at around −570 kA/m (~−7160 Oe). We observed that the magnetization of the whole structure rotates to the in-plane direction quite rapidly, then to the −z-direction, resulting in a high remanent magnetization and a large coercive field. In contrast, if we kept the size of the FePt core and increased the thickness of the iron shell to 7 nm, a switching loop similar to that of soft magnets (type C) was observed in our simulation. The magnetization in the iron shell phase started to rotate before the external field was completely withdrawn, whereas the 90-degree domain wall motion was taking place, leading to a decrease in magnetization. Note that a small jump of magnetization was observed at saturation in the opposite direction, which can be attributed to the switching of the hard magnetic FePt core.

For a composite of FePt(8 nm)/Fe(5 nm) structure, a hysteresis loop exhibited a two-step process (type D) with a small coercive field but a large remanent magnetization, which is typical for "exchange-coupled" magnetic composite. Interestingly, a vortex structure formed in the soft magnetic shell as a nucleus, further decreasing the external field; the formed nucleus propagated to the soft/hard interface where the magnetization became pinned. As the soft shell is strongly exchanged coupled to the hardcore, the whole structure is completely reversed to the opposite direction. The simulation results show that in this case, there is a presence of an exchange coupling effect, and the reduction in coercivity is much more significant, which is very close to the theoretical predictions of Goh et al. [17].

Based on the simulation results, a switching type diagram, i.e., a representation of switching loops as a function of core radius and shell thickness, is constructed and illustrated in Figure 3. The radius of the FePt core changes from 3 nm to 11 nm, and the nominal thickness of Fe varies from 3 nm to 7 nm. In the diagram, if the FePt core size is small, the switching hysteresis exhibits a constricted loop (type A); this behavior is due to the hardcore as a small single-domain particle is easy to be switched. For large FePt core size, the switching type is sensitive to the thickness of the iron layer. The switching diagrams indicate that for a relatively large core size (r > 6 nm), there are three different types of switching loops with increasing iron layer thickness from 3 nm to 7 nm. If the shell thickness is lower than 4 nm, the switching mechanism is dominated by the FePt core. The hysteresis loop exhibits a square-like switching loop (type B), which is analogous to the loops of hard magnets. In contrast, when the iron layer thickness increases to 6 nm, the magnetization in the soft layer starts to rotate before the external field is reduced to zero, which results in a decrease in remanent magnetization. Further increasing the iron layer thickness, the calculated hysteresis loop is analogous to the loops of soft magnets (type C), as the iron layer plays a major role in the switching process. It should be noted that a transition region can be observed between type B and type C regions, as can be seen in Figure 3. By controlling the thickness of the iron shell to ~5 nm, a hysteresis loop with relatively high remanent magnetization and a small coercive field is confirmed, which can be attributed to a typical exchange-coupled effect in the composite magnets.

r (nm) \ d (nm)	3	4	5	6	7
3	A	A	A	A	A
4	A	A	A	A	A
5	B	A	A	A	C
6	B	B	A	C	C
7	B	B	D	C	C
8	B	B	D	C	C
9	B	B	D	C	C
10	B	B	D	C	C

Figure 3. The switching type diagram of FePt/Fe core-shelled structure as a function of the thickness of the iron shell and the radius of FePt core. d and r in the figure represents the thickness of iron shell in the range of 3–7 nm and the radius of FePt core in the range of 3–10 nm, respectively. The ○ indicates the location of the switching loops shown in Figure 2. The symbols A, B, C, and D represent the switching type A, B, C, and D of the core-shelled FePt/Fe structure in Figure 2, respectively.

In order to probe the expected influence of the composite structures on the magnetic properties, the calculated remanent magnetization and the coercive field are summarized in Figure 4. The remanent magnetization and coercive field can be tuned by changing the core size and soft layer thickness of the composite. Obviously, the remanent magnetization and coercive field decrease with the thickness increase in the iron layer. In contrast, increasing the hardcore radius can increase the coercive field of the composite. It is also interesting to note that for the composites with the same shell thickness, the coercive field has minor changes with the radius of FePt core when the core radius is beyond 6 nm. This phenomenon is because the increase in hardcore radius also increases the volume of the soft layer. In general, the magnetization properties of the composite are strongly influenced by the switching type of the hysteresis loops.

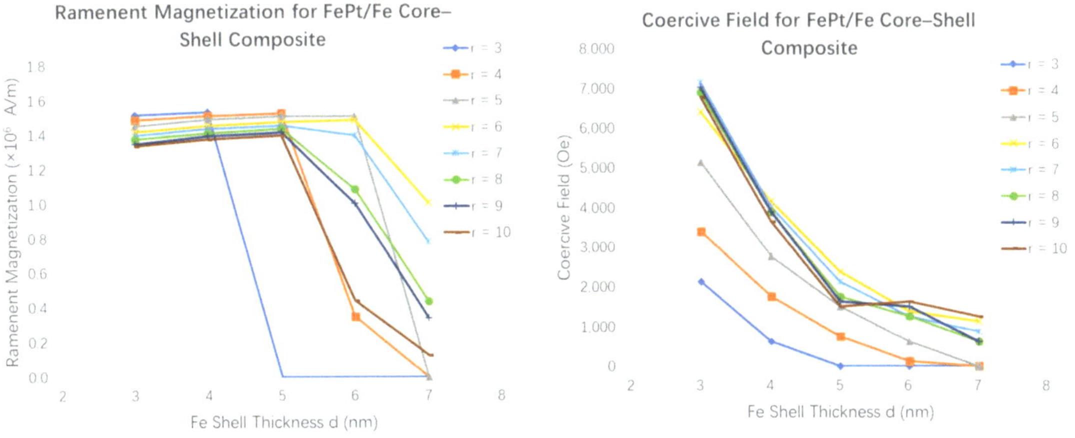

(a) **(b)**

Figure 4. The magnetic properties as a function of the thickness of the iron shelland the radius of FePt core. (**a**) The remanent magnetization (**b**) the coercivities. In this figure, r is the radius of FePt core.

For achieving high-density memory devices, bit patterned media with a lower coercive field, lower anisotropy field, smaller island sizes, and a high remanence is required [3,9,15,19]. Hereby, a remanent magnetization-coercive field (M_r–H_c) diagram was introduced to illustrate the shape of the hysteresis loops, as shown in Figure 5. It is easy to see the typical hard magnets with high coercive field and high remanence located in the upper right corner of the M_r–H_c diagram, while typical soft magnets with low coercive field and low remanence are located in the lower-left corner of this diagram. The magnetic composite with the geometric structures located in the upper left corner of Figure 5 is our aim material, which meets the requirements for high-density storage devices. As expected, type B hysteresis loops exhibit high remanent magnetization and high coercive field, while type C loops show a hysteresis similar to soft magnetic materials with relatively small remanent magnetization and small coercive field. Respectively, in the M_r–H_c diagram, the magnetic composites of type B and type C are distributed in the upper right and lower left corners of the diagram in a relatively large range. Thus, both type B and type C hysteresis loops are not suitable for the bit patterned memory devices. The candidate structure for bit patterned media should have a hysteresis loop of type A or type D. As shown in Figure 5, the magnetic properties of composites of type A and type D are concentrated in the upper left corner of the diagram with a relatively small range, which means that the exchange coupling effect of type A/type D loops can lead to the smaller coercive field and maintaining high remanent magnetization. If considering the size of the composite island, the type A hysteresis loop with a core radius around 3–4 nm and a shell thickness of 3–5 nm may potentially find uses in future memory devices. A disadvantage of the type A hysteresis loop is that a relatively high anisotropy field is observed, which results in a higher switching field and may limit its applications. Therefore, further studies are necessary to decrease the anisotropy field of the type A hysteresis loop.

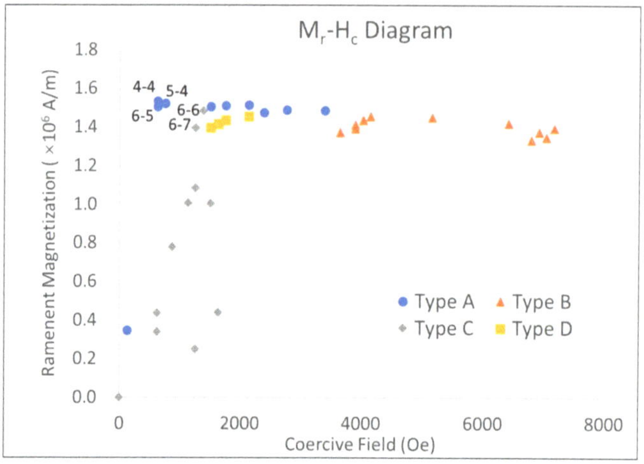

Figure 5. The remanent magnetization-coercive field (M_r–H_c) diagram for the core-shell FePt/Fe composite. Solid circles, triangles, diamonds, and squares denote the M_r and H_c of loops of type A–D in Figure 3, respectively. Some key structures with their geometric structure (FePt radius-Fe thickness, in the unit of nm, e.g., 4–4 represents a composite structure with a core of 4 nm radius and a shell of 4 nm thickness) are also noted in the diagram.

4. Conclusions

In this work, a switching type diagram as a function of the iron shell thickness and the FePt core radius was constructed for FePt/Fe core-shell composite structure based on a micromagnetic simulation. Four types of switching types were found in the core-shell type composite with different core sizes and shell thicknesses. This study demonstrated that the magnetic switching hysteresis loop could be tuned by controlling the core size and the shell thickness of the core-shell composite. Furthermore, the magnetic domain configuration during the switching process was analyzed to obtain insight into the mechanisms of the switching properties. The remanent magnetization and the coercive field for the core-shell type composite were predicted based on the switching type diagram. The possible composite structure for bit patterned media should have a hysteresis of type A or type D because the exchange coupling effect between the hard magnetic core and soft iron shell lowered the coercive field but maintained high remanent magnetization. These simulation results should motivate further exploration of magnetic composite structures with smaller sizes and better performance, and it will be beneficial in future design and development of bit patterned memory device applications.

Author Contributions: Conceptualization, Y.L., Z.W. and P.W.; methodology, P.W.; validation, P.W.; investigation, Y.W. and Z.Z.; data curation, Y.Z.; writing—original draft preparation, Y.W. and Z.Z.; writing—review and editing, Y.Z.; visualization, P.W.; supervision, P.W.; project administration, P.W.; funding acquisition, Y.L., Z.W. and P.W. All authors have read and agreed to the published version of the manuscript.

Funding: This work was supported by State Key Lab for Advanced Metals and Materials in the University of Science and Technology Beijing with the contract No. 2021-ZD02, the outstanding young research talents program (2016) in colleges and universities of Fujian province, and the Cultivating Advanced Functional Materials Research Group Program of Xiamen Institute of Technology with the contract No. KYTD202004. This work was also supported by the Key Research and Development Program of the Science and Technology Ministry of China (No. 2017YFA0403704) and the National Natural Science Foundation of China (Grant No. 12075320).

Institutional Review Board Statement: Not applicable.

Informed Consent Statement: Not applicable.

Data Availability Statement: Not applicable.

Conflicts of Interest: The authors declare no conflict of interest.

References

1. Albrecht, T.R.; Arora, H.; Ayanoor-Vitikkate, V.; Beaujour, J.-M.; Bedau, D.; Berman, D.; Bogdanov, A.L.; Chapuis, Y.-A.; Cushen, J.; Dobisz, E.E.; et al. Bit-Patterned Magnetic Recording: Theory, Media Fabrication, and Recording Performance. *IEEE Trans. Magn.* **2015**, *51*, 1–42. [CrossRef]
2. Kuo, D.; Xiao, S.; Terris, B. Bit Patterned Magnetic Recording Media. In *Reference Module in Materials Science and Materials Engineering*; Elsevier: Amsterdam, The Netherlands, 2016; pp. 1–7.
3. Goll, D.; Bublat, T. Large-area hard magnetic $L1_0$-FePt and composite $L1_0$-FePt based nanopatterns. *Phys. Status Solidi* **2013**, *210*, 1261–1271. [CrossRef]
4. Varvaro, G.; Laureti, S.; Fiorani, D. $L1_0$ FePt-based thin films for future perpendicular magnetic recording media. *J. Magn. Magn. Mater.* **2014**, *368*, 415–420. [CrossRef]
5. Varvaro, G.; Albertini, F.; Agostinelli, E.; Casoli, F.; Fiorani, D.; Laureti, S.; Lupo, P.; Ranzieri, P.; Astinchap, B.; Testa, A.M. Magnetization reversal mechanism in perpendicular exchange-coupled Fe/$L1_0$–FePt bilayers. *New J. Phys.* **2012**, *14*, 073008. [CrossRef]
6. Huang, L.S.; Hu, J.F.; Zong, B.Y.; Zeng, S.W.; Chen, J.S. Magnetic properties of $L1_0$-FePt/Fe exchange-coupled composite nanodots. *J. Phys. D Appl. Phys.* **2014**, *47*, 245001. [CrossRef]
7. Poudyal, N.; Gandha, K.H.; Liu, J.; Elkins, K.E.; Arami, H.; Liu, J.P. Ferromagnetic FePt/Au Core/Shell Nanoparticles Prepared by Solvothermal Annealing. *IEEE Magn. Lett.* **2016**, *7*, 4102105. [CrossRef]
8. Kurosu, R.; Sugawara, A.; Iwama, H.; Doi, M.; Shima, T. Improvement of Magnetic Properties in Microfabricated FePt and Fe Nano-Composites Upon Post-Annealing. *IEEE Magn. Lett.* **2017**, *8*, 5502804. [CrossRef]
9. Suess, D.; Schrefl, T.; Fähler, S.; Kirschner, M.; Hrkac, G.; Dorfbauer, F.; Fidler, J. Exchange spring media for perpendicular recording. *Appl. Phys. Lett.* **2005**, *87*, 012504. [CrossRef]
10. Suess, D. Multilayer exchange spring media for magnetic recording. *Appl. Phys. Lett.* **2006**, *89*, 113105. [CrossRef]
11. Wang, H.; Zhao, H.; Rahman, T.; Isowaki, Y.; Kamata, Y.; Maeda, T.; Hieda, H.; Kikitsu, A.; Wang, J.-P. Fabrication and Characterization of FePt Exchange Coupled Composite and Graded Bit Patterned Media. *IEEE Trans. Magn.* **2013**, *49*, 707–712. [CrossRef]
12. Goll, D.; Macke, S. Thermal stability of ledge-type $L1_0$-FePt/Fe exchange-spring nanocomposites for ultrahigh recording densities. *Appl. Phys. Lett.* **2008**, *93*, 152512. [CrossRef]
13. Lomakin, V.; Shaojing, L.; Livshitz, B.; Inomata, A.; Bertram, H.N. Patterned Media for 10 Tb/in^2 Utilizing Dual-Section "Ledge" Elements. *IEEE Trans. Magn.* **2008**, *44*, 3454–3459. [CrossRef]
14. Goll, D.; Breitling, A. Coercivity of ledge-type $L1_0$-FePt/Fe nanocomposites with perpendicular magnetization. *Appl. Phys. Lett.* **2009**, *94*, 052502. [CrossRef]
15. Speliotis, T.; Giannopoulos, G.; Niarchos, D.; Li, W.F.; Hadjipanayis, G.; Barucca, G.; Agostinelli, E.; Laureti, S.; Peddis, D.; Testa, A.M.; et al. Ledge-type Co/$L1_0$-FePt exchange-coupled composites. *J. Appl. Phys.* **2016**, *119*, 233904. [CrossRef]
16. Wang, Z.; Zheng, Y.; He, W.; Ma, X.; Wu, P. Vortex Domain Wall Structures and Exchange Coupled Magnetization Switching in Bit Patterned Media. *J. Magn.* **2021**, *26*, 32–40. [CrossRef]
17. Goh, C.-K.; Yuan, Z.-M.; Liu, B. Magnetization reversal in enclosed composite pattern media structure. *J. Appl. Phys.* **2009**, *105*, 083920. [CrossRef]
18. Ma, B.; Wang, H.; Zhao, H.; Sun, C.; Acharya, R.; Wang, J.-P. $L1_0$ FePt/Fe Exchange Coupled Composite Structure on MgO Substrates. *IEEE Trans. Magn.* **2010**, *46*, 2345–2348. [CrossRef]
19. Ma, B.; Wang, H.; Zhao, H.; Sun, C.; Acharya, R.; Wang, J.-P. Structural and magnetic properties of a core-shell type $L1_0$ FePt/Fe exchange coupled nanocomposite with tilted easy axis. *J. Appl. Phys.* **2011**, *109*, 083907. [CrossRef]
20. Büttner, F.; Moutafis, C.; Bisig, A.; Wohlhüter, P.; Günther, C.M.; Mohanty, J.; Geilhufe, J.; Schneider, M.; von Korff Schmising, C.; Schaffert, S.; et al. Magnetic states in low-pinning high-anisotropy material nanostructures suitable for dynamic imaging. *Phys. Rev. B* **2013**, *87*, 134422. [CrossRef]
21. Pfau, B.; Günther, C.M.; Guehrs, E.; Hauet, T.; Hennen, T.; Eisebitt, S.; Hellwig, O. Influence of stray fields on the switching-field distribution for bit-patterned media based on pre-patterned substrates. *Appl. Phys. Lett.* **2014**, *105*, 132407. [CrossRef]
22. Wu, P.; Liang, Y. Lattice Phase Field Model for Nanomaterials. *Materials* **2021**, *14*, 7317. [CrossRef] [PubMed]
23. Jubert, P.-O. Micromagnetic Simulations of Exchange-Coupled Core–Shell Particulate Media. *IEEE Trans. Magn.* **2014**, *50*, 3002004. [CrossRef]
24. Lin, J.Q.; Chen, J.P.; Tan, Z.Y.; Chen, Y.; Chen, Z.F.; Li, W.A.; Gao, X.S.; Liu, J.M. Manipulation of Skyrmion Motion Dynamics for Logical Device Application Mediated by Inhomogeneous Magnetic Anisotropy. *Nanomaterials* **2022**, *12*, 278. [CrossRef] [PubMed]
25. Sinnecker, E.; Garcia-Martin, J.M.; Altbir, D.; D'Albuquerque, E.C.J.; Sinnecker, J.P. A Magnetic Force Microscopy Study of Patterned T-Shaped Structures. *Materials* **2021**, *14*, 1567. [CrossRef] [PubMed]
26. Hu, C.; Zhang, Z.; Yang, T.; Li, W.; Chen, L. Phase-field simulation of magnetic microstructure and domain switching in $(Tb_{0.27}Dy_{0.73})Fe_2$ single crystal. *AIP Adv.* **2021**, *11*, 015207. [CrossRef]

27. Sheng, Q.; Liu, X.L.; Chen, W.J.; Xiong, W.M.; Jiang, G.L.; Zheng, Y. Domain stability and polar-vortex transformations controlled by mechanical loads in soft ferromagnetic nanodots. *AIP Adv.* **2016**, *6*, 035003. [CrossRef]
28. Schabes, M.; Aharoni, A. Magnetostatic interaction fields for a three-dimensional array of ferromagnetic cubes. *IEEE Trans. Magn.* **1987**, *23*, 3882–3888. [CrossRef]
29. Wang, X.P.; Garcıa-Cervera, C.J.; Weinan, E. A Gauss–Seidel Projection Method for Micromagnetics Simulations. *J. Comput. Phys.* **2001**, *171*, 357–372. [CrossRef]
30. Huang, L.S.; Hu, J.F.; Chen, J.S. Critical Fe thickness for effective coercivity reduction in FePt/Fe exchange-coupled bilayer. *J. Magn. Magn. Mater.* **2012**, *324*, 1242–1247. [CrossRef]
31. O'Handley, R.C. *Modern Magnetic Materials: Principles and Applications*; John Wiley & Sons: New York, NY, USA, 2000.
32. Kikuchi, N.; Okamoto, S.; Kitakami, O. The critical size between single domain and multidomain in $L1_0$-FePt particles. *J. Appl. Phys.* **2008**, *103*, 07D511. [CrossRef]

materials

Article

Investigation of Phase Transitions in Ferromagnetic Nanofilms on a Non-Magnetic Substrate by Computer Simulation

Sergey V. Belim

Physics Department, Omsk State Technical University, Omsk 644050, Russia; sbelim@mail.ru

Abstract: Magnetic properties of ferromagnetic nanofilms on non-magnetic substrate are examined by computer simulation. The substrate influence is modeled using the two-dimensional Frenkel-Kontorova potential. The film has a cubic crystal lattice. Cases of different ratio for substrate period and ferromagnetic film period are considered. The difference in film and substrate periods results in film deformations. These deformations result in a change in the magnetic properties of the film. The Ising model and the Metropolis algorithm are used for the study of magnetic properties. The dependence of Curie temperature on film thickness and substrate potential parameters is calculated. Cases of different values for the coverage factor are considered. The deformation of the film layers is reduced away from the substrate when it is compressed or stretched. The Curie temperature increases when the substrate is compressed and decreases when the substrate is stretched. This pattern is performed for films with different thicknesses. If the coating coefficient for the film is different from one, periodic structures with an increased or reduced concentration of atoms are formed in the film first layer. These structures are absent in higher layers.

Keywords: thin film; nonmagnetic substrate; phase transition; computer simulation; Frenkel-Kontorova potential

Citation: Belim, S.V. Investigation of Phase Transitions in Ferromagnetic Nanofilms on a Non-Magnetic Substrate by Computer Simulation. *Materials* **2022**, *15*, 2390. https://doi.org/10.3390/ma15072390

Academic Editor: Dana Popescu

Received: 11 February 2022
Accepted: 22 March 2022
Published: 24 March 2022

1. Introduction

Ferromagnetic film composite systems on a non-magnetic substrate are widely used in spintronics devices. A substrate of non-magnetic material has no effect on the giant magnetic resistance. The ability to control magnetization in such a system is important. Antiferromagnetic or ferromagnetic substrates change the state of the magnetic film, creating an exchange bias. This effect is nonlinear. Non-magnetic substrates can change the mechanical properties of the film. Magnetostrictive effects change the magnetization of the film during mechanical deformations. The substrate deformations can cause a ferromagnetic phase transition at temperatures below the Curie point for the free film.

The influence of the substrate on various physical properties (optical, electrical and magnetic) in epitaxial films has been experimentally studied [1–5]. The study of the substrate deformations effect on the magnetic properties in thin ferromagnetic films is of great interest. Film deformations may be caused by thermal expansion of the substrate [6,7]. The substrate may be made of ferroelectric material. In this case, compression or tension deformations can be caused by an electric field [8–10]. We're looking at non-magnetic substrates. The effects of ferromagnetic and antiferromagnetic substrates require consideration of additional exchange and dipole–dipole interactions.

Different types of interactions may occur on the film and substrate interface. Chemical bonds at the interface between the film and the substrate play a large role in the formation of the film by various techniques. These bonds can determine the type of crystal lattice for the film. One type of crystal lattice is studied in this work; therefore, the main type of interaction by Van der Waals forces is considered. Different types of interactions should be considered when studying structural phase transitions induced by the substrate in nanofilms.

The effect of a non-magnetic substrate on epitaxial films is due to the interaction of its atoms with the film atoms. A change in the mutual arrangement of atoms occurs when the substrate is deformed. The film atoms are displaced due to interaction with the substrate atoms. Deformation of the film occurs due to displacement of its atoms. Magnetostrictive effects lead to a change in the magnetic properties of the film. The dependence of the exchange integral for the spin–spin interaction on distance has a main influence on the change in magnetic properties. Studies of the substrate deformations effect on the magnetic properties of single-layer 2D films on a non-magnetic substrate [11–13] have shown that uniform substrate deformations can lead to non-uniform film deformations at the free ends of the film. Islands with increased concentration of atoms are formed in the film when the substrate is compressed. These islands are separated by a periodic structure with a reduced concentration of atoms. Islands with reduced concentration of atoms are formed when the substrate is stretched. The space between the islands is a periodic superstructure with an increased concentration of atoms. If it is limited to the case with the exponential law for changing the exchange interaction with distance, then in both cases the Curie temperature of the ferromagnetic film decreases. If the exchange interaction decreases with the distance according to a slower law, then with a small compression of the substrate, the Curie temperature can increase. These patterns are qualitatively consistent with the experiment's results [14–16].

We can expect a more complex dependence for magnetic properties when deforming a substrate for films with a thickness of more than one layer. Curie temperature depends on film thickness at any substrate [17,18]. The non-magnetic substrate interacts only with the film atoms at the boundary of their contact. The substrate does not directly affect the second and subsequent film layers due to the rapid decrease in intermolecular interaction with distance. The layer of film atoms directly contacting the substrate is influenced by the substrate and the subsequent atoms layer. Film deformations are complex. The deformation of the film layers is reduced when removed from the substrate. We can expect that an increase in film thickness will reduce the effect of the substrate on the film magnetic properties.

Computer simulation of a thin ferromagnetic film on a substrate with a periodic potential is performed in this article. The simulation was carried out in two stages. In the first step, the size change for the film layers is examined by compressing and stretching the substrate. In the second step, the change in the magnetic properties of the film is studied under the influence of these deformations.

2. Model and Methods

The substrate influences the arrangement of the ferromagnetic film atoms due to the interaction between their atoms. We are studying the case for a substrate with a cubic crystal lattice. Atoms in such a substrate create a periodic potential for film atoms. This potential can be written as a two-dimensional Frenkel–Kontorova potential [19].

$$U_{sub} = \frac{A}{2}\sum_{n}\left(1 - \cos\left(\frac{2\pi}{b}x_n\right) \times \cos\left(\frac{2\pi}{b}y_n\right)\right) \quad (1)$$

b is the substrate crystal lattice period, A is the substrate potential amplitude and (x_n, y_n) are the film atom coordinates at the border with the substrate.

Interatomic interaction rapidly decreases with distance. Van der Waals forces obey the power law r^{-6}. The substrate has a direct effect only on the film layer that contacts it. This first atomic film layer can deform and cause deformations in the following layers. Harmonic approximation is used to describe the interaction energy between atoms in a film.

$$U_{\text{int}} = \frac{g}{2}\sum_{n}\left((x_{n+1} - x_n - a)^2 + (y_{n+1} - y_n - a)^2 + (z_{n+1} - z_n - a)^2\right) \quad (2)$$

g is an elastic constant and a is the period of the crystal lattice for the undeformed film.

We are looking at a cubic lattice film. The substrate's potential depends only on the x and y coordinates. We do not consider film deformations perpendicular to the OXY plane.

Each ferromagnetic film atom is described by the position in space (x_n, y_n, z_n) and spin S_n. We use the Ising model to describe the film's magnetic properties. The spin can take one of two values (1/2 or −1/2) in this model. Magnetic phenomena in the film are described by Hamiltonian of the Ising model.

$$H = -\sum_{i,j} J\left(\left|\vec{r}_i - \vec{r}_j\right|\right) S_i S_j \tag{3}$$

The exchange integral J depends on the distance between the spins. The law of this dependence can be different for different substances. The exchange integral rapidly decreases with distance in most substances. This case can be described by exponential law.

$$J\left(\left|\vec{r}_i - \vec{r}_j\right|\right) = J_0 \exp\left(-\left(\left|\vec{r}_i - \vec{r}_j\right| - a\right)/r_0\right) \tag{4}$$

The parameter r_0 determines the speed of energy decrease for exchange interaction between spins with distance. Summing only on the nearest neighbors remains in Hamiltonian for the Ising model in this case. The descending of the exchange integral according to a slower law leads to long-range effects. Long-range effects correct critical behavior near phase transition line [20].

It is necessary to minimize the total energy for the system to find an equilibrium state.

$$U_{sub} + U_{\text{int}} + H \to \min \tag{5}$$

Spin dynamics are characterized by a much shorter relaxation time than the mechanical movement of atoms. Therefore, the task can be divided into two parts. The equilibrium mechanical state for film atoms may be determined in a first step. After that, the spin dynamics can be studied at an unchanged position for atoms. We first minimize the mechanical part of the potential energy.

$$U = U_{sub} + U_{\text{int}} \to \min \tag{6}$$

The atom's coordinates are used to calculate exchange integrals for the interaction between neighboring spins.

Changing the substrate potential period simulates substrate compression. Film edges are rigidly fixed at substrate boundaries in the considered model. Linear dimensions of the first film layer coincide with linear dimensions of the substrate. Higher layers can resize because they do not interact directly with the substrate.

The sequential approximation method is used to search for an equilibrium state. The film atoms are located at the nodes of the cubic crystal lattice with period a in zero approximation. All film atoms are bypassed at each iteration. Each atom is shifted by a random vector (dx, dy) in the OXY plane. The displacement of the atom does not exceed $0.01a$. If the new position is more energetically advantageous, then it is accepted, otherwise the atom returns to the original position. The total number of iterations is 10,000. As the computer experiment showed, subsequent iterations do not change the position of atoms. The distribution of atoms in the film in the basic state depends on the ratio of the substrate crystal lattice periods and the undisturbed film b/a and the coating coefficient θ.

$$\theta = L/M \tag{7}$$

L is the number of film atoms along one axe and M is the number of substrate minima along one axe.

The Wolf cluster algorithm [21] is used to study phase transitions in thin films. The magnetization in the film m is used as order parameter for the system.

$$m = \sum S_i / N \tag{8}$$

N is the number of atoms in the film.

Periodic boundary conditions are used in modeling magnetic phenomena. Finite dimensional scaling theory is used to calculate phase transition parameters [22]. Systems with different linear dimensions are modeled. The behavior for an infinite system is approximated based on the results for the endpoint systems. The fourth order Binder cumulants [23] are calculated to determine the phase transition temperature.

$$U = 1 - \frac{\langle m^4 \rangle}{3 \langle m^2 \rangle^2} \tag{9}$$

Angle brackets denote averaging by thermodynamic configurations. The dependence of Binder cumulants on temperature is calculated for systems with different linear dimensions. The Binder cumulant value does not depend on the size of the system at the phase transition temperature, so all plots must intersect at one point [22]. The Curie temperature Tc is calculated from the crossing point of the plots. The temperature of system T is measured in relative units J_0/k_B (k_B—Boltzmann constant) in computer modeling.

Crystal lattices with different geometry can be obtained by varying lattice periods along the OX and OY axes in potential (1). Examination of substrates with any crystal lattices can be carried out by adding periodic terms to the potential (1). The algorithm proposed in this paper can be used to model a system with any substrate potential. If the symmetry of the crystal lattice for the film and the substrate is different, then competition of the ordering types is observed in the system.

3. Results

Computer experiment performed for films with size $L \times L \times D$. The film thickness varies from $D = 2$ to $D = 8$ in increments $\Delta D = 2$. A film having a thickness one-layer $D = 1$ is also considered. The undeformed crystal lattice period in the film is $a = 1$ in a computer experiment. Values from $b = 0.9$ to $b = 1.10$ in increments of $\Delta b = 0.05$ are selected for the substrate period. The linear dimensions for the system vary from $L = 20$ to $L = 100$ in increments $\Delta L = 10$. The exchange integral parameter is $r = 0.1$. The value of r affects the rate at which the exchange interaction with distance descends. Increasing this parameter reduces the change in phase transition temperature during substrate deformation. The general patterns remain unchanged. Substrate potential amplitude $A = 0.5$ was used in the simulation. Increasing the substrate potential amplitude slightly changes the phase transition temperature. Decreasing the substrate potential amplitude brings the Curie temperature closer to the undeformed film temperature.

Systems with a coverage factor $\theta = 1$ are considered in the first stage. In this case, the first layer atoms interacting with the substrate are located in the minima of the potential U_{sub}. These atoms offset changes the linear dimensions of the layer. The relative change in layer size along one direction is indicated through the ε.

$$\varepsilon = \Delta L / L \tag{10}$$

The relationship of the relative increase in the layer's linear length ε depending on its number d at $b = 1.10$ is shown in Figure 1.

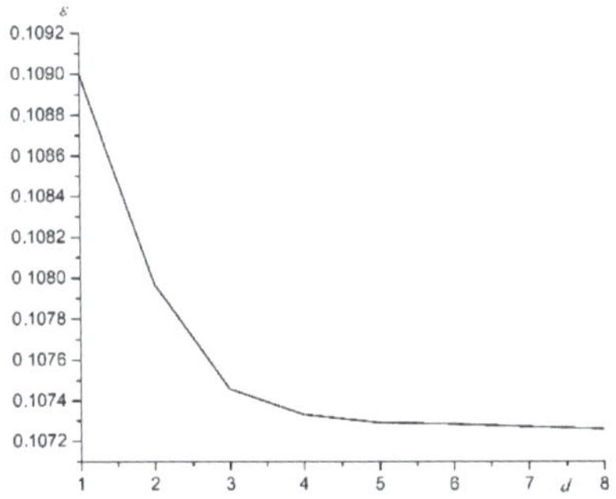

Figure 1. The relationship of the relative increase in the layer's linear length ε depending on its number d at b = 1.10.

As seen in Figure 1, substrate deformations affect all layers. The substrate influence decreases as the number of the atomic layer increases. The first film layer interacts most intensively with the substrate. The deformations for the first layer almost exactly repeat the deformations for the substrate. Deformations are transmitted to the following layers by elastic forces of interaction between the film atoms. For layers with numbers greater than the first, there is a competition of elastic forces. The first force tends to deform the layer under the influence of the substrate. The second force tends to return the atoms to their original equilibrium state. The first force is created by layers located closer to the substrate. The second force is created by higher layers. The first force decreases as the distance to the substrate increases. Substrate deformations affect each next layer less.

The deformation of the film changes the distances between the spins. This change affects the exchange integral and Curie temperature of the film. A similar pattern is observed for another values $b \neq 1$. System compression occurs at substrate period b < 1. The distance between the spins is reduced. The exchange integral increases. Curie's temperature is rising. System straining occurs at b > 1. The phase transition temperature decreases. A plot for Curie temperature versus substrate period is shown in Figure 2 for films with different thicknesses.

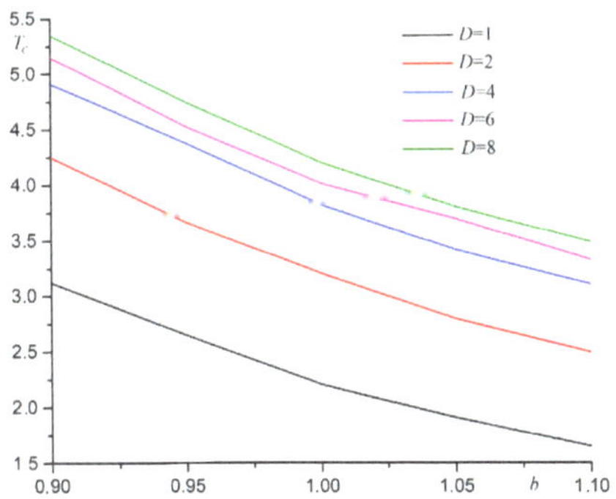

Figure 2. A plot of Curie temperature T_C versus substrate period b for films with different thicknesses D.

As shown in Figure 2, the phase transition temperature is lowered by stretching the substrate and increased by compressing the substrate. The phase transition temperature increases with increasing film thickness. The general pattern for change in phase transition temperature from substrate deformation is the same for films with different thicknesses.

The dependence of Curie temperature on film thickness at various substrate deformations is shown in Figure 3.

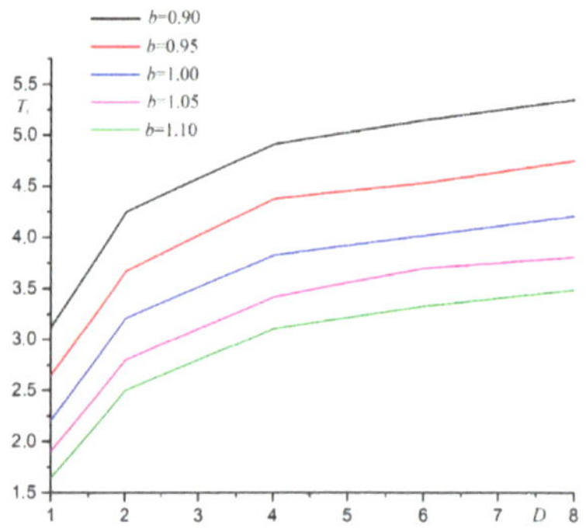

Figure 3. The dependence of Curie temperature T_C on film thickness D at various substrate deformations b.

As shown in Figure 3, the Curie temperature increases with increasing film thickness. Phase transition temperature reaches asymptotic value $T_0 = 4.51$ at $b = 1.00$. This asymptotic value corresponds to the Curie temperature for the three-dimensional Ising model. The asymptotic value increases when the substrate is deformed.

The change in Curie temperature is caused by a change in the exchange integral for the interaction between spins. The exchange integral changes due to a decrease or increase in the distance between atoms possessing spins. The change in Curie temperature can be considered as a manifestation of the magnetostrictive effect during film deformation.

Systems with a coverage factor not equal to one are considered in the second stage. If the coating coefficient is $\theta < 1$, then the atom density is less than for a free lattice. This atom arrangement leads to a tighter package. Curie temperature rises. A less dense package of atoms is realized for the coating factor $\theta > 1$. Curie temperature drops. The phase transition temperature is further varied by substrate deformation. The main patterns of change in Curie temperature when the substrate period changes remain the same as when the single coating coefficient $\theta = 1$ (Figure 4). The law for changing the phase transition temperature with increasing film thickness remains the same as for a single coating coefficient $\theta = 1$ (Figure 5).

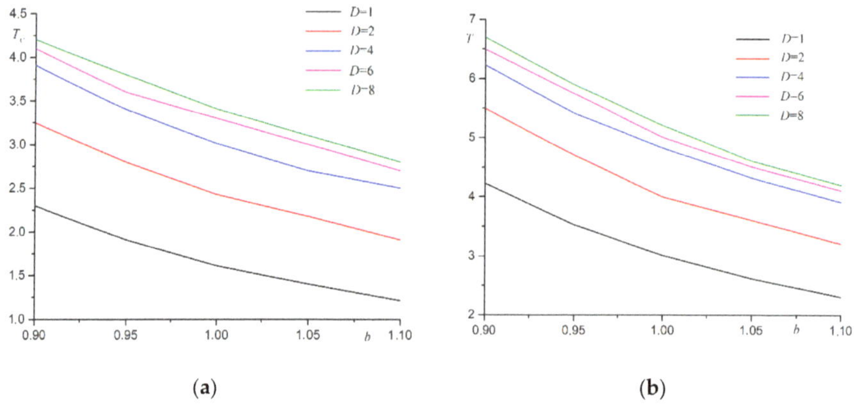

Figure 4. Dependence of Curie temperature T_C on substrate period b for films with different thicknesses D at different coating coefficients: (**a**) $\theta = 0.9$, (**b**) $\theta = 1.1$.

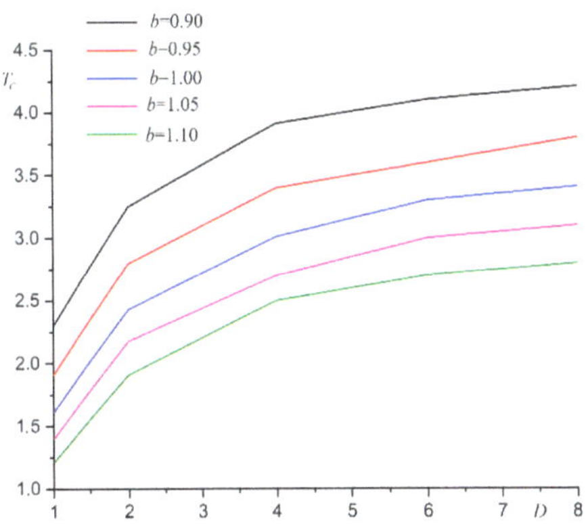

(a)

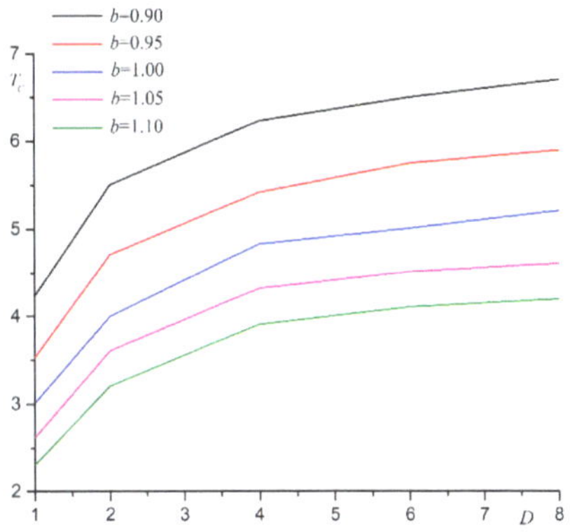

(b)

Figure 5. Dependence of phase transition temperature T_C on film thickness D, at different substrate periods b at different coating coefficients: (**a**) θ = 0.9, (**b**) θ = 1.1.

The deformation's nature changes with a coating coefficient not equal to one ($\theta \neq 1$). Periodic structures are formed in the first film layer under the influence of the substrate. The shape of these structures depends on the coating coefficient. If the coating coefficient is less than one ($\theta < 1$), square regions with an increased concentration of atoms are formed. These regions are separated by spaces with a reduced concentration of atoms (Figure 6a). These structures are weakly manifested in the second layer (Figure 6b). Atoms are located at the nodes of the square lattice in higher layers.

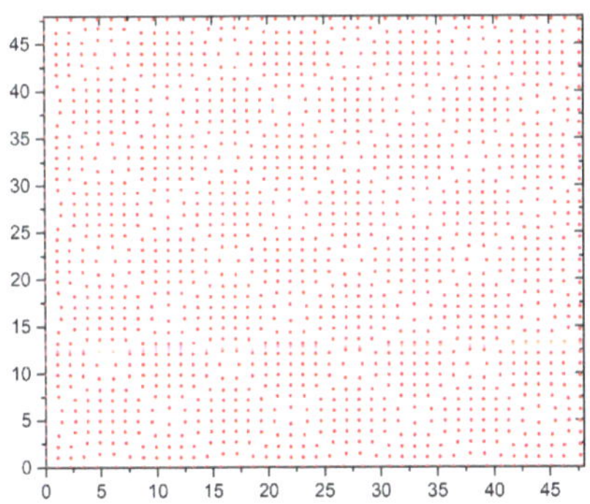

(a)

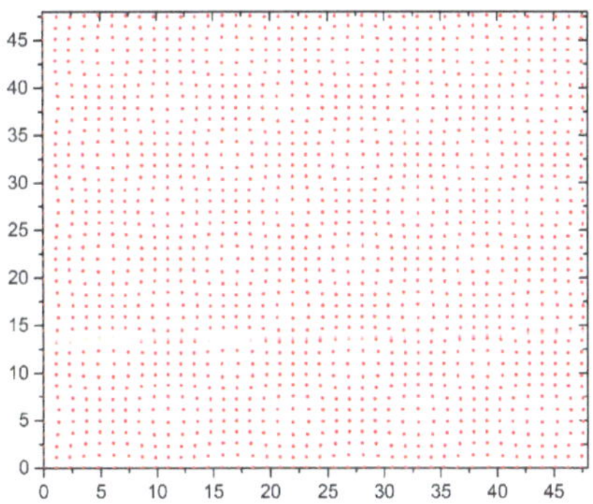

(b)

Figure 6. Placement of atoms in the first two layers at θ = 0.9 and b = 1.10: (**a**) the first layer, (**b**) the second layer.

If the coating coefficient is greater than one ($\theta > 1$), then square regions with a reduced concentration of atoms are formed in the first layer under the influence of the substrate (Figure 7a). These regions are separated by a space with an increased concentration of atoms. These structures are weakly manifested in the second layer (Figure 7b). These structures are practically untraceable in higher layers.

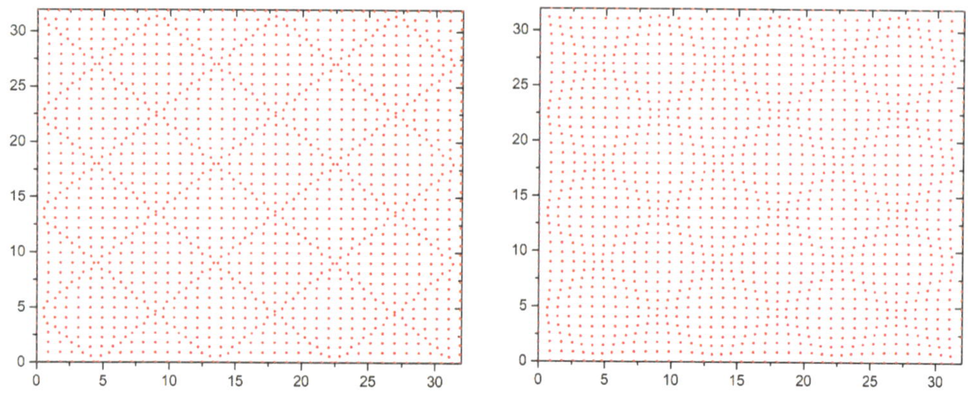

(**a**) (**b**)

Figure 7. Placement of atoms in the first two layers at $\theta = 1.1$ and $b = 0.90$: (**a**) the first layer, (**b**) the second layer.

Periodic structures in the first layer are preserved when the substrate is deformed. The distance between atoms changes. The relative arrangement of atoms remains unchanged.

4. Conclusions

Ferromagnetic phase transition in thin multilayer films on a deformable non-magnetic substrate was studied by computer modeling. Substrate deformations change phase transition temperature. The increase in film thickness also increases the phase transition temperature. These facts are consistent both with the experimental data [14–16] and with the results of modelling by other methods [17,18]. The film coating coefficient also affects the phase transition temperature. Deformation of the substrate causes general compression or stretching in the film without changing its structure at a single coating coefficient. The layer interacting with the substrate experiences the greatest deformation. The deformations of the remaining layers decrease when removed from the substrate. If the coating coefficient is not one, periodic structures are formed in the layer interacting with the substrate. These structures are islands with increased or decreased concentrations of atoms. These structures are most pronounced in monoatomic films. These structures are weakly expressed in higher layers. The formation of periodic structures has little effect on the phase transition temperature in the multilayer films. These structures can influence the functioning of spintronic elements implemented based on film structures on the substrate.

We will compare the results with some experimental facts. For thin films CrN on the Si substrate, a structural phase transition is observed with an increase in film thickness [24]. This phase transition is associated with a decrease in the effect of deformations on the film/substrate interface with an increase in film thickness. Experimental studies of a thin VO_2 film on a sapphire substrate showed that its deformation depends on thickness [25]. The phase transition temperature also increases with increasing film thickness. An experimental study of highly deformed thin films $BiFeO_3$ showed a significant effect of the substrate interface on the structure and structural phase transitions in them [26]. Increasing the film thickness reduces the dependence on the substrate. The state of the film becomes more stable.

Experimental studies of thin films $Sm_{0.35}Pr_{0.15}Sr_{0.5}MnO_3$ on various single-crystal substrates confirm the conclusions about the change in Curie temperature [27]. The film undergoes compression deformation when deposited on the substrate $LaAlO_3$. The temperature of the ferromagnetic phase transition rises to 165 K. The substrates $SrTiO_3$ and $La_{0.18}Sr_{0.82}$ stretch the film. The Curie temperature drops to 120 K and 130 K, respectively. Thin films $Sm_{0.53}Sr_{0.47}MnO_3$ on a single-crystal substrate $LaAlO_3$ undergo compression deformations [28]. This compression leads to an increase in the phase transition temperature. An experimental study of thin films FeRh on substrates MgO, $SrTiO_3$ and $KTaO_3$ demonstrated the dependence of Curie temperature on deformations and film thickness [29]. Compression strains shift the Curie point above room temperature. Reducing the film thickness to less than 15 nm significantly reduces the phase transition temperature.

Funding: The reported study was funded by RFBR, project number 20-07-00053.

Institutional Review Board Statement: Not applicable.

Informed Consent Statement: Not applicable.

Data Availability Statement: Not applicable.

Conflicts of Interest: The authors declare no conflict of interest.

References

1. Ghidini, M.; Dhesi, S.S.; Mathur, N.D. Nanoscale magnetoelectric effects revealed by imaging. *J. Magn. Magn. Mater.* **2021**, *520*, 167016.
2. Averyanov, D.V.; Sokolov, I.S.; Tokmachev, A.M.; Parfenov, O.E.; Karateev, I.A.; Taldenkov, A.N.; Storchak, V.G. High-temperature magnetism in graphene induced by proximity to EuO. *ACS Appl. Mater. Interfaces* **2018**, *10*, 20767–20774.
3. Kim, J.; Akinaga, H.; Kim, J. Observation of magnetic domain structures in epitaxial MnAs film on GaAs(001) with temperature hysteresis. *Appl. Phys. Lett.* **2011**, *98*, 102511.
4. Paes, V.Z.C.; Graff, I.L.; Varalda, J.; Etgens, V.H.; Mosca, D.H. The role of magnetoelastic and magnetostrictive energies in the magnetization process of MnAs/GaAs epilayers. *J. Phys. Condens. Matter* **2012**, *25*, 046003. [PubMed]
5. Fu, X.; Warot-Fonrose, B.; Arras, R.; Seine, G.; Demaille, D.; Eddrief, M.; Etgens, V.; Serin, V. In situ observation of ferromagnetic order breaking in MnAs/GaAs(001) and magnetocrystalline anisotropy of α-MnAs by electron magnetic chiral dichroism. *Phys. Rev. B* **2016**, *93*, 104410.
6. Shi, M.; Xu, Y.; Zhang, Q.; Yu, Q.; Gu, C.; Zhao, Z.; Guo, L. Impact of heat-treatment conditions on ferroelectric, ferromagnetic and magnetoelectric properties of multi-layered composite films of $Ba_{0.9}Ca_{0.1}TiO_3/CoFe_2O_4$. *J. Mater. Sci. Mater. Electron.* **2019**, *30*, 19343–19352.
7. Ivanov, M.S.; Buryakov, A.M.; Vilarinho, P.M.; Mishina, E.D. Impact of compressive and tensile epitaxial strain on transport and nonlinear optical properties of magnetoelectric $BaTiO_3-(LaCa)MnO_3$ tunnel junction. *J. Phys. D Appl. Phys.* **2021**, *54*, 275302.
8. Du, W.; Liu, M.; Su, H.; Zhang, H.; Liu, B.; Meng, H.; Xu, G.; Peng, R.; Tang, X. Wide range voltage-impulse-controlled non-volatile magnetic memory in magnetoelectric heterostructure. *Appl. Phys. Lett.* **2020**, *117*, 222401.
9. Chen, A.; Huang, H.; Wen, Y.; Liu, W.; Zhang, S.; Kosel, J.; Sun, W.; Zhao, Y.; Lu, Y.; Zhang, X.-X. Giant magnetoelectric effect in perpendicularly magnetized Pt/Co/Ta ultrathin films on a ferroelectric substrate. *Mater. Horiz.* **2020**, *7*, 2328–2335.
10. Wang, A.-p.; Song, G.; Zhou, F.-p.; Zhao, L.N.; Jin, M.; Liu, M.; Zhang, Y.; Hu, L.-l.; Qi, J.; Xu, H.; et al. Strengthened magnetoelectric coupling in $Pb(Zr_{0.52}Ti_{0.48})O_3$/Ni composite through interface modification with $LaNiO_3$ buffer layer. *J. Mater. Sci. Mater. Electron.* **2021**, *32*, 5920–5927.
11. Belim, S.V.; Tikhomirov, I.V. Composition of the Frenkel-Kontorova and Ising models for investigation the magnetic properties of a ferromagnetic monolayer on a stretching substrate. *Sci. Rep.* **2021**, *11*, 21428. [PubMed]
12. Byckov, I.V.; Belim, S.V.; Maltsev, I.V.; Shavrov, V.G. Phase Transition and Magnetoelectric Effect in 2D Ferromagnetic Films on a Ferroelectric Substrate. *Coatings* **2021**, *11*, 1325.
13. Belim, S.V.; Bychkov, I.V.; Kuzmin, D.A.; Shavrov, V.G. Tuning of 2D magnets Curie temperature via substrate. *J. Magn. Magn. Mater.* **2022**, *541*, 168553.
14. Apostolova, I.N.; Apostolov, A.T.; Wesselinowa, J.M. Room temperature ferromagnetism in multiferroic $BaCoF_4$ thin films due to surface, substrate and ion doping effects. *J. Thin Solid Films* **2021**, *722*, 138567.
15. Tongue Magne, G.E.; Keumo Tsiaze, R.M.; Fotué, A.J.; Hounkonnou, N.M.; Fai, L.C. Cumulative effects of fluctuations and magnetoelectric coupling in two-dimensional $RMnO_3$ (R = Tb, Lu and Y) multiferroics. *Phys. Lett. A* **2021**, *400*, 127305.
16. Sun, Q.; Mahfouzi, F.; Velev, J.P.; Tsymbal, E.Y.; Kioussis, N. Ferroelectric-driven tunable magnetism in ultrathin platinum films. *Phys. Rev. Mater.* **2020**, *4*, 124401.
17. Lang, X.Y.; Zheng, W.T.; Jiang, Q. Size and interface effects on ferromagnetic and antiferromagnetic transition temperatures. *Phys. Rev. B* **2006**, *73*, 224444.
18. Velasquez, E.A.; Altbir, D.; Mazo-Zuluaga, J.; Duque, L.F.; Mejia-Lopez, J. Searching for the nanoscopic-macroscopic boundary. *J. Magn. Magn. Mater.* **2013**, *348*, 154–159.
19. Kontorova, T.A.; Frenkel, Y.I. To the theory of plastic deformation and twinning. *Z. Eksp. Teor. Fiz.* **1938**, *8*, 89–95.
20. Belim, S.V. Effect of elastic deformations on the critical behavior of disordered systems with long-range interactions. *J. Exp. Theor. Phys.* **2004**, *98*, 316–321.
21. Wolff, U. Collective Monte Carlo Updating for Spin Systems. *Phys. Rev. Lett.* **1989**, *62*, 361. [PubMed]

22. Landau, D.P.; Binder, K. Phase Diagrams and Multicritical Behavior of a Three-Dimensional Anisotropic Heisenberg Antiferromagnet. *Phys. Rev. B* **1978**, *17*, 2328.
23. Binder, K. Critical Properties from Monte-Carlo Coarse-Graining and Renormalization. *Phys. Rev. Lett.* **1981**, *47*, 693.
24. Akkera, H.S.; Reddy, N.N.K.; Sekhar, M.C. Thickness dependent structural and electrical properties of magnetron sputtered nanostructured CrN thin films. *Mater. Res.* **2017**, *20*, 712–717.
25. Jian, J.; Wang, X.; Li, L.; Fan, M.; Zhang, W.; Huang, J.; Qi, Z.; Wang, H. Continuous tuning of phase transition temperature in VO_2 thin films on c-cut sapphire substrates via strain variation. *ACS Appl. Mater. Interfaces* **2017**, *9*, 5319–5327.
26. Yang, Y.; Beekman, C.; Siemons, W.; Schlepütz, C.M.; Senabulya, N.; Clarke, R.; Christen, H.M. Origin of thickness dependence of structural phase transition temperatures in highly strained $BiFeO_3$ thin films. *APL Mater.* **2016**, *4*, 036106.
27. Giri, S.K.; Nath, T.K. Strain-induced first-order magnetic phase transition in epitaxial $Sm_{0.35}Pr_{0.15}Sr_{0.5}MnO_3$ thin film. *J. Supercond. Nov. Magn.* **2015**, *28*, 895–900.
28. Srivastava, M.K.; Singh, M.P.; Kaur, A.; Singh, H.K. First order phase transition in $Sm_{0.53}Sr_{0.47}MnO_3$ films. *AIP Conf. Proc.* **2012**, *1447*, 125.
29. Ceballos, A.; Chen, Z.; Schneider, O.; Bordel, C.; Wang, L.-W.; Hellman, F. Effect of strain and thickness on the transition temperature of epitaxial FeRh thin-films. *Appl. Phys. Lett.* **2017**, *111*, 172401.

materials MDPI

Article

Grain Growth Behavior and Electrical Properties of $0.96(K_{0.46-x}Na_{0.54-x})Nb_{0.95}Sb_{0.05}O_3$–$0.04Bi_{0.5}(Na_{0.82}K_{0.18})_{0.5}ZrO_3$ Ceramics

Yeon-Ju Park [1], Il-Ryeol Yoo [1], Seong-Hui Choi [1], Jiung Cho [2,*] and Kyung-Hoon Cho [1,*]

1 School of Materials Science and Engineering, Kumoh National Institute of Technology, Gumi 39177, Korea; pyjuuu0902@kumoh.ac.kr (Y.-J.P.); yiy1204@kumoh.ac.kr (I.-R.Y.); shchoi7@kumoh.ac.kr (S.-H.C.)
2 Western Seoul Center, Korea Basic Science Institute, Seoul 03579, Korea
* Correspondence: jiung14@kbsi.re.kr (J.C.); khcho@kumoh.ac.kr (K.-H.C.)

Abstract: This study investigated the causes of microstructural changes and the resultant electrical properties according to the sintering temperature of $0.96(K_{0.46-x}Na_{0.54-x})Nb_{0.95}Sb_{0.05}O_3$-$0.04Bi_{0.5}(Na_{0.82}K_{0.18})_{0.5}ZrO_3$ lead-free ceramics by analyzing the correlation between vacancy concentrations and 2D nucleation. When sintered for 4 h, no grain growth occurred for the x = 0.000 composition over a wide temperature range, demonstrating that the existence of initial vacancies is essential for grain growth. As x increased, that is, as the vacancy concentration increased, the critical driving force (ΔG_C) for 2D nucleation decreased, and abnormal grain growth was promoted. The number and size of these abnormal grains increased as the sintering temperature increased, but at sintering temperatures above 1100 °C, they decreased again owing to a large drop in ΔG_C. The x = 0.005 specimen sintered at 1085 °C exhibited excellent piezoelectric properties of d_{33} = 498 pC/N and k_p = 0.45 due to the large number of large abnormal grains with an 83% tetragonal phase fraction. The x = 0.000 specimen sintered at 1130 °C with suppressed grain growth exhibited good energy storage properties because of its very high relative density and small grain size of 300 to 400 nm.

Keywords: abnormal grain growth; lead-free; piezoelectric; 2D nucleation; vacancy

Citation: Park, Y.-J.; Yoo, I.-R.; Choi, S.-H.; Cho, J.; Cho, K.-H. Grain Growth Behavior and Electrical Properties of $0.96(K_{0.46-x}Na_{0.54-x})Nb_{0.95}Sb_{0.05}O_3$–$0.04Bi_{0.5}(Na_{0.82}K_{0.18})_{0.5}ZrO_3$ Ceramics. *Materials* **2022**, *15*, 2357. https://doi.org/10.3390/ma15072357

Academic Editor: Dana Popescu

Received: 28 February 2022
Accepted: 20 March 2022
Published: 22 March 2022

1. Introduction

Piezoelectric ceramics are smart materials that enable the interconversion between mechanical and electrical energy and are used in various electronic devices such as actuators, transducers, sensors, and motors [1–5]. Most piezoelectric ceramic materials have been commercialized, mainly for $Pb(Zr,Ti)O_3$ (PZT)-based ceramics with a high lead content. However, recently tightened environmental regulations around the world have urged the development of lead-free piezoelectric ceramics that do not contain lead oxide, which leads to harmful effects such as high toxicity and high vapor pressure in manufacturing and disposal processes [2,6–8]. $(K,Na)NbO_3$ (KNN)-based ceramics are ferroelectric materials with ABO_3–type perovskite structures and are actively being studied as eco-friendly lead-free piezoelectric materials because of their high Curie temperatures and good piezoelectric properties [6–10]. High piezoelectric coefficient values can be obtained compared to pure KNN by modifying KNN with various elements (A-site: Li, Bi, etc., B-site: Ta, Sb, etc.) to move orthorhombic-tetragonal (O-T) and rhombohedral-orthorhombic (R-O) phase transition temperatures (T_{O-T} and T_{R-O}) to near room temperature [11–14]. In addition, when T_{O-T} and T_{R-O} meet near room temperature by doping Li, Bi, Ag, etc. at the A-site and Sb, Zr, Hf, etc. in the B-site of KNN, excellent piezoelectric properties comparable to those of PZT-based ceramics are obtained because of the coexistence of rhombohedral and tetragonal phases [15–23].

For the commercialization of KNN-based ceramics, not only excellent piezoelectric properties but also reproducibility of microstructures directly related to the piezoelectric properties are very important. The microstructure of KNN-based ceramics can be sensitively changed depending on the sintering conditions, and various microstructures and piezoelectric properties with large deviations have been reported, even in materials with the same composition [24–27]. This is a major obstacle in ensuring the reproducibility of piezoelectric characteristics.

The grains of KNN-based ceramics exhibit a faceted cube shape, signifying their very high interfacial energy anisotropy [28–30]. The faceted grain boundary makes normal grain growth difficult, owing to the absence of kink sites for the adsorption of atoms, which is associated with the low sinterability of KNN-based ceramics. The grain growth of ceramics with faceted grain boundaries can be enabled by 2D nucleation and growth mechanisms [28,29,31]. The critical driving force (ΔG_C) for the formation of 2D nuclei on the grain surface is expressed as

$$\Delta G_C = \frac{\Omega \varepsilon^2}{3hkT}, \tag{1}$$

where ε is the step free energy of the 2D nucleus, Ω is the molar volume, k is Boltzmann's constant, T is the temperature, and h is the height of the 2D nucleus. The driving force (ΔG) for grain growth of the grains is given as follows:

$$\Delta G = 2\gamma_{sl}\Omega\left(\frac{1}{r^*} - \frac{1}{r}\right), \tag{2}$$

where γ_{sl} is the solid-liquid interfacial energy, r is the radius of the growing grain, and r^* is the average radius of the grains around the growing grain. Only grains that satisfy the condition $\Delta G \geq \Delta G_C$ can grow via 2D nucleation. The variable ε in Equation (1) is inversely proportional to the configurational entropy; therefore, it can be decreased as the vacancy increases [29,30].

In this study, $0.96(K_{0.46-x}Na_{0.54-x})Nb_{0.95}Sb_{0.05}O_3$-$0.04Bi_{0.5}(Na_{0.82}K_{0.18})_{0.5}ZrO_3$ compositions with intentionally reduced alkali element content were designed, and the effect of vacancy concentration on the microstructure and electrical properties of sintered ceramics was investigated. It was found that grain growth was suppressed in compositions in which the alkali element content was not controlled, and there were specific compositions and sintering temperatures to obtain a microstructure exhibiting excellent piezoelectric properties. In addition, we report that when a highly dense microstructure with suppressed grain growth is secured, the energy storage performance is enhanced, even if the ferroelectricity is weakened owing to the small grain size.

2. Experimental

Scheme 1 shows the experimental procedure including sample preparation and analysis methods.

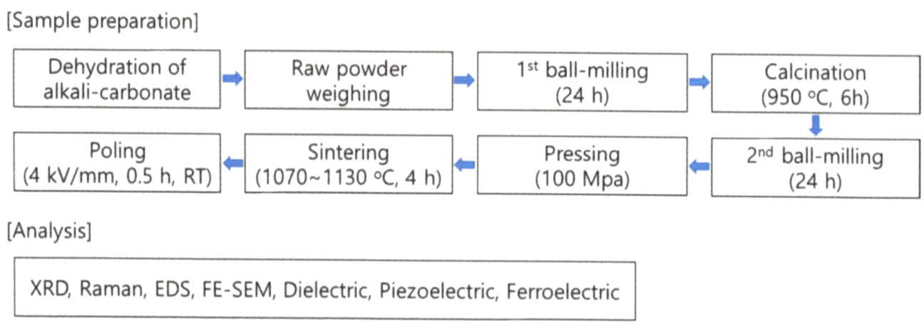

Scheme 1. Schematic diagram of the experimental procedure in this study.

The $0.96(K_{0.46-x}Na_{0.54-x})Nb_{0.95}Sb_{0.05}O_3$-$0.04Bi_{0.5}(Na_{0.82}K_{0.18})_{0.5}ZrO_3$ compositions (x = 0.000, 0.005, 0.010, 0.015) were prepared using K_2CO_3 (99% purity), Na_2CO_3 (99%

purity), Nb_2O_5 (99.9% purity), Sb_2O_3 (99.9% purity), Bi_2O_3 (99.9% purity), and ZrO_2 (98% purity) raw powders (all from Kojundo Korea Co., Uiwang-si, Korea) by solid state synthesis. K_2CO_3 and Na_2CO_3 powders were heated at 400 °C for 24 h under a vacuum to achieve complete dehydration. The raw powders were mixed according to their compositions and milled with ethanol and yttria-stabilized zirconia balls in a polyethylene jar for 24 h. The powder mixtures were calcined at 950 °C for 6 h with a heating rate of 4 °C/min and naturally cooled down to room temperature in a furnace (furnace-cooling). The calcined powders were re-milled for 24 h and pressed into disk-shaped powder compacts at 100 MPa. The powder compacts were sintered for 4 h at temperatures ranging from 1070 °C to 1130 °C with a heating rate of 4 °C/min and naturally cooled down to room temperature in a furnace.

An Ag paste was printed on both surfaces of the sintered specimens and fired at 600 °C for 30 min with a heating rate of 4 °C/min and naturally cooled down to room temperature in a furnace. The specimens were electrically poled by applying a DC electric field of 4 kV/mm for 30 min at room temperature in a silicone oil bath. Structural analyses were performed using X-ray diffraction (XRD, D-MAX/2500, Rigaku, Japan), Raman spectroscopy (System 1000, Renishaw, UK), energy-dispersive X-ray spectroscopy (EDS; XFlash 630, Bruker Nano GmbH, Berlin, Germany), and scanning electron microscopy (SEM, JSM-6500F, JEOL, Akishima, Japan). The piezoelectric and dielectric properties of the sintered specimens were measured using a d_{33}-m (YE2730A, Sinocera Piezotronics, Yangzhou, China) and impedance analyzer (IM3570, Hioki, Japan). Polarization vs. electric field (P-E) curves were measured using a ferroelectric test system (PK-CPE1801, PolyK Technologies, Philipsburg, PA, USA).

3. Results and Discussion

Figure 1a shows the ratio of alkali ion content to Nb ion content in $0.96(K_{0.46-x}Na_{0.54-x})Nb_{0.95}Sb_{0.05}O_3$-$0.04Bi_{0.5}(Na_{0.82}K_{0.18})_{0.5}ZrO_3$ powders calcined at 950 °C, and the ratio decreased as x increased. As all of the compositions exhibit perovskite structures without a secondary phase (inset in Figure 1a), the concentration of alkali ion vacancies is expected to increase as x increases. As shown in Figure 1b, the v_1 and v_5 mode peaks of the Raman spectra shifted toward higher wavenumbers as x increased.

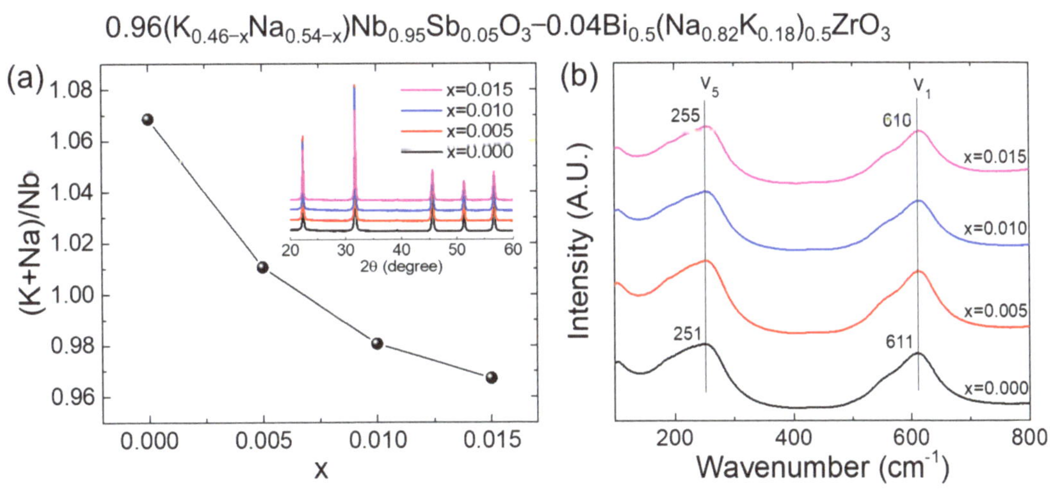

Figure 1. (**a**) Ratio of alkali-ion content to niobium ion content of $0.96(K_{0.46-x}Na_{0.54-x})Nb_{0.95}Sb_{0.05}O_3$-$0.04Bi_{0.5}(Na_{0.82}K_{0.18})_{0.5}ZrO_3$ calcined powders obtained by EDS analysis. Inset shows their XRD patterns. (**b**) Raman spectra of $0.96(K_{0.46-x}Na_{0.54-x})Nb_{0.95}Sb_{0.05}O_3$-$0.04Bi_{0.5}(Na_{0.82}K_{0.18})_{0.5}ZrO_3$ calcined powders.

The v_1 and v_5 modes correspond to the stretching and bending modes of the B-site octahedra (NbO_6), respectively, and the v_1 peak shifts to a higher wavenumber when an alkali ion deficiency (or vacancy) exists or when ions with a small ionic radius, such as lithium, are substituted for the A-site [32–34]. This also demonstrates that the concentration of alkali ion vacancies present in the $0.96(K_{0.46-x}Na_{0.54-x})Nb_{0.95}Sb_{0.05}O_3$-$0.04Bi_{0.5}(Na_{0.82}K_{0.18})_{0.5}ZrO_3$ grains increases with an increase in x.

Figure 2 shows microstructures of $0.96(K_{0.46-x}Na_{0.54-x})Nb_{0.95}Sb_{0.05}O_3$-$0.04Bi_{0.5}(Na_{0.82}K_{0.18})_{0.5}ZrO_3$ ceramics sintered for 4 h at temperatures ranging from 1070 to 1085 °C.

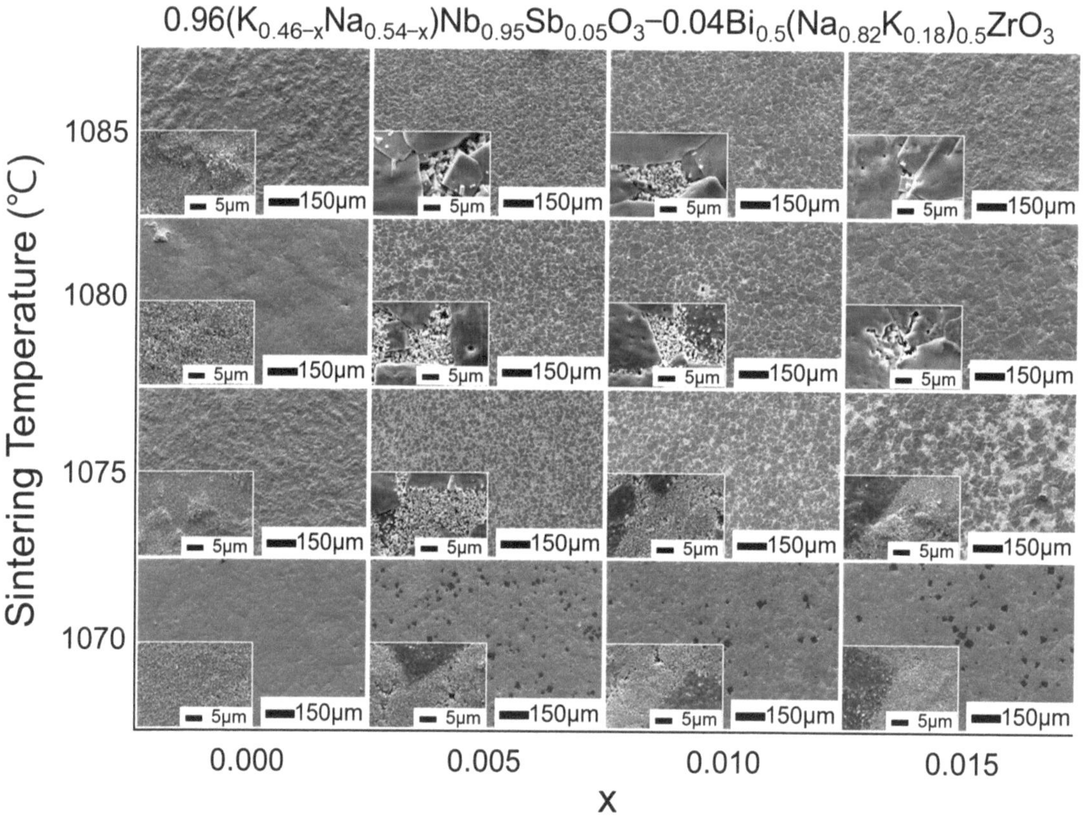

Figure 2. SEM images showing the microstructural evolution of $0.96(K_{0.46-x}Na_{0.54-x})Nb_{0.95}Sb_{0.05}O_3$-$0.04Bi_{0.5}(Na_{0.82}K_{0.18})_{0.5}ZrO_3$ ceramics sintered in the temperature range from 1070 to 1085 °C for 4 h. Insets show magnified SEM images for each specimen.

Unusually, no grain growth was observed in the x = 0 specimen under all sintering temperatures, signifying that 2D nucleation was suppressed in all grains owing to the high ΔG_C because the vacancy concentration of grains was too low. However, for x = 0.005, 0.010, and 0.015 specimens sintered at 1070 °C, abnormal grains of several tens of micrometers in size were observed in the small matrix grains that did not grow. This can be interpreted as 2D nucleation and growth progressing in some grains, satisfying the $\Delta G \geq \Delta G_C$ condition as ε decreases owing to the increased vacancy concentration. Moreover, the number and size of these abnormal grains increased as the sintering temperature increased. This implies that the higher the sintering temperature, the more active the alkali element volatilization, which increases the number of grains satisfying the $\Delta G \geq \Delta G_C$ condition and advances the onset of 2D nucleation during sintering. Matrix grains of the x = 0.005, 0.010, and 0.015 specimens sintered at temperatures below 1080 °C did not grow. However, the x = 0.005 specimen sintered at 1085 °C exhibited a microstructure composed of large grains of several tens of micrometers and small grains of several micrometers, suggesting that the ΔG_C continued to decrease with time at 1085 °C, and the growth of small-sized grains was also initiated.

Figure 3 shows the piezoelectric properties of $0.96(K_{0.46-x}Na_{0.54-x})Nb_{0.95}Sb_{0.05}O_3$-$0.04Bi_{0.5}(Na_{0.82}K_{0.18})_{0.5}ZrO_3$ ceramics sintered at 1085 °C for 4 h. The x = 0.000 specimen, in which grain growth did not occur, exhibited poor piezoelectric properties of $d_{33} < 100$ pC/N and $k_p < 0.2$, whereas the x = 0.005 specimen, in which large grain growth was observed, exhibited excellent piezoelectric properties of $d_{33} = 498$ pC/N and $k_p = 0.45$. A large grain size makes domain rotation easier, thereby increasing the piezoelectric properties [26,35]. However, in the case of the x = 0.010 and 0.015 specimens, the piezoelectric properties were reduced compared with those of the x = 0.005 specimen, although they displayed a microstructure composed of large grains similar to that of the x = 0.005 specimen.

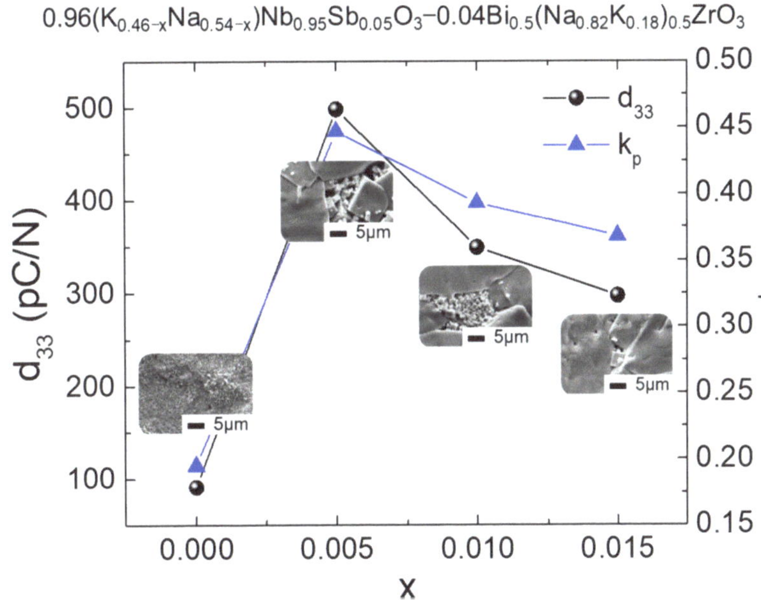

Figure 3. Piezoelectric charge coefficient (d_{33}) and electromechanical coupling coefficient (k_p) of $0.96(K_{0.46-x}Na_{0.54-x})Nb_{0.95}Sb_{0.05}O_3$-$0.04Bi_{0.5}(Na_{0.82}K_{0.18})_{0.5}ZrO_3$ ceramics sintered at 1085 °C for 4 h.

The $0.96(K_{0.46-x}Na_{0.54-x})Nb_{0.95}Sb_{0.05}O_3$-$0.04Bi_{0.5}(Na_{0.82}K_{0.18})_{0.5}ZrO_3$ ceramics sintered at 1085 °C for 4 h showed a pure perovskite structure without a secondary phase, as shown in Figure 4a. The x = 0.000 specimen exhibited a pseudo-cubic structure owing to its very small grain size. However, clear separation of the $(200)_C$ peak was observed for the x = 0.005, 0.010, and 0.015 specimens composed of large grains. Deconvolution analysis of the $(200)_C$ peaks in Figure 4b confirmed that x = 0.005, 0.010, and 0.015 compositions were in a tetragonal-rich rhombohedral-tetragonal (R-T) phase coexistence (or phase transition) state at room temperature [36,37]. The rhombohedral and tetragonal phase fractions are shown in Table 1. It can be seen that the fraction of the rhombohedral phase decreases as x increases. Furthermore, as x increased, both the R-T transition temperature ($T_{R\text{-}T}$) and Curie temperature (T_C) shifted toward lower temperatures (Figure 4c), which supports the results in Table 1. The piezoelectric properties of the R-T phase coexistence compositions are maximized when the fraction of the tetragonal phase is approximately 80% [38,39]. The deterioration of the piezoelectric properties of the x = 0.010 and 0.015 specimens observed in Figure 3 might be due to the high tetragonal phase fraction of more than 90%. Therefore, even for specimens with similar microstructures, their phase status and piezoelectric properties may vary significantly depending on the initial vacancy concentration. For the compositions used in this study, an excessively high vacancy concentration seems unsuitable for securing excellent piezoelectric properties.

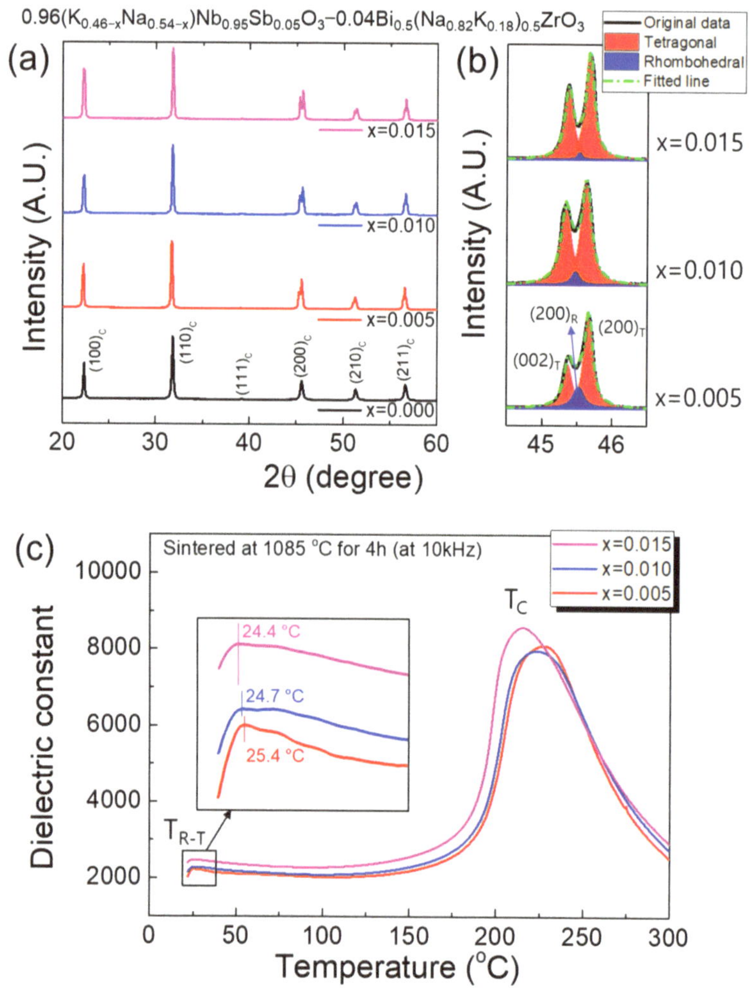

Figure 4. (**a**) XRD patterns of $0.96(K_{0.46-x}Na_{0.54-x})Nb_{0.95}Sb_{0.05}O_3$-$0.04Bi_{0.5}(Na_{0.82}K_{0.18})_{0.5}ZrO_3$ ceramics sintered at 1085 °C for 4 h. (**b**) Finely scanned XRD patterns for $(200)_C$ peaks of x = 0.005, 0.010, and 0.015 specimens. (**c**) Dielectric constant vs. temperature curves of x = 0.005, 0.010, and 0.015 specimens.

Table 1. Phase fraction of the rhombohedral and tetragonal phases in x = 0.005, 0.010, and 0.015 specimens sintered at 1085 °C for 4 h.

Composition	Phase Fraction (%)	
	Rhombohedral	Tetragonal
x = 0.005	16.74	83.26
x = 0.010	5.79	94.21
x = 0.015	1.65	98.35

Figure 5 shows the microstructures of $0.96(K_{0.46-x}Na_{0.54-x})Nb_{0.95}Sb_{0.05}O_3$-$0.04Bi_{0.5}(Na_{0.82}K_{0.18})_{0.5}ZrO_3$ ceramics sintered at temperatures higher than 1085 °C. For the x = 0.000 specimen, grain growth was not observed even at a very high sintering temperature of 1130 °C, which indicates that the ΔG_C value of the stoichiometric composition, in which the content of the initial alkali element was not reduced, could be maintained at a very high level even at a high sintering temperature. In the case of the x = 0.005 specimen, as the sintering temperature increased, the number of large abnormal grains decreased, and the matrix grains grew to a size of several micrometers. This phenomenon was also observed in the x = 0.010 specimen, but the number of abnormal grains began to decrease at a lower sintering temperature (1100 °C) than that of the x = 0.005 specimen (1110 °C). The x = 0.010 specimen sintered at 1130 °C displayed a microstructure in which all grains were grown to a size of several micrometers without large abnormal grains.

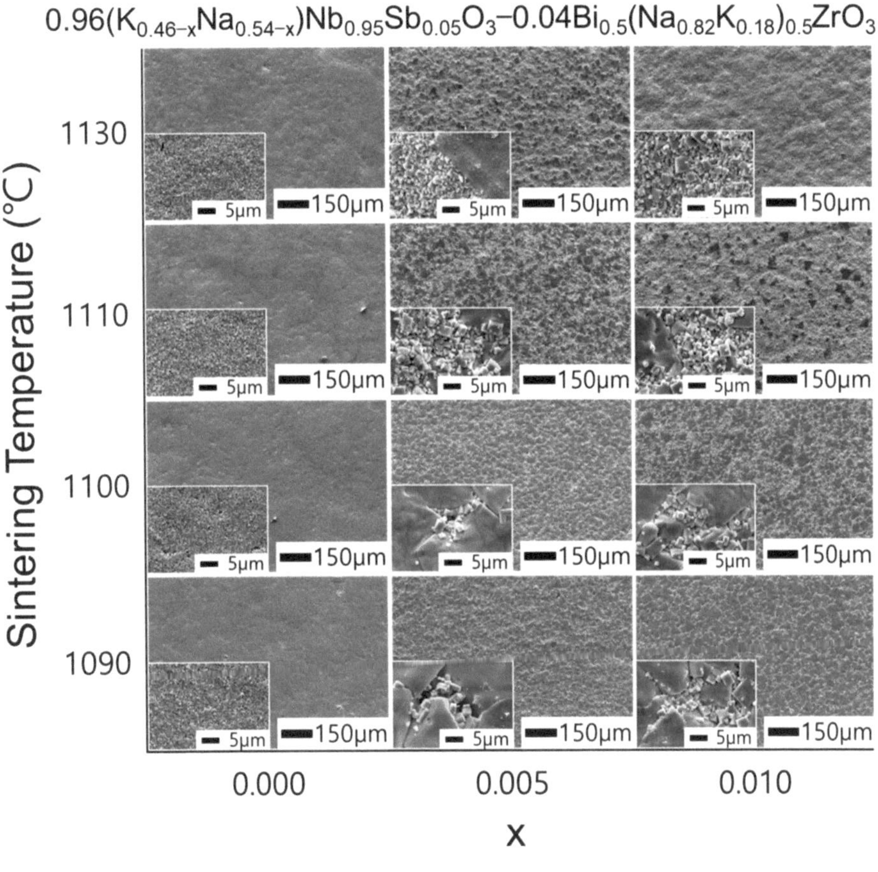

Figure 5. SEM images showing the microstructural evolution of $0.96(K_{0.46-x}Na_{0.54-x})Nb_{0.95}Sb_{0.05}O_3$-$0.04Bi_{0.5}(Na_{0.82}K_{0.18})_{0.5}ZrO_3$ ceramics sintered in the temperature range from 1090 to 1130 °C for 4 h. Insets show magnified SEM images for each specimen.

The decrease in the number of abnormal grains as the sintering temperature increased is considered to be due to the increase in the vacancy concentration due to the enhanced volatilization of the alkali element. ΔG_C decreases as the concentration of vacancies increases, and if the number of grains satisfying the $\Delta G \geq \Delta G_C$ condition rapidly increases during the sintering process, the growth of large abnormal grains is inevitably suppressed, and the number of abnormal grains is also limited. In the case of the x = 0.010 specimen, because the initial vacancy concentration was higher than that of the x = 0.005 specimen,

ΔG_C decreased more rapidly, and the abnormal grain growth started to be suppressed at lower temperatures.

Figure 6 shows the piezoelectric properties of x = 0.005 specimens sintered at various temperatures. Both d_{33} and k_p sharply increased as the sintering temperature increased up to 1085 °C but decreased again at higher sintering temperatures. By correlating the results of Figure 2, Figure 5, and Figure 6, we can conclude that the piezoelectric properties were enhanced as the content of large grains increased. The reduction in the piezoelectric properties of the x = 0.005 specimens sintered at temperatures higher than 1085 °C is believed to be attributed to the decrease in $T_{R\text{-}T}$ due to the increased vacancy concentration (similar to the explanation in Figure 4) and the decrease in the number of large abnormal grains.

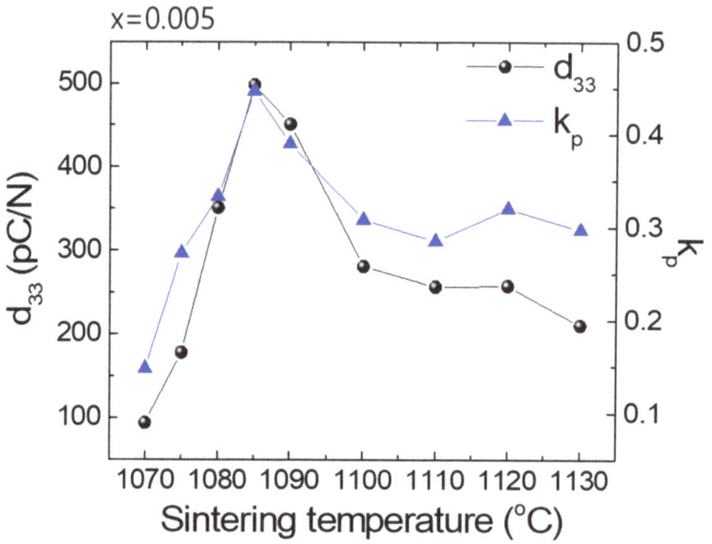

Figure 6. Piezoelectric charge coefficient (d_{33}) and electromechanical coupling coefficient (k_p) of the x = 0.005 specimen sintered at various temperatures for 4 h.

As shown in Figures 2 and 5, the x = 0.000 specimen exhibited a microstructure in which grain growth was suppressed over a wide sintering temperature range from 1070 °C to 1130 °C. However, as the sintering temperature increased, the density of the sintered specimen rapidly increased, and the x = 0.000 specimen sintered at 1130 °C exhibited a high relative density of 98%, as shown in Figure 7a. In addition, owing to the fine grain size of 300–400 nm, the x = 0.000 specimens sintered at high temperatures maintained a pseudo-cubic structure (inset of Figure 7a). Figure 7b shows the unipolar P-E curve of the x = 0.000 specimen sintered at 1130 °C. Owing to the very small grain size and high relative density, a high polarization value was stably obtained without dielectric breakdown, even under a high electric field of 100 kV/cm.

Furthermore, the weakening of ferroelectricity owing to the fine grains with a pseudo-cubic structure led to a slim P-E curve with very little hysteresis loss. The stored energy density (W_{st}), recoverable energy density (W_{rec}), and efficiency (η) of the dielectric capacitor are expressed as follows:

$$W_{st} = \int_0^{P_{max}} EdP \quad (3)$$

$$W_{rec} = \int_{P_r}^{P_{max}} EdP \quad (4)$$

$$\eta = \frac{W_{rec}}{W_{st}} \times 100\% \quad (5)$$

Due to the high maximum polarization (P_{max}) and low remnant polarization (P_r) of the P-E curve in Figure 7b, the x = 0.000 specimen sintered at 1130 °C exhibited a good energy storage performance of W_{st} = 0.66 J/cm^3, W_{rec} = 0.52 J/cm^3, and η = 79%. Therefore, by designing a dense microstructure composed of small grains with a low vacancy concentration, excellent energy storage characteristics can be obtained, even if the piezoelectric characteristics deteriorate.

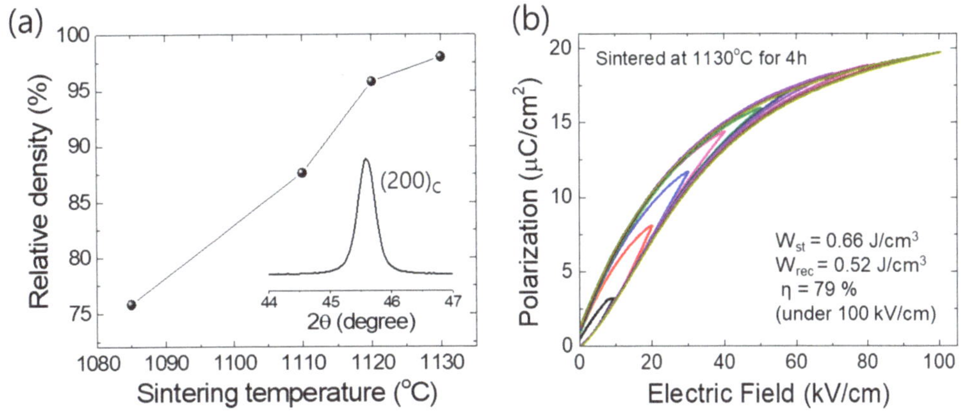

Figure 7. (**a**) Relative density of the x = 0.000 specimen sintered at various temperatures for 4 h. Inset of (**a**) shows the XRD pattern of the $(200)_C$ peak of the x = 0.000 specimen sintered at 1120 °C. (**b**) Unipolar P-E curves of the x = 0.000 specimen sintered at 1130 °C for 4 h.

4. Conclusions

In this study, the effects of alkali element deficiency and sintering temperature on the microstructure and electrical properties of $0.96(K_{0.46-x}Na_{0.54-x})Nb_{0.95}Sb_{0.05}O_3$-$0.04Bi_{0.5}(Na_{0.82}K_{0.18})_{0.5}ZrO_3$ ceramics were investigated. For the x = 0.000 specimen, grain growth was not observed at any sintering temperature from 1070 °C to 1130 °C because the high ΔG_C owing to the very low vacancy concentration was maintained even at high sintering temperatures. In the case of the x = 0.005, 0.010, and 0.015 specimens, the grains satisfying the $\Delta G \geq \Delta G_C$ condition owing to the increased vacancy concentration grew into abnormal grains by 2D nucleation. The x = 0.005 specimen sintered at 1085 °C exhibited a microstructure composed of large grains of several tens of micrometers and small grains of several micrometers with excellent piezoelectric properties of d_{33} = 498 pC/N and k_p = 0.45. For the x = 0.010 and 0.015 specimens sintered at 1085 °C, the piezoelectric properties deteriorated compared to those of the x = 0.005 specimen because the $T_{R\text{-}T}$ of x = 0.010 and 0.015 specimens moved to lower temperatures than that of the x = 0.005 specimen. For the x = 0.005 specimen sintered at temperatures higher than 1085 °C, the number of large abnormal grains decreased again as the number of grains satisfying the $\Delta G \geq \Delta G_C$ condition rapidly increased, owing to the further increased vacancy concentration; accordingly, the values of the piezoelectric coefficients also decreased. The relative density of the x = 0.000 specimen rapidly increased with an increase in the sintering temperature, and the x = 0.000 specimen sintered at 1130 °C with a high relative density of 98% and a fine grain size of 300–400 nm exhibited good energy storage performance of W_{st} = 0.66 J/cm^3, W_{rec} = 0.52 J/cm^3, and η = 79%. The results of this study demonstrate that the microstructure of $0.96(K_{0.46-x}Na_{0.54-x})Nb_{0.95}Sb_{0.05}O_3$-$0.04Bi_{0.5}(Na_{0.82}K_{0.18})_{0.5}ZrO_3$ ceramics highly depends on the vacancy concentration, and their applications can be diversified depending on the microstructure. These findings are expected to be applied to the study of microstructure control of other KNN-based ceramics.

Author Contributions: Conceptualization, Y.-J.P. and K.-H.C.; experiment, Y.-J.P. and I.-R.Y.; analysis, J.C. and S.-H.C.; writing—original draft preparation, Y.-J.P. and K.-H.C.; writing—review and editing, J.C. and K.-H.C.; supervision, K.-H.C.; funding acquisition, K.-H.C. All authors have read and agreed to the published version of the manuscript.

Funding: This research was supported by Kumoh National Institute of Technology (2018104100).

Institutional Review Board Statement: Not applicable.

Informed Consent Statement: Not applicable.

Conflicts of Interest: The authors declare no conflict of interest.

References

1. Rödig, T.; Schönecker, A.; Gerlach, G. A Survey on Piezoelectric Ceramics for Generator Applications. *J. Am. Ceram. Soc.* **2010**, *93*, 901–912. [CrossRef]
2. Zheng, T.; Wu, J.; Xiao, D.; Zhu, J. Recent development in lead-free perovskite piezoelectric bulk materials. *Prog. Mater. Sci.* **2018**, *98*, 552–624. [CrossRef]
3. Damjanovic, D. Ferroelectric, dielectric and piezoelectric properties of ferroelectric thin films and ceramics. *Pep. Prog. Phys.* **1998**, *61*, 1267–1324. [CrossRef]
4. Bellaiche, L.; García, A.; Vanderbilt, D. Finite-Temperature Properties of $Pb(Zr_{1-x}Ti_x)O_3$ Alloys from First Principles. *Phys. Rev. Lett.* **2000**, *84*, 5427–5430. [CrossRef]
5. Cho, K.-H.; Park, H.-Y.; Heo, J.S.; Priya, S. Structure-performance relationships for cantilever-type piezoelectric energy harvesters. *J. Appl. Phys.* **2014**, *115*, 204108. [CrossRef]
6. Panda, P.K. Review: Environmental friendly lead-free piezoelectric materials. *J. Mater. Sci.* **2009**, *44*, 5049–5062. [CrossRef]
7. Wu, J.; Xiao, D.; Zhu, J. Potassium-Sodium Niobate Lead-Free Piezoelectric Materials: Past, Present, and Future of Phase Boundaries. *Chem. Rev.* **2015**, *115*, 2559–2595. [CrossRef]
8. Rödel, J.; Jo, W.; Seifert, K.T.P.; Anton, E.-M.; Granzow, T.; Damjanovic, D. Perspective on the Development of Lead-free Piezoceramics. *J. Am. Ceram. Soc.* **2009**, *92*, 1153–1177. [CrossRef]
9. Saito, Y.; Takao, H.; Tani, T.; Nonoyama, T.; Takatori, K.; Homma, T.; Nagaya, T.; Nakamura, M. Lead-free piezoceramics. *Nature* **2004**, *432*, 84–87. [CrossRef]
10. Zhang, S.; Xia, R.; Shrout, T.R. Lead-free piezoelectric ceramics vs. PZT? *J. Electroceram.* **2007**, *19*, 251–257. [CrossRef]
11. Guo, Y.; Kakimoto, K.; Ohsato, H. Phase transitional behavior and piezoelectric properties of $(Na_{0.5}K_{0.5})NbO_3$-$LiNbO_3$ ceramics. *Appl. Phys. Lett.* **2004**, *85*, 4121. [CrossRef]
12. Guo, Y.; Kakimoto, K.-i.; Ohsato, H. $(Na_{0.5}K_{0.5})NbO_3$–$LiTaO_3$ lead-free piezoelectric ceramics. *Mater. Lett.* **2005**, *59*, 241–244. [CrossRef]
13. Hollenstein, E.; Davis, M.; Damjanovic, D.; Setter, N. Piezoelectric properties of Li- and Ta-modified $(K_{0.5}Na_{0.5})NbO_3$ ceramics. *Appl. Phys. Lett.* **2005**, *87*, 182905. [CrossRef]
14. Zang, G.Z.; Wang, J.F.; Chen, H.C.; Su, W.B.; Wang, C.M.; Qi, P.; Ming, B.Q.; Du, J.; Zheng, L.M.; Zhang, S.; et al. Perovskite $(Na_{0.5}K_{0.5})_{1-x}(LiSb)_xNb_{1-x}O_3$ lead-free piezoceramics. *Appl. Phys. Lett.* **2006**, *88*, 212908. [CrossRef]
15. Wang, X.; Wu, J.; Xiao, D.; Zhu, J.; Cheng, X.; Zheng, T.; Zhang, B.; Lou, X.; Wang, X. Giant Piezoelectricity in Potassium–Sodium Niobate Lead-Free Ceramics. *J. Am. Chem. Soc.* **2014**, *136*, 2905–2910. [CrossRef]
16. Wu, B.; Wu, H.; Wu, J.; Xiao, D.; Zhu, J.; Pennycook, S.J. Giant Piezoelectricity and High Curie Temperature in Nanostructured Alkali Niobate Lead-Free Piezoceramics through Phase Coexistence. *J. Am. Chem. Soc.* **2016**, *138*, 15459–15464. [CrossRef]
17. Xu, K.; Li, J.; Lv, X.; Wu, J.; Zhang, X.; Xiao, D.; Zhu, J. Superior Piezoelectric Properties in Potassium-Sodium Niobate Lead-Free Ceramics. *Adv. Mater.* **2016**, *28*, 8519–8523. [CrossRef]
18. Wu, J.; Wang, X.; Cheng, X.; Zheng, T.; Zhang, B.; Xiao, D.; Zhu, J.; Lou, X. New potassium-sodium niobate lead-free piezoceramic: Giant-d33 vs. sintering temperature. *J. Appl. Phys.* **2014**, *115*, 114104. [CrossRef]
19. He, B.; Du, Y.; Liu, Y.; Wang, J.; Liu, W.; Xu, H. Phase boundary design and enhanced electrical properties in $(Bi_{0.5}Li_{0.45}Ag_{0.05})(Zr_{0.98}Hf_{0.02})O_3$-modified KNN-based lead-free piezoceramic. *J. Mater. Sci. Mater. Electron.* **2021**, *32*, 18240–18250. [CrossRef]
20. Liu, W.; Wang, H.; Hu, W.; Du, Y.; Cheng, C. Understanding the origin of the high piezoelectric performance of KNN-based ceramics from the perspective of lattice distortion. *Ceram. Int.* **2022**, *48*, 9731–9738. [CrossRef]
21. Shi, C.; Ma, J.; Wu, J.; Chen, K.; Wu, B. Enhanced electrical properties in KNN-based ternary ceramics with high Curie temperature. *J. Mater. Sci.: Mater. Electon.* **2020**, *31*, 2809–2816. [CrossRef]
22. Wang, J.; He, B.; Du, Y.; Cheng, C.; Liu, Y.; Liu, W.; Ma, J.; Xu, H. Improved electrical properties and luminescence properties of lead-free KNN ceramics via phase transition. *J. Mater. Sci. Mater. Electon.* **2021**, *32*, 28819–28829. [CrossRef]
23. Li, J.W.; Liu, Y.X.; Thong, H.C.; Du, Z.; Li, Z.; Zhu, Z.X.; Nie, J.K.; Geng, J.F.; Gong, W.; Wang, K. Effect of ZnO doping on $(K,Na)NbO_3$-based lead-free piezoceramics: Enhanced ferroelectric and piezoelectric performance. *J. Alloys Compd.* **2020**, *847*, 155936. [CrossRef]
24. Park, H.-Y.; Ahn, C.-W.; Song, H.-C.; Lee, J.-H.; Nahm, S. Microstructure and piezoelectric properties of $0.95(Na_{0.5}K_{0.5})NbO_3$-$0.05BaTiO_3$ ceramics. *Appl. Phys. Lett.* **2006**, *89*, 062906. [CrossRef]

25. Wang, R.; Xie, R.J.; Hanada, K.; Matsusaki, K.; Kawanaka, H.; Bando, H.; Sekiya, T.; Itoh, M. Enhanced piezoelectricity around the tetragonal/orthorhombic morphotropic phase boundary in (Na,K)NbO_3-$ATiO_3$ solid solutions. *J. Electroceram.* **2008**, *21*, 263–266. [CrossRef]
26. Cho, K.-H.; Park, H.-Y.; Ahn, C.-W.; Nahm, S.; Uchino, K.; Park, S.-H.; Lee, H.-G.; Lee, H.-J. Microstructure and Piezoelectric Properties of 0.95($Na_{0.5}K_{0.5}$)NbO_3-0.05$SrTiO_3$ Ceramics. *J. Am. Ceram. Soc.* **2007**, *90*, 1946–1949. [CrossRef]
27. Guo, Y.; Kakimoto, K.; Ohsato, H. Dielectric and piezoelectric properties of lead-free ($Na_{0.5}K_{0.5}$)NbO_3-$SrTiO_3$ ceramics. *Solid State Commun.* **2004**, *129*, 279–284. [CrossRef]
28. Thong, H.C.; Xu, Z.; Zhao, C.; Lou, L.Y.; Chen, S.; Zuo, S.Q.; Li, J.F.; Wang, K. Abnormal grain growth in (K, Na)NbO_3-based lead-free piezoceramic powders. *J. Am. Ceram. Soc.* **2019**, *102*, 836–844.
29. Kabakov, P.; Dean, C.; Kurusingal, V.; Cheng, Z.; Lee, H.-Y.; Zhang, S. Solid-state cystal growth of lead-free ferroelectrics. *J. Mater. Chem. C.* **2020**, *8*, 7606–7649. [CrossRef]
30. Fisher, J.G.; Kang, S.-J.L. Microstructural changes in ($K_{0.5}Na_{0.5}$)NbO_3 ceramics sintered in various atmospheres. *J. Eur. Ceram. Soc.* **2009**, *29*, 2581–2588. [CrossRef]
31. Yang, J.; Yang, Q.; Li, Y.; Liu, Y. Growth mechanism and enhanced electrical properties of $K_{0.5}Na_{0.5}NbO_3$-based lead-free piezoelectric single crystals grown by a solid-state crystal growth method. *J. Eur. Ceram. Soc.* **2016**, *36*, 541–550. [CrossRef]
32. Kakimoto, K.-i.; Akao, K.; Guo, Y.; Ohsato, H. Raman Sattering Study of Piezoelectric ($Na_{0.5}K_{0.5}$)NbO_3-$LiNbO_3$ Ceramics. *Jpn. J. Appl. Phys.* **2005**, *44*, 7064–7067. [CrossRef]
33. Ahn, C.W.; Hwang, H.I.; Lee, K.S.; Jin, B.M.; Park, S.; Park, G.; Yoon, D.; Cheong, H.; Lee, H.J.; Kim, I.W. Raman Spectra Study of $K_{0.5}Na_{0.5}NbO_3$ Ferroelectric Thin Films. *Jpn. J. Appl. Phys.* **2010**, *49*, 095801. [CrossRef]
34. Hreščak, J.; Dražić, G.; Deluca, M.; Arčon, I.; Kodre, A.; Dapiaggi, M.; Rojac, T.; Malič, B.; Bencan, A. Donor doping of $K_{0.5}Na_{0.5}NbO_3$ ceramics with strontium and its implications to grain size, phase composition and crystal structure. *J. Eur. Ceram. Soc.* **2017**, *37*, 2073–2082. [CrossRef]
35. Gio, P.D.; Vuong, L.D.; Tu, L.T.U. Enhanced piezoelectric and energy storage performance of 0.96($K_{0.48}Na_{0.48}Li_{0.04}$)($Nb_{0.95}Sb_{0.05}$)$O_3-0.04Bi_{0.05}$($Na_{0.82}K_{0.18}$)$_{0.5}ZrO_3$ ceramics using two-step sintering method. *J. Mater. Sci. Mater. Electron.* **2021**, *32*, 13738–13747.
36. Rubio-Marcos, F.; López-Juárez, R.; Rojas-Hamandez, R.E.; Campo, A.; Razo-Pérez, N.; Fernandez, J.F. Lead-Free Piezoceramics: Revealing the Role of the Rhombohedral-Tetragonal Phase Coexistence in Enhancement of the Piezoelectric Properties. *ACS Appl. Mater. Interfaces.* **2015**, *7*, 23080–23088. [CrossRef]
37. Yang, S.A.; Kim, B.H.; Lee, M.K.; Bu, S.D.; Lee, G.J. The effects of the phase structure of the polymorphic phase boundary on the piezoelectric properties of (K,Na)NbO_3-based ceramics. *Curr. Appl. Phys.* **2016**, *16*, 593–598. [CrossRef]
38. Lee, M.K.; Yang, S.A.; Park, J.J.; Lee, G.J. Proposal of a rhombohedral-tetragonal phase composition for maximizing piezoelectricity of (K,Na)NbO_3 ceramics. *Sci. Rep.* **2019**, *9*, 4195. [CrossRef]
39. Lv, X.; Wu, J.; Zhu, J.; Xiao, D.; Zhang, X. A new method to improve the electrical properties of KNN-based ceramics: Tailoring phase fraction. *J. Eur. Ceram. Soc.* **2018**, *38*, 85–94. [CrossRef]

MDPI AG
Grosspeteranlage 5
4052 Basel
Switzerland
Tel.: +41 61 683 77 34

Materials Editorial Office
E-mail: materials@mdpi.com
www.mdpi.com/journal/materials

www.ingramcontent.com/pod-product-compliance
Lightning Source LLC
Chambersburg PA
CBHW041355210926
51850CB00013B/3086
* 9 7 8 3 7 2 5 8 3 7 3 9 7 *